Toward Quality Assurance and Excellence in Higher Education

RIVER PUBLISHERS SERIES IN INNOVATION AND CHANGE IN EDUCATION - CROSS-CULTURAL PERSPECTIVE

Series Editor:

XIANGYUN DU
Department of Learning and Philosophy
Aalborg University, Denmark
and
College of Education, Qatar University, Qatar

Guest Editor:

NIAN ZHIYING
Beijing Normal University
China

Indexing: All books published in this series are submitted to the Web of Science Book Citation Index (BkCI), to CrossRef and to Google Scholar.

Nowadays, educational institutions are being challenged as professional competences and expertise become progressively more complex. This is mainly because problems are more technology-bounded, unstable and ill-defined with the involvement of various integrated issues. Solving these problems requires interdisciplinary knowledge, collaboration skills, and innovative thinking, among other competences. In order to facilitate students with the competences expected in their future professions, educational institutions worldwide are implementing innovations and changes in many respects.

This book series includes a list of research projects that document innovation and change in education. The topics range from organizational change, curriculum design and innovation, and pedagogy development to the role of teaching staff in the change process, students' performance in the areas of not only academic scores, but also learning processes and skills development such as problem solving creativity, communication, and quality issues, among others. An inter- or cross-cultural perspective is studied in this book series that includes three layers. First, research contexts in these books include different countries/regions with various educational traditions, systems, and societal backgrounds in a global context. Second, the impact of professional and institutional cultures such as language, engineering, medicine and health, and teachers' education are also taken into consideration in these research projects. The third layer incorporates individual beliefs, perceptions, identity development and skills development in the learning processes, and inter-personal interaction and communication within the cultural contexts in the first two layers.

We strongly encourage you as an expert within this field to contribute with your research and help create an international awareness of this scientific subject.

For a list of other books in this series, visit www.riverpublishers.com

Toward Quality Assurance and Excellence in Higher Education

Ahmed Odeh Al Jaber

Secretary General of the Ministry of Public Sector Development
Jordan
Vice President of Al-Balqa Applied University
Jordan
Dean of Information Technology, University of Jordan
Jordan

Haifaa Omar Elayyan

Middle East University (MEU)
Jordan
Arab Open University
KSA

LONDON AND NEW YORK

Published 2018 by River Publishers
River Publishers
Alsbjergvej 10, 9260 Gistrup, Denmark
www.riverpublishers.com

Distributed exclusively by Routledge
4 Park Square, Milton Park, Abingdon, Oxon OX14 4RN
605 Third Avenue, New York, NY 10158

First published in paperback 2024

Toward Quality Assurance and Excellence in Higher Education / by Ahmed Odeh Al Jaber, Haifaa Omar Elayyan.

Routledge is an imprint of the Taylor & Francis Group, an informa business

Publisher's Note
The publisher has gone to great lengths to ensure the quality of this reprint but points out that some imperfections in the original copies may be apparent.

While every effort is made to provide dependable information, the publisher, authors, and editors cannot be held responsible for any errors or omissions.

ISBN: 978-87-93609-55-6 (hbk)
ISBN: 978-87-7004-396-0 (pbk)
ISBN: 978-1-003-33983-0 (ebk)

DOI: 10.1201/9781003339830

Contents

Preface   xi

Acknowledgment   xiii

List of Figures   xv

List of Tables   xix

List of Abbreviations   xxi

1   Concept of Higher Education, Quality Assurance, and E-learning   **1**
    1.1   Introduction   1
        1.1.1   Challenges of Higher Education   5
    1.2   Concept of Quality Assurance in Higher Education   6
        1.2.1   Definition of Quality and Quality Assurance   6
        1.2.2   Challenges of Adopting Quality Assurance   12
    1.3   Concept of E-learning in Higher Education   14
        1.3.1   Introduction   14
        1.3.2   E-learning Timeline   15
        1.3.3   Standards of E-learning   16
        1.3.4   Quality-oriented Concepts of E-learning Development Process   20
    1.4   Concept of Blended Learning in Terms of Quality Assurance   23
    1.5   E-government: The Missing Factor in Quality Assurance Concept   28
      References and Suggested Literature   33

2 The Role of Higher Education (Quality Assurance-based System) in Promoting Lifelong Learning   **39**

2.1   Introduction   39

2.2   The Role of Higher Education in Sustainable Development   43

2.3   Quality Assurance for Higher Education   47

     References and Suggested Literature   57

3 Quality-oriented Strategic Planning for Higher Education   **61**

3.1   Introduction   61

3.2   Definition of Strategic Planning   62

     3.2.1   Definition   62

3.3   Components of Strategic Planning   67

3.4   Integration of Strategic Planning and Quality Assurance . .   81

3.5   Strategic Planning and Assessments   86

     References and Suggested Literature   87

4 Quality Assurance Management and Control   **91**

4.1   Introduction   91

4.2   Quality Assurance in Higher Education   93

     4.2.1   Concept of Quality Assurance   93

     4.2.2   Mechanism of Quality Assurance in Higher Education Systems   99

     4.2.3   Quality Enhancement Cells   100

         4.2.3.1   Academic Standards   100

     4.2.4   Quality Assurance and Management: Extending the Framework   102

4.3   Total Quality Management and Higher Education   108

     4.3.1   Definition of Total Quality Management (TQM) . .   108

     4.3.2   Total Quality Management: Framework Fits Higher Education Institutions   110

     4.3.3   Advantages and Disadvantages of Definition of Total Quality Management (TQM) Strategies   114

4.4   Accreditation   116

4.5   Globalization and Its Impact on Higher Education's Quality Assurance System   121

     4.5.1   General Understanding   121

     4.5.2   Facts About Higher Education Globalization   121

     4.5.3   Quality Assurance and Globalization   123

4.5.4   How to Get to Globalization?   125
References and Suggested Literature   125

5   Quality Assurance for Universities   129
5.1   University Quality System   129
     5.1.1   Definitions   129
5.2   Quality Assurance General Framework for Universities . . .   131
     5.2.1   Standards and Guidelines   132
     5.2.2   Two Levels of Quality Assurance   135
     5.2.3   Evaluation and Assessment System for Quality
         Assessment   140
     5.2.4   University Assessments   141
     5.2.5   Universities and the Fitting for Globalization   144
5.3   Design a Manual for University Quality Assurance   145
     5.3.1   Quality at All Stages   145
     5.3.2   Main Keys in University Quality Assurance   148
References and Suggested Literature   150

6   Quality Assurance Piloting the E-Learning Sector   153
6.1   Introduction   153
6.2   E-Learning on the Cloud   156
     6.2.1   Cloud Computing   157
     6.2.2   Virtual Learning and Cloud Computing   160
6.3   The Future of Artificial Intelligence in e-Learning
     Systems   167
References and Suggested Literature   182

**7   Teaching–Learning Process and Quality Assurance in Higher
Education   191**
7.1   Overview   191
     7.1.1   Introduction   191
7.2   Quality of Teaching–Learning Process   192
     7.2.1   Quality of Teaching–Learning Process   192
     7.2.2   Quality Assurance System Is a Personal Touch
         When it Comes to Teaching–Learning Process . . .   196
     7.2.3   Quality Assurance Leads to Quality of Teaching–
         Learning Process   199
     7.2.4   Responsibilities for Quality Assurance
         in Teaching and Learning   203

7.3   Quality of Teaching–Learning Effectiveness and
Challenges 206
   7.3.1   Quality of Teaching–Learning as Indicator 206
   7.3.2   Quality Teaching and Research 208
7.4   How Quality Assurance Should Communicate about Quality
Teaching? 209
   References and Suggested Literature 211

8   Assessments in a Quality Assurance Suit   217
8.1   Introduction 217
8.2   Types of Assessments Defined 224
   8.2.1   Definition and Principles 224
   8.2.2   Educational Assessment Types 229
8.3   Designing a Quality-oriented Assessment 230
   8.3.1   Quality-oriented Assessment Framework 231
8.4   Importance and Impact 236
   References and Suggested Literature 246

9   Quality-oriented learning Resources   249
9.1   Introduction 249
9.2   Aligning Learning Resources with Teaching–learning
Process for Quality Assurance Purposes 251
   9.2.1   Quality Assurance Process for Learning Resource
Design and Managing 254
9.3   Quality Ideologies for Digital Learning Resources 262
   9.3.1   Pedagogy in Quality Framework 264
   9.3.2   Learning Principles to Guide Pedagogy and the
Design of Learning Environments and Resources . . 265
   9.3.3   Quality Procedures for Pedagogic and Design
Principles 267
9.4   Semantic Connection of Learning Objects 268
   References and Suggested Literature 273

**10 Crystalize the Bond between University Ranking Systems
and University Quality Assurance Systems   277**
10.1 Introduction 277
10.2 Ranking Systems 280
   10.2.1 Definition 280
   10.2.2 General Correlation Coefficient 282

10.3 Global Academic Ranking Systems 292
    10.3.1 Academic Ranking of World Universities (ARWU) 295
    10.3.2 QS World University Rankings 304
10.4 Positives, Negatives, and Impact of University Rankings . . 313
    10.4.1 Positives and Negatives of University Ranking . . . 313
    10.4.2 Impact of Internal and Global University Ranking
        Systems 323
    10.4.3 Impact of Global Rankings on Higher Education
        Research and the Production of Knowledge 325
        10.4.3.1 How Rankings Impact Research 328
        10.4.3.2 Institutional Responses to Rankings . . . 328
    10.4.4 Impact on Graduates 331
        10.4.4.1 QS Graduate Employability Rankings
            2018 Methodology 332
    10.4.5 Proposing a New Impact: Impact on University
        Selection for Higher Education Affordable
        Scholarships — The Passive Motivator 334
10.5 Data-mine-based Data Pool of Students' Learning
    Experience on Social Network for Ranking Purposes 338
    10.5.1 Universities Create their Own Social Networks
        for Students 338
10.6 Quality-oriented Crystallization Process of University
    Ranking Systems 340
    10.6.1 How Ranking Helps Quality Assurance
        and Vice Versa 345
    References and Suggested Literature 346

11 Toward Excellence of Higher Education   **351**
11.1 Introduction 351
11.2 Excellence in Higher Education 352
    11.2.1 Introduction 352
        11.2.1.1 Definitions 353
    11.2.2 Characteristics of Excellence for Professional
        Practice in Higher Education 359
        11.2.2.1 General Knowledge and Skills 360
        11.2.2.2 Interactive Competencies 361
        11.2.2.3 Self-mastery 362
    11.2.3 Quality Assurance and Excellence of
        Higher Education 363

       11.2.4 Accreditation and Excellence of Higher Education . 367
  11.3 The Never Ending Journey in Pursuit of Excellence
        of Higher Education 368
        11.3.1 Journey Just Started 368
        11.3.2 Excellence in Teaching 372
        11.3.3 Excellence in Student Performance 372
        11.3.4 Excellence in Research 373
        11.3.5 International Peer-review Method 373
        References and Suggested Literature 375

12 **Quality Assurance for Higher Education in Japan**   **377**
  12.1 Quality Assurance in Japan Higher Education 377
        12.1.1 Quality-oriented Education System in Higher
                Education of Japan 377
                12.1.1.1 System in order to realize quality
                          assurance 381
        12.1.2 The Future of Higher Education in Japan 390
        12.1.3 Challenges of Higher Education in Japan 391
  12.2 Japanese Universities and Ranking System:
        University of Tokyo 394
        12.2.1 About the University of Tokyo 394
        12.2.2 Explore Rankings' Data for University of Tokyo . . 394
        References and Suggested Literature 395

13 **Quality Assurance for Higher Education in Finland**   **397**
  13.1 The European Education System 397
  13.2 Quality Assurance in Finnish Higher Education 402
        13.2.1 Education System 403
        13.2.2 The Quality Assurance of Finnish Higher Education 409
        13.2.3 Strategic planning in Finnish Higher Education . . . 411
        13.2.4 Internationalization of Finnish Higher Education . . 411
  13.3 Examples of Finnish Universities: University of Helsinki . . 413
        13.3.1 University of Helsinki 413
        13.3.2 University of Helsinki Rankings 414
        References and Suggested Literature 415

Index   **419**

About the Authors   **425**

Preface

Recently, several academic calls have been sprouted for changes in teaching at the educational levels reflecting, reelecting and exploiting the ongoing growth in our life in general and education process in specific. Taken into our consideration the unlimited adopting of technologies and being more satisfied with their involvedness and complexity. The adopting of technology, submitting and being influenced by marketplace extreme requirements sometimes unwillingly drive us to step down and roll back our standards denning and impacting the quality in our education system making excuses of quality being inflexible that restricts our open-mind motivation toward new teaching-learning technologies. These justifications have molded the enthusiasm to be unrestricted trying to re-form the higher education system without an acceptable infrastructure of the education system ignoring how this can shape the future of the outcomes.

Despite of the fact that; Quality Assurance not being a new concept in the education sector in general, and higher education in particular, though it is becoming increasingly more relevant and important. And despite the fact as well that there is no simple definition of the concept of quality in education, though numerous models and theories have been devised. The higher education have been always contributing and helping to improve an individual's quality of life by enabling them to inflate their knowledge and expertise, to grasp abstract concepts and theories, and to raise their awareness of the world and their community, and as such the assurance of quality is becoming more pivotal in the whole education process.

"Toward Quality Assurance and Excellence of Higher Education" book is inscribed with a new-fangled vision to improve and expand the Higher education theoretical and practical progression using a plenty of quality assurance oriented methods and procedures blended in a combination of original strategies that are expected to alternate old plans and promote the intellectual for a better teaching-learning environment. Toward Quality Assurance and Excellence of Higher Education is a new episode of the Quality Assurance perception in higher education, which identifies the quality culture and orientation from the beginning, integrating crucial factors to build a "pyramid" of higher education excellence. The book compares concepts from the main theories of Quality Assurance, management and control when they are applied to educational systems in higher education. The book also presents a new model of excellence in higher education. Excellence is an architecture of building blocks that includes process performance, effectiveness, harmony and collaboration, and these bocks should be incorporated in a quality-oriented concept for sustainable excellence of higher education. The model integrates four main facets: the Educational System, Quality Assurance Managing and Control, Strategic Planning and Globalization. Also presented are international "best-practices" in quality assurance in higher education, from Japan and Finland.

Authors confirm that there is no single way leading to adopt the quality in our education, but there is always a place and challenge for presenting a best way to orient the educational process on the ultimate quality we can form and perform. Our book presents a new challenge on how the integration of many essential factors and available requirements can easily grasp efforts toward the excellence in education progress in Higher education Imitations.

Sincerely;
Prof. Dr. Ahmed Odeh Al Jaber
Haifaa Omar Elayyan

Acknowledgment

It is our pleasure to contribute to the Academic field with a book such as "Toward Excellence and Quality Assurance" for higher education institution. As we really hope it will be a remarkable contribution and will benefit all who are interested in this field. We would acknowledge the support from our families to afford what writing needed and the outstanding help that we could not do it without their dedication as well.

We also would like to extend our gratitudes and deep appreciation to everyone who have dedicated time, material and experience to form this book as well, everyone who has added value information by their available academic researches and proposals for better academic environment. We would like also to thank all institutions, universities and higher education systems mentioned in our references whom helped to raise the awareness of a better form of higher education systems.

Our grateful thanks also to our publishers and editors who helped in our book revision.

Special thanks for the following institutions:

UNESCO
EQNA
UNICEF
THE ranking system
QS ranking system
ARWU ranking system
E4 group
ESIB
NIAD-UE
EHEA
Ministry Of education and culture, Finland
MEXT, Japan
NCEE, Japan
The University of Tokyo, Japan
University of Helsinki, Finland

List of Figures

Figure 1.1   Age at which cumulative earnings of college graduates exceed those of high-school graduates, by degree and college cost. 3

Figure 1.2   Concept of blended learning. 23

Figure 1.3   Concepts related to the quality of learning at university. 28

Figure 1.4   Chart of U.S. households without the Internet in rural areas, 2010. 30

Figure 1.5   Percentage of U.S. population with Internet access, 2015. 31

Figure 1.6   Internet usage statistics, the Internet big picture world Internet users, and 2017 population stats. . . 32

Figure 1.7   Conceptual model of e-Government role in quality assurance concept. 32

Figure 2.1   Categories of "Quality." 49

Figure 2.2   Quality system. 56

Figure 3.1   Types of possible responses to sustainability in strategic plans. 66

Figure 3.2   Components of a strategic plan. 68

Figure 3.3   Sample of a charge letter. 74

Figure 3.4   SWOT (strengths, weakness, opportunities, and threats). 79

Figure 3.5   Quality map of a higher education institution. . . . 83

Figure 3.6   Quality cycle of continuous improvement in higher education. 84

Figure 3.7   "The cyclical process and when the process should include reflection on how it worked and what changes might make it better". 86

Figure 4.1   Conceptual model of quality-oriented definition for higher education institutions. 96

Figure 4.2   Bilateral relations between educational institution
             elements and society. 106
Figure 4.3   Model of total quality management. 114
Figure 5.1   P–D–C–A Mechanism. 131
Figure 5.2   Quality assurance developing system – University
             vision. 138
Figure 5.3   Assessment for excellence – feedback process. . . . 141
Figure 5.4   Assessment for accountability. 142
Figure 5.5   University of Eastern Finland Main Quality Manual. 146
Figure 6.1   Cloud computing and big data. 158
Figure 6.2   Cloud computing compared to traditional. 159
Figure 6.3   Architecture of a simplified learning system. 161
Figure 6.4   Modified cloud based e-learning system architecture. 163
Figure 6.5   Detailed e-learning cloud architecture. 164
Figure 6.6   Connectivity scenario of the institutions
             in the proposed architecture. 165
Figure 6.7   Interactive mode of the proposed architecture. . . . 166
Figure 6.8   Potential system architecture of the personalized
             e-learning system with ontology-based
             concept maps. 177
Figure 6.9   Quality-oriented artificial intelligence unit. 182
Figure 7.1   Quality assurance system to build and control the
             quality of teaching–learning and teacher quality. . . 200
Figure 7.2   Quality of teaching–learning measures quality and
             quantity of teaching that have been transferred to
             the learner and vice versa in terms of feedback and
             satisfaction. 200
Figure 7.3   Students' orientation, teaching method, and level of
             engagement. 202
Figure 7.4   Relationship between indicators and teaching–
             learning quality. 207
Figure 8.1   Assessment conceptual model. 232
Figure 8.2   Charts of student's result with no-coordination
             system. 239
Figure 8.3   A power of feedback to enhance learning. 243
Figure 10.1   A Spearman correlation of 1 results. 285
Figure 10.2   Elliptically distributed data. 285
Figure 10.3   Correlation is less sensitive. 286

Figure 10.4 A positive Spearman correlation coefficient corresponds to an increasing monotonic trend between X and Y 287

Figure 10.5 A negative Spearman correlation coefficient corresponds to a decreasing monotonic trend between X and Y 287

Figure 10.6 Chart of the data presented. It can be seen that there might be a negative correlation, but that the relationship does not appear definitive. 289

Figure 10.7 The ARWU first appeared in 2003. Although there have been some changes in the criteria used to rank universities, the ranking methodology has been fairly consistent since 2005. 297

Figure 10.8 Example of university ranking changing over years. 306

Figure 10.9 Importance of ranking. 314

Figure 10.10 QS Graduate Employment ranking. 335

Figure 10.11 Scholarship and partnerships passive and influential impact. 337

Figure 11.1 Fundamental concepts of excellence. 356

Figure 11.2 EFQM Excellence Model®, The EFQM Excellence Model is a registered trademark ©Sheffield Hallam University. 357

Figure 11.3 Baldrige framework. 358

Figure 11.4 Excellence in Higher Education: Framework, Categories, and Point Structure. 359

Figure 11.5 The evolution of quality and excellence. 364

Figure 11.6 Building blocks of excellence of higher education pyramid. 371

Figure 12.1 Organization of universities in Japan. 378

Figure 12.2 Illustration of quality assurance framework in Japan education system. 381

Figure 12.3 World university ranking. 392

Figure 12.4 Research criteria. 392

Figure 12.5 History of University of Tokyo ranking. 392

Figure 12.6 Other criteria of ranking. 393

Figure 12.7 Adjacently ranked institutions against international students. 394

Figure 13.1 Specificities of the Finnish education system. . . . 405

Figure 13.2   FINEEC develops, experiments, reforms, involves, and serves. 412

Figure 13.3   Performance in academic ranking of world universities. 414

Figure 13.4   Performance in academic ranking of world universities by broad subject fields. 414

Figure 13.5   Performance in academic ranking of world universities by subject fields. 415

List of Tables

Table 2.1   A compilation of quality assurance definitions   51
Table 4.1   Categories of quality indicators   95
Table 4.2   ISO standards   104
Table 8.1   AFL study 10 principles   228
Table 10.1   Types of rankings   279
Table 10.2   IQ per TV watching   288
Table 10.3   IQ per TV watching and correlation coefficients. . .   288
Table 10.4   ARWU indicators   297
Table 10.5   ARWU subject area   298
Table 10.6   Indicators for ARWU subject ranking   298
Table 10.7   QS ranking indicators   307
Table 10.8   QS ranking indicators' data   307
Table 10.9   Mapping institutions actions against rankings.   330
Table 10.10 Summary of data collection from academic SNS . .   339
Table 10.11 Inventory of Likert scale items and the theme each
             draws upon   341
Table 11.1   Elements that constitute the definition of excellence
             in higher education   354

List of Abbreviations

AFL	Assessment for learning
AI	Artificial intelligence
APC	Academic Programs Committee
ARWU	Academic Ranking of World Universities
DLR	Digital learning resources
E4 group	EQNA; EUA; EURASHE; AND ESU
EAS	Established approval system
e-Content	Electronic content
e-Course	Electronic course
e-design	Electronic design
e-Government	Electronic government
EHEA	European higher education area
e-Learning	Electronic learning
EQAR	European quality assurance register for higher education
EQNA	European Association for Quality Assurance in Higher Education
ESG	Standards and guidelines for quality assurance in European higher education
ESIB	European students union (national unions of students)
e-technology	Electronic technology
EUA	European Universities Association
EURASHE	European association of institutions in higher education
HE	Higher education
HEIs	Higher education institutions
IA	Intelligent Agents
ICT	Information communication technology
ISO	International Organization for Standardization
ITA	Intelligent Tutoring Applet

ITS	Intelligent tutoring system
ITTs	Intelligent Tutoring Tools
LMS	Learning Management Systems
LQM	learning object metadata standards
MEXT	Ministry of education, culture, sport, science and technology
MI	Machine intelligence
NI	Natural intelligence
NIAD-UE	National institution for academic degrees and university evaluation
OECD	Organization for Economic Co-operation and Development
OOLR	Quality oriented learning resources
PDCA	Plan-do-check-action
PEST	Political, economic, socio-cultural and technological
QA	Quality Assurance
QAA	Quality Assurance Agency
QAC	Quality Assurance Control
QAC	Quality Assurance Council
QAC	Quality assurance committee
QAM	Quality Assurance management
QAS	Quality Assurance System
QS	Quacquarelli Symonds ranking system
SEU	Standard for establishing universities
SPC	Selection Procedures Committee
SWOT	Strengths, weakness, opportunities and threats
THE	Times Higher Education ranking system
TLDC	Teaching and Learning Development Committee
TLQAC	Teaching and Learning Quality Assurance Committee
TQM	Total quality management
UNESCO	The United Nations Educational, Scientific and cultural Organization
Wi-Fi	Wireless networking, Wireless fidelity

1

Concept of Higher Education, Quality Assurance, and E-learning

1.1 Introduction

Education is a vital sector attracting millions of people at different ages and interests, employing thousands of teachers and engaging millions of students in their life-long learning. It is a fundamental human right that promotes individual's life and ambitions. Education is a crucial tool to survive and succeed in contemporary world. The knowledge being attained by the education systems, helps to mitigate most of the life challenges, and helps to open doors to more opportunities for better prospective life and career. It is actually a life support for a promising present and future. Educational system is a term to refer to any academic institution that affords learning process and allows learners to attain education in different fields and sciences and accumulate practices and experiences with diverse learning models and methodologies, which means that these education systems offer education policies and services. This diversity generates a competent environment among these academic institutions to perform the best. Internationally, The United Nations Educational, Scientific, and Cultural Organization (UNESCO) is a specialized agency that endorses education as an essential human right and contributes in education systems to promote a global participation and collaboration to assure the right for every person to be educated. UNESCO's mission is expanded to progress the quality of education and to enable a common dialog for all educational systems.

Every country, by now, established a hierarchical education structure with a sequential order to allow citizens to practice their right for education at any life stage. The education structure takes a pyramid shape and requires the climbing to be a level by level which means that learners should accomplish first level to be promoted to the next level. Each level is expected to promote the learning and increase the attained knowledge. A climb of this pyramid is a significant achievement and has become a compulsory form for individual attainment, social suppleness, and commercial success.

The environments of individual success are determined by one's ability to sustain in the learning pyramid, given that the size of the pyramid at the top is small, and usually called the higher education. Higher education occupies the top position in the educational hierarchy. By quality of its location in the hierarchy, it has a responsibility and makes an involvement and impact to the lower levels of learning. This influence can be either direct and perceptible or indirect and invisible. Likewise, the lower levels of education are spontaneously accompanying to the higher education segment and standards.

Higher education is an education, a training, and a research guidance that takes place after the postsecondary level. It comes as a final stage to formalize learning process and get the higher outcome. The structure of higher education is a multi-layered organized in a different way from other education systems at the prime levels. The higher education subdivision in most of the countries consists of a university and a non-university institution. Universities are academic institutions with the ability to grant academic degrees. The universities can be private or public in terms of tenure and sources of sponsoring. The non-university institutions are technical institutions that present courses in professional subject areas, training colleges, and specialized institutions proposing programs of study and carrying out inquiries and academic research, as well as institutions involved in policy and development support to the learning sector.

The role and importance of higher education in individual's life go beyond affording education. It is actually a golden ticket to a better life. Studies show that, linked to high-school alumni, college graduates have longer life extents, better access to health care, better nutritional and health performs, greater financial constancy and security, more prominent occupation, and greater job fulfillment. Furthermore, they are less reliance on government backing and greater acquaintance of supervision. Higher education, theoretically, will also enable individuals to increase their awareness and develop their skills. It helps to promote their understanding and get involved in dialogs that require intellectual backgrounds. Higher education endorses person's constructive contribution to society and encourages self-development progress. For example, educated people are more likely to involve and contribute in volunteering, they are able to elect rationally, and more have motivation to donate blood when a blood donation campaign is launched, for instance. The advantage of high level of education can be credentials for next generations as well. Children of educated people are recited to more habitually and more likely to perform much better than other children which in turn impact the social outflows positively. Obviously, and as time goes on, the relationship between the higher education and the success will be with no doubt.

Higher education presents more advantages to individual students and their relatives. In marketspace, typical wages increase significantly with higher levels of education. Over their occupied lives, standard college ex-students receive much more than classic high-school graduates and those with progressive degrees gross two to three times as much as high-school graduates. More educated people are less likely to be without a job and less likely to live in scarcity. These economic profits make sponsoring a college education a profitable project. The following chart presents indicators and advantages we have been detailing. Figure 1.1 is itemizing earnings premium relative to price of education.

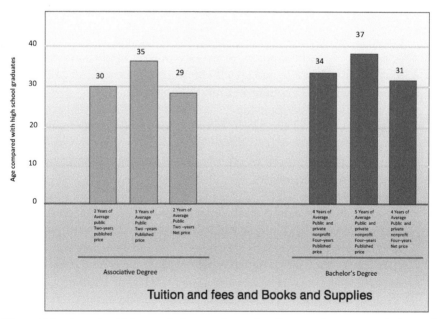

Figure 1.1 Age at which cumulative earnings of college graduates exceed those of high-school graduates, by degree and college cost.

Source: [http://www.unesco.org/new/en/education/themes/leading-the-international-agenda/education-for-all/resources/statistics/]

In most cases, when we talk about the benefits of having a higher level of education, we refer to the higher education proposal for long-term and sustainable strategies with commitment, reliability, and accountability features in order to promote learning process and endorse quality culture in each layer of higher education pyramid. Strategies and quality-oriented plans come with an evaluation system to check and validate the main goal of quality, "fit to purpose." The purpose of the "evaluation system" is to manage and maintain the education pyramid and control any violations. Diverse of purposes and benefits from attaining to higher education are well known and not so well known to young learners. Some are detailed but with the intention of embedding the concept and benefit of having a quality-oriented higher education system.

1. **Career Groundwork**

   Learners have goals and objectives to achieve while learning. But most of the times, learners, especially young, are not sure of these goals and objectives as their belief changes and their experience expands with each knowledge they learn. Higher education plays the role of a revenue for skills' acquisition, training, and knowledge and its academic frames offer one of the best places to explore others' options and make that high-quality choice to end up with prestige and long career life term.

2. **Comprehensive Practical Reimbursements**

   People who pursue their higher education studies are expected to be superior in terms of life practices and to society. They are more capable of taming their behavior and training brain to be out-of-box. They are also able to run processes with required intellectual background and quality-oriented awareness.

3. **Personal Development**

   Personal development is a challenge of higher education and a benefit at the same time. Promoting individual's life to the level improvement and enhancement on quality basis is not a simple task. But it comes with excellence in self-development. Quality cycle in each education system impacts learner's performance and promotes their learning process to achieve their goals. An effective higher education environment can demonstrate to be the most important experience to any one's life. The following are examples of how quality-oriented cycles of improvement and enactment in higher education systems can make difference and add values to the learner:

a. Better communication (written and verbal): Learners get a chance to enhance their communication skills. The cycles of progressive writing and speaking assignments train students to express themselves clearly and interconnect more effectively with others.

b. Critical thinking skills: Learners will develop the capability of thinking and acquiring and then being able to analyze and act rationally.

c. Identification of skills: Learners will explore skills that were not known to have before being exposed to new things and new ideas in a higher education environment.

d. Passions' realization: Practicing an educational setting can dabble with diverse disciplines to bring up passions to reality.

e. Greater sense of self-control: Learning is not an easy task as most think. It requires lots of commitment and accountability. Learners will be going through life change and transform; therefore, they will be trained for self-control and managing an unexpected situation. They will be experiencing different settings that sometimes necessitate that students do lots of learning by themselves, for example. This is a real self-development process and these skills can easily transfer to all other areas of life.

f. Intellect of success: The choice to pursue higher education and go above and beyond others gives the sense of achievement and develops confidence to pursue whatever ambition in life.

4. Pursuing a Passion

Pursuing higher education is in fact pursuing a passion. Pursuing a passion means that students select the strategy that drives to achieve their goals and objectives. Learners will learn the right framework to plan for their goals and how they put actions to attain them.

1.1.1 Challenges of Higher Education

Simplicity is not a found in higher education systems, and seeking the perfection always comes with a suit of challenges. In recent years, considerable attention has been paid to determine these challenges and try to reduce the impacts. Around 20 challenges are mentioned in previous literature and some will be listed below:

1. Curriculum alignment and design
2. Students' employability

3. Widening participation
4. Quality of teaching – learning process
5. Effective academic research
6. Accreditation polices and standards
7. Global collaboration
8. Students' involvement and retention
9. Emergency of technology transfer
10. Evaluations and assessments
11. Quality of teachers and quality of teaching
12. Funding
13. Self-learning and self-development terminologies
14. Impact on society
15. Quality management and development.

Most of these challenges will be discussed later through the book's dedicated chapters. It is worthy to end this section with the agreement that higher education institutions are under the stress of rapid change and the need to respond to environment's needs. On the other hand, higher education is supposed to act on reputation and prestige strong foundation. Balancing in between is neither a simple task nor a check list to meet. Excellence is a repetitive cycle's dedication and recognition of challenges that can impact the higher education main role which is affording education systems with the ultimate quality presented. The next section is a brief discussion on "quality" in higher education.

1.2 Concept of Quality Assurance in Higher Education

1.2.1 Definition of Quality and Quality Assurance

The role of higher education as a crucial factor of the productive economy has been intensely identified and recognized in the 21st century, resulting in a continuous need of higher education institutions to demonstrate quality, standards, and excellence to stakeholders and investors. To achieve this, quality-oriented strategies and action procedures have been endorsed, emerged, and then kept in a cycle of evaluation and development. Therefore, the concept of quality assurance has been presented and implemented by many systematic definitions to guarantee the quality assurance system within the higher education pyramid. The quality system has also been promoted and determined by special interventions that were designed to monitor and follow the quality standards for higher education. The need for quality has proven to be the decisive factor in determining the success or failure of many services in society and most important in higher education.

The definition of quality concept in educational systems, has varied from being simple and straightforward as "Fit for Purpose" to other terms as quality, equity, effectiveness, and efficiency which have been synchronously used not only in academic field but also in all working fields that need standards (Adam, 1993), to the most complicated as a nested integrated cycles of procedures and policies that perform to assure a quality-oriented building blocks endorsed in each system, process, program and communication occur in an educational system. The urgent need to emerge the concept of excellence and practices into higher education systems was necessitated to conceptually distill previous notions of "quality." Education systems consider quality and quality culture as elusive to define in fixed terms; therefore, universities used to take quality with multiple meanings depending on the context within which it is discussed (Annis et al., 2014; Dagles, 2008). According to Kontogeorgos (2012), "quality" refers to a set of features that offer a complete representation of a person, a system, a process, or a product. In correspond to the same context, "quality" is a multi-dimensional concept that takes different models and definitions according to the desired prospective feature or service (Zavlanos, 2003). The multi-dimensional concept does not promise supremacy or excellence; it promotes cycles of methods to measure and evaluate the suitability of strategy process rather than the perfection of the strategy.

Defining "quality" necessitates to define a strategy to foster the quality definition, and in turn defining a strategy is worthless without developing procedures and action processes to implement this strategy properly with full satisfaction of the main goal "Fitness for Purpose." In general, there is an agreement that the definition of "quality" should be flexible and open for change based on the nature of new challenges in education. Quality has to allow for complexity

> *"Where there are differentiated systems and diverse higher education institutions with different traditions and identities, it is necessary to allow for different interpretations of quality and different responses to standards..."* (Jose' Diad Sobrinho, 2008).

The importance of quality is not that hard as defying it. Higher education systems all over the word have agreed on the necessity of "qualifying" educational system and presented this integrity in different methods. Some of these experiments are remarkable experiments that have been adopted in other countries as they achieved the ultimate. Higher education systems have been reconstructed for quality assurance purposes. According to Harvey (2008), these quality purposes pushed the higher education to adopt and achieve the goal of the following definition as:

1. Distinction which means ultimate exclusivity and excellence in higher education systems.
2. Perfection and integrity away from any defect or impacts to help rending successfully with the first trial.
3. Fitness for purposes which is the main mission of integrating quality into any academic system. It assesses whether the academic system is able to achieve its mission or not.
4. Adapting with the idea of being an investment, or a value for money. The quality role here is assessing if the net outcomes deserve the efforts and expenses.
5. Worth the transformation, which means that the academic systems have added values and positive contributions to students' performance and society.

There have been many other definitions and goals of integrating quality concept into higher education systems, many as the number of interested people in quality and who came up with new definitions, strategies and concept of integration (Conseil Supérieur de l'Education de Québec [CSE], 2012). As a result, we can all agree that university education quality assurance is polysemous with multi-occurrences. And quality strategies will be polysomic as well.

> Other main factor is who is going to evaluate this quality? *"Quality is seen differently by different people. It is not the job of the evaluator to find a consensus but to weight the evidence, make judgments, and report the different ways that merits and shortcomings are seen. Observations and interpretations that do not agree necessary indicate a failing of evaluation but perhaps the complexity of the program and it contexts. It is problematic to assume that there is a simpler world behind the world that people see."* *(Stake, 2004, p. 286)*

Harvey and Green (1993) talked about quality concept and cleared that it was accompanied to excellence in the 1980s and early of the 1990s; Kilowski (2006) confirmed. Over years in association with Internet revolution, concepts of quality have evolved to be determined by adapting to "Fitness for Purpose." This transformation was also stated by Trembly and

Kish (2008) in the Organization for Economic Co-operation and Development (OECD) report. The transformation continued toward complexity and using it for accreditation purposes. Accreditation is a system being used to certify any educational system in terms of meeting the higher education's regulations, rules, and standards. Quality assurance procedures are cycles integrated inside the educational system to measure the strategic and action procedures for evaluation purposes that can lead latter to accreditation level, which is crucial for any university. Then concept of quality and quality assurance is collectively accredited factor in successful commerce taking into consideration that one purpose is investing.

Endearing universities are the ones that could endorse the excellency of quality criteria using unusual terminologies and methodologies, which apparently helped to define, deploy, and develop inner cycle's quality assurance in order to enhance the associated educational system and consequently meet the requirement of higher education. Globalization thinking pushes toward an accumulative demand for a quality-oriented building block inside the pyramid of higher education. Globalization is a golden ticket for international partnerships and students' exchange programs; therefore, quality is actually an international demand for higher education and stakeholders who invest to achieve this high level of "quality orientation" leading to a stable investment indeed.

It is very challenging to attach "quality-orientation" educational system, quality-based strategic plans, quality-based evaluations' cycles, and globalization purposes in higher education systems. This book is written to present this compilation that inspects how numerous quality assessments in higher education and how we can combine the combination with an evocative and synergistic quality custom. Integration is occurring horizontally as well as vertically ranging from external evaluations to self-assessment aiming to meet the excellence at the top. Stakeholders, who are students, staff, faculty, and administrative staff, are part of this integration to attain an advance harmony of work, learning, and investing. The pervasive concept quality-based building blocks of the excellence pyramid based on this integration can be seen as a linking factor in integration, distinction, and an excellence award model that presented higher education.

Wolf (2007) in the article "Evaluating and Enhancing Outcomes Assessment Quality in Higher Education Programs" wrote that achieving academic system accreditation is a "mark of distinction" if the academic institution met the high standards set by the higher education. Wolf's (2007) writing was mainly to focus on the important feature of the accreditation process which is "outcomes' assessment," a cycle of quality-oriented evaluation procedure is each academic institution longing for accreditation.

Seven years later, Doley (2014) inspected and investigated the "Role of Assessment in Improving Quality in Higher Education." The study presented and compared different innovative assessment methods aimed to promote and improve the quality learning, teaching, and required skills for attaining and teaching substantive knowledge. The finding of the study implied the shortage of strategic plans designed and enhanced for assessment procedures, and therefore, the study recommends that educators and instructors should plan the assessment procedure methodically for the development based on the requirements of quality standards endorsed by higher education.

The Bologna Process Stocktaking Report (2009) further confirms such observations:

> *"The national reports demonstrate that Higher Education Institutions (HEIs) in most countries are aggressively working to found coherent interior QA systems and bring into line with the external assessment procedures. A number of countries state that they do not prescribe particular mechanisms for internal quality assurance in HEIs but rather require that HEIs create them as they see fitting, on condition that the interior QA of each HEI is intelligible, operative and fits its purposes. Some countries use ISO, Total Quality Management or EFQM excellence model methodologies for internal quality assurance in HEIs." (Bologna Process Stocktaking Report, 2009)*

Quality assurance system is actually a supervised cycle with tactics plans to ensure that developing and enhancing are going on a continuous basis. At the Conference of the European Ministers Responsible for Higher Education, in Leuven and Louvain-la-Neuve, 28–29 April, 2009, the deduction which mentions to the development of quality assurance was:

> *"Higher education is being modernized with the adoption of the European Standards and Guidelines for quality assurance. We have also seen the creation of a European register for quality assurance agencies and the establishment of national qualifications frameworks linked to the overarching European Higher Education Area Framework, based on learning outcomes and workload." (Leuven and Louvain-la-Neuve Communiqué, 2009, p. 2)*

This drives us to inquire how the institutions can endorse the quality assurance concepts and work toward achieving quality goals. Higher education plays a great role in guiding more than restricting and punishing, so different and elastic designed tactics have been recommended and released by higher education to end up with the quality assurance development. The following argues some of these tactics.

- **Policy and procedures for quality assurance:**
  The academic institution is expected to develop an education policy allied with evaluations and measures that declare "quality" standards. Measurements are to help pledging the system explicitly to the development of a quality culture and quality assurance system. The approach, policy, and procedures should have a prescribed status and be publicly obtainable. They should also take account the involvements of students and other stakeholders.

- **Approval, monitoring, and periodic review of programs and awards:**
  The academic institutions should have official instruments for the approval, periodical review, and monitoring of their programs and rewards.

- **Assessment of students:**
  Students' evaluation should be executed with pre-published criteria, pre-posted regulations, and procedures which are applied consistently. This raise the awareness of students on which basis their performance will be evaluated, which is considered fair and attaining the goal of improvement more than grading tasks.

- **Quality assurance of teaching staff:**
  The academic institution should be keen enough to quality assuring its quality of teaching and learning processes. This should involve attaining the quality at the level of staff, students, and administrative as well. The academic system is expected to develop ways of recruiting teaching staff based on relevance experiences. Teaching experts help to promote the culture of quality within the system and contribute constructively to the quality cycle for enhancing purposes.

- **Learning resources and student support:**
  The academic institution should afford internal learning resources and make them available for teachers and students. These learning resources are sustenance of student learning and expected to be suitable and appropriate for each program offered.

- **Information systems:**
  The academic institution should guarantee that they collect, analyze, and use pertinent information for the operative management of their study programs and other accomplishments.
- **Public information:**
  The academic institutions should regularly distribute impartial, neutral, and up-to-date material and information, both quantitative and qualitative, on the programs and awards they are subscripting.

Excellence model of higher education means a continuous cycle of this excellency. It should persist as a key objective of higher education institutions to mature and conserve the concept of this excellency. Achieving this, most universities have established interior quality assurance measures, but nevertheless some are much up to quality and more restrict than others. While systems for countersignature of programs and qualifications are well developed, it is clear that linking programs with learning outcomes and designing assessment measures to quantize achievement of the envisioned learning outcomes are the most problematic chunks and will take lengthier to implement. Also student contribution in quality assurance is not taking seriously when students contribute in reviews only as viewers, they are not always allowed to elaborate in concocting self-assessment reports, and they are very rarely involved in follow-up measures (Bologna Process Stocktaking Report, 2009, p. 14). Interior quality assurance is not the only concern for higher education institutions but also the internationalization dream has widened the quality assurance concepts and standards. More challenges can be discussed next.

1.2.2 Challenges of Adopting Quality Assurance

Defining and adapting to "quality" in educational systems are with obstacles and challenges that most of times add further burdens to higher education institutions. Higher education cannot run independently from frequent international policies which up to some level are directing the future application of quality assurance process across academic institutions. The adoption of the framework law on HE is the most challenging. Add to this, also enhancing capacities for self-evaluation in the academic institution is still a critical point to approve, adopt, and apply.

Guidance for concrete further enhancement of quality assurance in higher education should be defined and well formed to identify the aims and recommend the actions required to implement internal quality assurance systems along with internal approval of programs. Publishing results of linking programs with learning outcomes and designing assessment procedures to measure the achievement of the intended learning outcomes is also a challenge for the academic institution as they test how far the institution is authentic and rational in measuring and evaluating its own quality system.

Establishing comprehensible internal quality assurance systems to cope the external assessment' s procedure, curricular review, modification, and modernization should also cope in line with upcoming national assessments and accreditation of programs. Internal quality assurance should comfort to further describing clear and concert indicators, to support the internal evaluation and quality control processes as always recommended by Higher education On other hand, Higher education is also facing obstacles of having any "quality-oriented" potential action plans that can help to identify the gabs and shortening of quality orientation in any educational system. University is also responsible as being a partner in the education procedure and required to develop processes of self-assessment, external review, and publishing of assessment results and follow-up measures of international peer review of programs. Universities are also required to mature the integrity and consistency of enhancing cooperation and creating conditions for collaboration between the university and higher education in terms of "quality" and quality assurance system.

We believe here that the accreditation itself is the most trending challenge for quality assurance system. Accreditation is the process used in higher education systems to guarantee that any of higher education institutions and other education providers meet, and maintain, minimum principles of quality and integrity concerning academics, administration, and related services. The idea behind considering accreditation as a challenge, because it is an intentional process based on the principle of academic self-governance. That means institution system and lineups (faculties) within institutions are asked to participate in accreditation. The entities which conduct accreditation are associations encompassed of institutions and academic experts in specific subjects, who create and enforce standards of membership and measures for conducting the accreditation process. This is not a comfortable task for all. Usually, any government recognizes accreditation as the mechanism by which institutional and programmatic legitimacy are ensured. In worldwide terms, accreditation by a recognized accrediting authority is recognized as the equivalent of other countries' ministerial recognition of institutions belonging to national edification systems.

The last challenge we can mention here is the "ambition of internal and external recognition." Being a crucial for any institution seeking the internal and external recognition, higher education is responsible to provide problem solving for improved measurement practice, covers aspects of related education and training, and fosters cooperation between national and international bodies operating in accreditation and quality assurance of measurements in different educational systems.

Higher education has to extend its flexibility to cope with technology transfer and for the exchange of views, especially on contentious topics. Any discussion in terms of accreditation should focus on accreditation–certification–quality assurance–tools for quality control–quality management–measurement uncertainty–comparability and equivalence of measurement results–verification and validation, and many others. These topics are clear to be related to quality-oriented concepts.

1.3 Concept of E-learning in Higher Education

1.3.1 Introduction

Long before the Internet was propelled, distance courses were introduced to provide students with learning on particular topics or expertise. In the 1840s, Isaac Pitman educated his pupil's shorthand via written letters with symbols. This system of symbolic writing was intended to improve writing speed and was popular among secretaries, journalists, and other individuals who did a great work of note taking or writing. Pitman, who was a qualified teacher, used to send completed assignments by his students via the mail system and he would then send them more work to be completed. In 1924, the first testing machine was invented. This device allowed students to test themselves. Then, in 1954, BF Skinner, a Harvard Professor, invented the "teaching machine," which empowered schools to control programmed instructions to their students. It was not until 1960, when the first computer-based training program was presented to the world. This computer-based training program (or CBT program) was known as PLATO – Programmed Logic for Automated Teaching Operations. It was originally designed for students attending the University of Illinois, but ended up being used in schools throughout the area.

The term "e-learning" has only been in existence since 1999, when the word was first formed at a CBT system's seminar. Other words also began to spring up in search of an accurate description such as "online learning" and "virtual learning." However, the philosophies behind e-learning have been well predictable throughout history, and there is even evidence which suggests that early forms of e-learning existed as far back as the 19th century.

1.3.2 E-learning Timeline

- *Pre-1920s:* The shift from phrenology to an empirical knowledge base for education based on Thorndike's laws of learning and the introduction of educational measurement.
- *1920s:* Matching of society needs to education and connecting outcomes and instruction.
- *1930s:* Great depression affected education in terms of funding and other respects; the 8-year-study plan (Tyler) was a major milestone in specifying general objectives for education and behavioral objectives were being shaped. Also formative evaluation was recognized.
- *1940s:* Mediated strategies such as the use of films for instruction were dominant and the term for instructional technologist was coined by Finn. The idea of an instructional development team was also introduced.
- *1950s:* Large group independent study instruction was a milestone and Sputnik which initiated federal funds to education was another milestone. The period also marked the birth of programmed instruction (PI) from behaviorism.
- *1960s:* The system's approach to designing instruction was introduced (Finn). A shift from norm-referenced testing to criterion-based testing was noted. The focus was on the development of instructional materials.
- *1970s:* In the history of e-learning, the work of Ausubel et al. on instructional strategies dominated this decade. The birth of AECT and the proliferation of models of instructional design were noted as well as the development of needs assessment procedures by Kauffman et al.
- *1980s:* Performance technology (Gilbert) and the focus on needs' assessment. Microcomputer instruction (CBI/CBT) flourished in this decade with the emphasis on design for interactivity and learner control.
- *1990s:* Focus on designing learning environments based on a constructivist approach to learning and multimedia development. Hypertext and hypermedia influence the field and cross-cultural issues are bridged using the Internet.
- *2000s:* Internet technologies are more and more integrated with personal, academic, and professional lives, making e-learning and online learning more accessible.

Understanding the history of e-learning and how it has evolved helps to make assembly about conceivable imminent directions and options for keeping up with current research and technological novelties. When it comes to education in general, the model is simple and direct until the early 2000s when education was still in a classroom of students with an educator who led the process. Then the computer evolution happened and it completely changed the learning scope. The adoption and use of e-learning concept in higher education took the instructional adaption as teachers like to undertake dialog processes in their teaching, and students are actively involved in the construction of awareness. These innovative customs are possibly accompanied with pedagogical wide conceptions and explicit and constructivist educational projects.

With the introduction of the computer and Internet in the late 20th century, e-learning delivery methods were developed and spread. The first MAC in the 1980s enabled individuals to have computers in their homes, making it easier for them to learn about particular subjects and develop certain skill sets. Then, in the following decade, virtual learning environments began to truly thrive, with people gaining access to a wealth of online information and e-learning prospects. In the 2000s, work market began using e-learning to train their employees. New and experienced workers alike now had the opportunity to improve upon their industry knowledge base and expand their skill sets. At home, individuals were granted access to programs that offered them the ability to earn online degrees and change lives through attained knowledge. Today, e-learning is more popular than ever, with countless individuals realizing the benefits that online learning can offer.

1.3.3 Standards of E-learning

E-learning has more than one description and more than one specific type. It can be distant learning; it can be synchronous or asynchronous learning. Learners and teachers may not be in the same definition or in the same vision of e-learning; they actually might never meet. In some of the computer-based learning, the teacher is actually a computer or an artificial intelligent unit that interacts to student's query. Worth to mention here that recently the second life concept is recently deployed into education using the computerized characters or "Avatars" to help in teaching and learning process, not to forget the game-based learning theory that uses games to teach claiming the concept "Fun to learn" is the best way of teaching. Then many forms of e-learning and more to forms to set standards that fits to this learning purpose.

The brave new pioneers who choose to experiment in e-learning are often those who are challenged by constraints of time and space. In addition, various authors have drawn conclusions that those who will be most successful in online learning have certain personality traits or characteristics, such as independence, assertiveness, persistence, and a reflective attitude (see Gibson, 1998; Gilbert, 2001; Wahlstrom et al., 2003).

To set standards, we need to understand the situation of users, mainly learners and teachers. E-learning should integrate all forms of e-learning and come with the dominate version that satisfies all users and meets their requirements and acquiring passions. E-learning or web-based learning, a relatively new form of distance education, is only now being integrated into distance learning research literature. It is, however, rapidly becoming the dominant form of distance learning delivery in developed countries.

To formalize and focus on e-learning, higher education makes available to ever-growing apprentice bodies. The purpose here is to present a matrix within which to examine, compare, contrast, and synthesize the standards of e-learning quality for quality assurance purposes. Higher education institutions in general have a long antiquity of liberation from governmental control. To debate quality assurance for e-learning, we need the following main troupes:

1. Professional faculty
2. Accrediting agencies to guide and control
3. University faculty and administrators.

However, learners' criteria for excellence in their e-learning experiences are largely not well understood yet as they have few challenges being actually an investment. E-learning is most frequently offered by institutions at prices representing "full cost recovery" which means programs that are not supported by government are required to be fully self-financing, including all overhead costs. This provides more enquiry relics to be done to fully sightsee consumer standards for e-learning and to fit in those standards with academic institutional concerns.

Quality standers of e-learning from the perspective view of the educationalist can be summarized by Jia Frydenberg Irvine. Irvine stated in his research quality standards in e-learning entitled "A matrix of analysis" Distance Learning Center – University of California (2002):

1. **Institutional commitment**

   Institutional commitment to the education and the provision of learning is an important standard. There is considerable variation in the understanding of commitment and how to develop it. Aspects such as financial commitment, the physical plant, articulation and other policies, technical support, and legal compliance are pledges that need to be emphasized and developed. Technological aspects of institutional support in terms of having a technology plan, security, redundancy in the delivery structure, and systems for maintenance of the technological infrastructure are important commitments as well (IHE, 2000).

2. **Technology**

   The technological infrastructure necessary for the delivery of a quality e-learning program that is often described separately is an important standard for e-learning. While none of the evaluators can set absolute description for this standard, yet they generally include such aspects as security and privacy of data and communication as well as the need for interactivity.

3. **Student services**

   E-learning standard does not concern neither teaching nor learning, but is instead centered on student services. It is subdivided into the services needed before students' entrance to a virtual classroom, support during the learning experience, and the continued connection between learners and the institution after the particular course or program has been completed.

4. **Instructional design and course development**

   Many parties are also offering standards for the design and development of e-learning programs. A report from The Pennsylvania State University titled *"An Emerging Set of Guidelines for the Design and Development of Distance Education"* (IDE, 1998) presented the following five aspects of course design with specific principles for each:

   - Learning goals and content presentation;
   - teaching–learning interactions;
   - Assessments and measurements;
   - Instructional tools;
   - Learner services and support.

5. **Instruction and Instructors**

   This standard is about instruction and instructor services. Instruction in a face-to-face setting such as depth of knowledge of the instructor, presentation and organizational skills, encouraging attitudes toward student dialog, feedback, and guidance is very important to accomplish the goal of learning. One issue that crops up in most of the standards documents that the distance learner is mostly solitary. The first principle of good practice from the American Association for Higher Education, for example, states that "Good practice encourages contacts between students and faculty" (AAHE, 2000).

6. **Delivery**

   This may or may not include student and instructor services. Delivery context is placed under "institutional context," noting such aspects as coordination, oversight, and articulation as parts of this domain. The Southern Regional Education Board (SREB, 2000) provides several principles of program management, which include such aspects as monitoring students to ensure academic morality, content revision and oversight, technical requirements for acceptable access, recourse for appeal if a web-based course is not delivered as described, and so on. Good program delivery depends on two aspects: (1) defined policies, procedures, and responsibilities and (2) communication, and fair and impartial management. In the best of all possible worlds, program delivery and program administration should be transparent to learners, just as the existence of the power plant on campus generally is.

7. **Finances**

   The headlong rush to develop courses and programs for e-learning sometimes fails to note standard business and accounting practices. The cost of one instructor at regular salary could be spread over hundreds of full tuition students and enables a rapid economy of scale. The financial business planning of e-learning programs is a necessary and is supposed to be the addressed domain of evaluation standards.

8. **Regulatory and legal compliance**

   Regulations governing e-learning are changing. The primary source of regulation in most countries in terms of quality assurance in higher education is the regional accreditation agencies and they have their charge to set standards of quality practice and to guide institutions toward achieving those standards. In addition, regional accreditation agencies must follow government laws.

   The institution seeks to understand the legal and regulatory requirements of the jurisdictions in which it operates, e.g., requirements for service to those with disabilities, copyright

law, state and national requirements for institutions offer-
ing educational programs, international restrictions such as
export of sensitive information or technologies, etc. (WICHE,
2000, p. 4)

9. Evaluation

E-learning programs' evaluation is frequently listed as a separate domain
as the assessment of student achievement is normally described as part
of instructional design and tied to specific course objectives; therefore,
program evaluation is a meta-activity that incorporates all the aspects
of the e-learning experience. Higher education identified the following
three criteria:

- The program's educational effectiveness and teaching/learning process are assessed through an evaluation process that uses several
  methods and applies specific standards.
- Data on enrollment, costs, and successful/innovative uses of technology are used to evaluate program effectiveness.
- Intended learning outcomes are reviewed regularly to ensure clarity, utility, and appropriateness (IHE, 2000).

1.3.4 Quality-oriented Concepts of E-learning Development Process

Quality assurance concept in e-learning is not a new, but yet not comfortably
applied and facing problem to trace and assure. Simple idea of tags on any
product can tell the customer the quality he/she is buying. Hiding irrelative
details from customer raises questions about the quality applied inside and
how this customer can assure the quality obtained. If this is the case for a
regular product, then how can we imagine it for an important sector like
e-learning? Especially, if this e-learning is a high way of education these
days. The answer will not take that much long to be identified as we have
referred that quality assurance concept is not new to e-learning sector, but
what will take time is explaining how this quality is taking place and how we
can assure the quality control and access.

Evaluation is a primary key that cannot be skipped or over taken in a
quality assurance of e-learning. Evaluation is the right framework of quality assurance in terms of e-learning that comes with lots of standards and
guidelines before actually deciding to evaluate the e-learning program. So,
logically we should stop by some standards and guidelines of e-learning
based on quality before divining into evaluation.

Let us define shortly, what is e-learning with simpler words. E-learning which actually stands for electronic learning is a system of education which integrates computer technology with a digitally networked infrastructure as a method of educational delivery. The system presents a way of informational and communication technology (ICT) that deploys all technologies available to provide access to the information that this e-learning center offers. To make the concept easier for non-computer-related customers, the technology of learning management systems (LMSs) has trended in the e-learning sector. The LMS is a software application or web-based technology used to deliver educational content and facilitate e-learning teaching and learning processes. Lots of new terms have been floated to the surfaces of e-learning to make the picture and frame of work much easier. E-content, e-course, e-design, virtual environment, and so on, which all lead to information, have been delivered to the user by using these platforms to present the knowledge via e-learning. Designing e-learning itself as a concept of education is a complicated procedure. So, to rephase what have been said, we can give more practical description of e-learning. E-learning is a range of several skill circles and often involves the exertions of a team of designers, developers, project managers, and subject-matter experts. The tag attached is "Quality is our first priority," and therefore, cycles of estimation and evaluation are a must. Here, we come to the concept of quality assurance in e-learning that requires the coordination side to design an e-learning product.

Adequate e-learning quality assurance is the motive why e-learning professionals have created frameworks that can be used to evaluate elements that subsidize to the factors influencing effective learning outcomes. This mainly includes four elements within e-learning development.

1. **The curriculum**

   A strong curriculum should be created for learner's possession in mind its relevance in the place of work. It should be modernized and reflects the best practices that would help learners perform better – within the workplace and beyond.

2. **The learning design**

   E-learning designers have to generate an effective learning atmosphere that provides not only required information but also engagement that ensures effective learning.

3. **The course content**

   The content provides required perspectives and should have conceptual strength to create acquaintance.

4. The delivery processes

Technology-aided learning can create a strong support system to sustain long-term learning.

While the overhead is the framework that forms the e-learning quality assurance, there are countless checks and procedures that developers can take to ensure that the framework is strong and delivers operative e-learning.

1. Course content

The e-learning content needs to be well thought-out, concise, and consistent. It is a known fact that learners do not prefer to read too much when they learn in a self-learning cycle. Therefore, e-learning is not too complicated. On contrary, there should be a constant effort to make difficult concepts easy to understand and learn.

2. Interactivity for engagement

Trainings can be uninteresting, especially in the context of "pure" technology-aided learning. Interactivities built within e-learning can effectively provide relief as well as challenge the learners to think and apply assimilated learning.

3. Visual design

E-learning developers should pay attentions to details such as font, space, color, graphics, and multimedia. Images and graphics have always been a part of learning and e-learning is also no exception.

4. Logical structure

The structure of the e-course should not be limiting to learning process. Navigation must be simple and consistent throughout the course. E-learning content should be logically structured and made searchable so that learners can access it better.

5. Adequate technology support

There is a lot in terms of delivery of e-learning that one can employ. E-learning delivery should be created intelligently keeping in mind the needs of the learners as well as available technology.

E-learning is an effective way of knowledge, but it definitely has allied costs and makes many demands. It is significant to be able to please investors and guarantee that e-learning is providing the supreme possible return on investment and that further speculation is warranted. With well-established e-learning quality assurance, this can be very well accomplished.

1.4 Concept of Blended Learning in Terms of Quality Assurance

Quality assurance in education is very greedy process that tries to control and promote the quality of education to the level being blended with each layer of education systems, if necessary. Moving from traditional ways of teaching to the most updated methods has been a very risky mission, especially when we talk about restrutting and reforming the whole system. In the presence of e-learning and facilities, it affords to education sector, education has become more enjoyable and opening more learning opportunities for students, more vacancies for faculty, and remarkable contribution to the society. But yet, considering the framework of quality and its limitations preferring the traditional ways, blended learning has raised up. Blended learning is a disruptive innovation in education that can take many forms that are mostly electronic forms. Blended learning is an education program (formal or non-formal) that combines online digital media with traditional classroom methods. It requires the physical presence of both the teacher and the student, with some element of student control over time, place, path, or pace. While students still attend a regular class with a teacher present, face-to-face classroom practices are combined with computer-mediated activities regarding content and delivery. "Blended learning" is sometimes used in the same breath as "personalized learning" and differentiated instruction (Figure 1.2).

Figure 1.2 Concept of blended learning.

https://da.wikipedia.org/wiki/Fil:Blended-learning.webm
https://coursefindr.co.uk/articles/studying-an-online-pgce-distance-learning-teacher-training/
http://laoblogger.com/english-teacher-clip-art.html#gal_post_77371_english-teacher-clipart-6.jpg

The terms "blended learning," "hybrid learning," "technology-mediated instruction," "web-enhanced instruction," and "mixed-mode instruction" are often used interchangeably in research literature. Although the concepts behind blended learning first developed in the 1960s, the formal terminology to describe it did not take its current form until the late 1990s. One of the earliest uses of the term appears in a 1999 press release, in which the Interactive Learning Centers, an Atlanta-based education business, announced a change of name to EPIC Learning. The release mentions that "The company currently operates 220 online courses, but will begin offering its Internet courseware using the company's blended learning methodology." The term "blended learning" was initially indefinite, encircling a wide diversity of technologies and pedagogical approaches in varying amalgamations. In 2006, the term became more tangible with the publication of the first *Handbook of Blended Learning* by Bonk and Graham. Graham confronted the breadth and haziness of the term's definition, and defined "blended learning systems" as learning systems that "combine face-to-face instruction with computer mediated instruction" (Bonk and Graham, 2006). In a report titled "Defining Blended Learning," researcher Norm Friesen suggests that, in its current form, blended learning "designates the range of possibilities presented by combining the Internet and digital media with established classroom forms that require the physical copresence of teacher and students."

So basically, blended learning presents a new methodology of teaching and learning to manipulate and flip the old traditional education system on its head. As changing the old concept related to education, blended learning is not about credits or grades; it is about mastery through tailored learning. Students typically have no predetermined evaluation level, and more to add, that there are no predetermined course completion dates. Blended learning combines face-to-face and digital lessons in an adult-supervised environment to maximize the benefits of e-learning tools and technologies. It fluctuates from one application to another, but a common denominator is that the location, equipment, and tools of instruction are imprinted out for each student's learning needs.

Blended learning has proven the authentication and importance of being able to promote learning processes and opening more opportunities for education, students, teachers, graduates, and also for the whole system as it paves the way to the globalization with quality assurance roots. Blended learning is becoming more and more widespread scheme of instruction in education, training, and self-promoting as well as knowledge and experience transfer. Many of higher education institutions have adopted blended learning programs for a variety of reasons, mainly to integrate the quality assurance access and control to education by following procedures and being able to evaluate and assess the educational process instead of completely depending on e-learning platforms. Other advantages can be as follows.

1. **Blended learning improves efficiency**

   Blended learning allows teachers to use a combination of digital instruction face-to-face interaction to improve efficiency in the classroom. It permits students to use adaptive learning technologies to work on their understanding of new concepts while teachers can use the additional class time to give struggling students the individualized attention that they need.

2. **Blended learning reduces the cost**

   Education can be extremely expensive. From continually reordering textbooks so that they are up to date to having school supplies on hand, the costs add up. Blended learning can help school districts save money by allowing students to bring their own technology devices to class to take full advantage of the adaptive learning software. The use of e-textbooks, which can also be accessed on computers, tablets, and e-readers, may also help drive down costs.

3. **Blended learning allows self-learning process**

   Blended learning is able to personalize education in a way that a more traditional classroom setup simply cannot. When teachers stand in front of a sea of desks and educate students about a new concept, kids of varying academic skill levels are not able to take extra time or work ahead as they may need. Blended learning allows students to work at their own pace, making sure that they fully understand new concepts before moving on.

4. **Blended learning affords a better student data**

   Software used with blended learning programs is able to collect student data that measure academic progress. In this way, teachers can clearly see the areas in which each individual student is excelling, and where he or she may need a little more guidance, and act accordingly.

5. **Blended learning affords Common Core State Standards**

   Digital fluency is a goal of the Common Core State Standards, and states that working to implement the new academic guidelines will need to make sure that technology is integrated into a variety of subject areas. Blended learning can easily accomplish this task.

6. Blended learning can take many models as:

- Face-to-face interaction where the teacher drives the instruction and augments with digital tools.
- Enhancement cycles where students can study and assure their attaining knowledge by cycles of online study and get the support during the face-to-face interaction.
- Most of the curriculum is delivered via a digital platform and teachers are available for face-to-face consultation and support.
- Blended learning can be in Labs. All of the curriculum is delivered via a digital platform but in a consistent physical location. Students usually take traditional classes in this model as well.
- Self-blend where students choose to augment their traditional learning with online course work.
- Online driver as students complete an entire course through an online platform with possible teacher check-ins. All curriculum and teaching are delivered via a digital platform and face-to-face meetings are scheduled or made available if necessary.

In case blended learning is effectively planned and executed, it makes a real difference in learning process. Otherwise, blended learning can bring more disadvantages in technical aspects since it has a strong dependence on the technical resources or tools with which the blended learning experience is delivered. These tools need to be reliable, easy to use, and up to date, for them to have a meaningful impact on the learning experience. Information technology knowledge can serve as a significant barricade for students struggling to get access to the course materials, making the accessibility of high-quality technical support overriding. Other phases of blended learning that can be challenging are the group work concept because of difficulties with online supervision.

From an educator's perspective, most recently, it has been noted that providing effective feedback is more time consuming – and therefore more expensive – when electronic media are used, in comparison to traditional as paper-based assessments. Using e-learning podiums can be overwhelming than traditional approaches and can also come with new costs as e-learning platforms and package providers may charge user fees to educators. Another critical issue is access to network infrastructure. Although the digital divide (digital divide: the division between individual with regard to the access and use of technology in society) is narrowing as the Internet becomes more inescapable, many students do not have pervasive and permeating access to the Internet – even in their classrooms. Any attempt to incorporate blended learning strategies into an organization's pedagogical scheme needs to account for this. This is why learning centers are built with good Wi-Fi connections to make sure that this issue is addressed.

Although quality concept is forming and taking a space in blended learning, yet the concept is not pragmatic faultlessly. Teachers anxious about the quality of learning in universities are facing a number of challenges correlated to information and communication technologies (ICT). High on the list of these contests is ascertaining appropriate ways of appraising the extent of their contribution to quality learning involvements. There is no room to sidestep this issue as there are a number of drivers encouraging the integration of ICT into the student experience. These include the litheness they bring to students with increasing work and familial obligations; the skills that they may help to develop such as modern communication and collaboration methods if they are used well, the immediate access they provide to an increasing amount of knowledge, both in the disciplinary and future qualified areas of students; and the thoughtful they stimulate if they are to upkeep learning properly. These drivers are being increasingly recognized by universities, not only those teaching students at a distance, but by universities offering a predominantly campus-based education (Higher Education Funding Council for England, 2004). From a student perspective, this most universally results in education processes that are banquet across face-to-face and online contexts. This type of learning, often discussed to as blended learning, creates challenges for the assessment process (Figure 1.3).

Evaluating the quality of blended learning involvements is no stress-free matter as technologies stereotypically support only part of the learning developments that the students engage in. Consequently, evaluating the contribution of the technologies in blended learning experiences involves research practices sufficiently penetrating so that they can recognize and acknowledge the relational nature of the know-hows methodology concept to the quality of learning. This book seeks to offer some useful ways into the debate of how to evaluate the quality of learning arising from learning experiences, in general and in e-learning and blended learning is specific later. It does so by complementing well-known and recent academic researchers in e-learning technologies used to evaluate the quality of learning in higher education and proposing a general concept of building a well-supported backbone for quality concepts in higher education institutions, e-learning, and blended learning as well. The next section is presenting this proposal in brief as it needs further experiments and studies which are open for researchers and mentors who are interested in the topic.

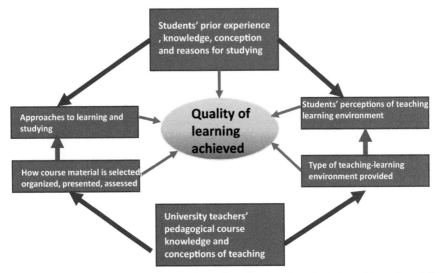

Figure 1.3   Concepts related to the quality of learning at university (adapted from Entwistle et al., 2002).

1.5 E-government: The Missing Factor in Quality Assurance Concept

Effective integration of information technology into educational processes has become increasingly crucial. The processes are not limited to teaching or delivering information as it should be seen as integrated educational community based on strong and active IT infrastructure. IT includes system software, application software, computer hardware, networks, and databases allied with managing an institution's information. However, when it comes to implementing quality standards in the IT monarchy, most face pressures to deliver systems and technologies which meet the organization's changing needs which consequently fail down the quality by wayside. It is not only institution; quality lack may expand to involve teachers, learners, and stakeholder's margins.

The industry as a whole has fallen short of conveying technology that people apprehend and can use. Several of the problems befall because of the complexity of technology and the rapid skip of change. Neither of these environments is likely to end; in fact, they are hastening at a disquieting ratio. If flawless execution was an elusive goal in the past, it is even more so today. To elaborate more about this side, we need to explain an important issue about ICT.

ICT is the infrastructure and components that enable modern computing. ICT is a term that is generally used to mean all devices, networking components, applications, and systems that combined allow people and organizations to interact in the digital world. ICT encompasses both the Internet-enabled sphere as well as the mobile one powered by wireless networks. It also includes antiquated technologies, such as landline telephones, radio, and television broadcast – all of which are still widely used today alongside cutting-edge ICT pieces such as artificial intelligence and robotics. For businesses, advances within ICT have brought a slew of cost savings, opportunities, and conveniences. They range from highly automated businesses processes that have cut costs, to the big data revolution where organizations are turning the vast trove of data generated by ICT into insights that drive new products and services, to ICT-enabled transactions such as Internet shopping and telemedicine and social media that give customers more choices in how they shop, communicate, and interact.

E-learning and blending learning are influenced sectors by the theory of ICT and digital divide as well as higher education institutions are influenced when they desire to adopt the quality assurance control and procedure as they also need a consistence ICT level to deliver, assess, and monitor the educational process. Not to mention that developing also needs a convenient concept of ICT. Then, and obviously there is a strong impact of ICT on quality assurance as a result while the ICT is actually meant to support and facilitate the purpose of quality assurance in higher education. It is worthy to mention that the relation should be clearly stated as the ICT is costly and divided parties into Have and Have-not categories. Those parties are mainly the motor of quality assurance system: institution, teacher, learners, and stakeholders.

In this section, we investigate the role of e-Government in education process – the learning process which is centered on ICT. Actually, e-learning is now embedded in most of the educational forms regardless of being traditional or non-traditional. Universities that yet are not adopted the e-learning in teaching/learning process completely are still established on ICT technology for internal and external communications. E-government is an electronic form to facilitate the concept of government line of work in order to integrate all government's institutions into one government body that manages, controls, and monitors the government workspace and to overcome all obstacles and manual work that cause the undesired delay in many work forms. The e-Government concept is actually well established and well adopted by many developed and developing countries. The form of e-Government is an architecture of e-learning centers that acquire knowledge and build-up procedures and processes to complete the mission they are designed for. The goal is to help the government to perform on high level with effectiveness that goes beyond the typical ways. Then, we conclude that e-learning concept is actually a requirement to make the e-Government project goes as expected. The e-learning centers built by different sectors for this government and accumulated data and information about citizens with amount can be discussed as big data. Big data require consistency of ICT levels and stable Internet connection with maximum convenience to scale the mission of the procedures and processes of this government. The fund of such a huge project is based on the governments' funding, contribution, and willing to moderate the work frame that catches with international race. Yet international statistics show that although many developed and developing countries have applied the concept of e-Government, they have yet been suffering a digital divide and rural areas with lack of Internet connections. Figure 1.4 shows the number of U.S. households without the Internet in rural areas in 2010, by householders' educational attainment.

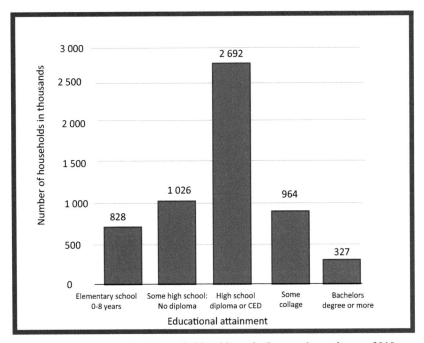

Figure 1.4   Chart of U.S. households without the Internet in rural areas, 2010.

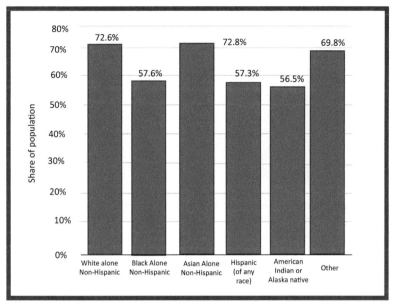

Figure 1.5 Percentage of U.S. population with Internet access, 2015.

Figure 1.5 shows a recent statistics; it shows the percentage of U.S. population with Internet access at home as of July 2015, by race and Hispanic origin.

Figure 1.6 shows the general statistics of Internet usage in different countries, although most of these countries have applied the concept of e-Government.

Countries have been aware of digital divide and ICT differences as many rural areas in different developing and developed countries are being penalized by a lack of the Internet and the so-called digital gap between remote and urban areas is widening, and more to tell that even if they have the Internet, they have at very low speed which affected business sector and investments which rely on the consistency of the Internet. Internet connection is also offered by providers with high cost, and most of the times, people pay without getting the quality they pay for because simply the government itself has not yet invested properly to have the right infrastructure for any kind of telecommunication system. Therefore, most of countries supported with researchers have planned to overcome the digital divide and bridge the gap. Lately, for example, U.S. Department of Agriculture (USDA) announced new rules to fund broadband services in unserved rural communities which are expected to refresh the ICT level investments.

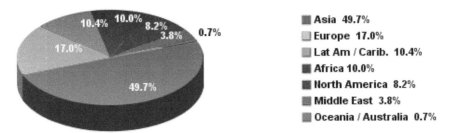

**Internet Users in the World
by Regions - June 30, 2017**

- Asia 49.7%
- Europe 17.0%
- Lat Am / Carib. 10.4%
- Africa 10.0%
- North America 8.2%
- Middle East 3.8%
- Oceania / Australia 0.7%

Source: Internet World Stats - www.internetworldstats.com/stats.htm
Basis: 3,885,567,619 Internet users in June 30, 2017
Copyright © 2017, Miniwatts Marketing Group

Figure 1.6   Internet usage statistics, the Internet big picture world Internet users, and 2017 population stats.

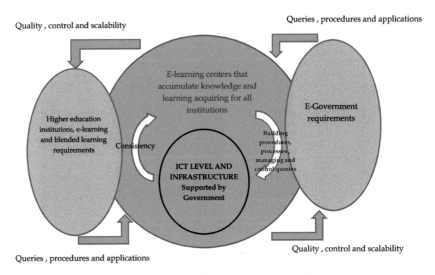

Figure 1.7   Conceptual model of e-Government role in quality assurance concept.

Then it is not only a matter of higher education institutions to decide whether to follow up with quality assurance or not if they are located in these rural areas. This section is a call for e-Government countries to play roles to support the model and need of education quality as they already have the equipment and have already installed the infrastructure to facilitate the work with e-Government applications. The following model is a kind of a brainstorming how the supported ICT level by e-Government can help greatly for quality assurance purposes of education (Figure 1.7).

The above model is as mentioned before opened for further investigations and studying, but in general, there is nothing to limit this theory as the e-learning centers are common in between e-Government and quality assurance. Not to forget that e-Government needs lots of quality assurance systems and application to assure the quality of delivering the information. Higher education itself is a government institution and all institutions underneath this higher education should be considered in this model. The government funding will help a lot to overcome the obstacle and bridge the gap caused by ICT differences.

References and Suggested Literature

[1] Available at: http://www.universityworldnews.com/article.php?story=20140508162141672

[2] Barrett, A. M., Chawla-Duggan, R., Lowe, J., Nikel, J., Ukpo, E., *The concept of quality in education: a review of the 'international' literature on the concept of quality in education*, EdQual Working Paper No. 3, UK, 2006.

[3] Kigotho, W., *Higher education challenges post 2015–UNESCO*, issue no: 319, 2014.

[4] MEXT, *"International Exchange and Cooperation,"* available at: http://www.mext.go.jp/english/org/eshisaku/ekokusai.htm, 120, 2000.

[5] JSPS, *Approaches for Systematic Planning of Development Projects/ Higher Education*, available at: http://www.jsps.go.jp/english/index.html.

[6] MEXT, *"International Cooperation,"* available at: http://www.mext.go.jp/english/org/exchange/65.htm, *The Importance of University Education in Developing Countries*, available at: http://www.educationalpathwaysinternational.org/?page_id=99.

[7] UNESCO, *The Positive Impact of eLearning—2012 UPDATE, WHITE PAPER Education Transformation The Role of Higher Education in Society: Quality and Pertinence, 2nd UNESCO-Non-Governmental organizations Collective Consultation on Higher Education*, Paris, 1991.

[8] e-Book, *E-learning concepts, trends, applications*, V 1.1, 2014.

[9] Nicholson, P., "*A history of e-learning*," in: B. Fernández-Manján B et al. *(eds) Computers and Education: E-Learning, From Theory to Practice. Springer, Netherlands, 1–11,* 2007.

[10] Hite, P., *An experimental study of the effectiveness of group exams in an individual income tax class. Issues Account. Educ.* 11(1), 61–75, 1996.

[11] Hite, P., Hasseldine, J., *A primer on tax education in the United States of America. Account. Educ.* 10(1), 3–13, 2001.

[12] Craner, J., Lymer, A., *Tax education in the UK: a survey of tax courses in undergraduate accounting degrees. Account. Educ.* 8(2), 127–156, 1999.

[13] Engelbrecht, E., *Adapting to changing expectations: post-graduate students' experience of an e-learning tax program. Comput. Educ.* 45(2), 217–229, 2005.

[14] Rosenberg, M., *E-Learning: Strategies for Delivering Knowledge in the Digital Age. McGraw Hill Companies,* Columbus, OH, 2001.

[15] Henderson, A. J., *The E-Learning Question and Answer Book: A Survival Guide for Trainers and Business Managers. AMACOM, New York, NY,* 2003.

[16] Garnham, C., Kaleta, R. *Introduction to hybrid courses. Teaching with Technology Today* 8(6), 2002.

[17] Singh, H. *Building effective blended learning programs. Educ. Technol.* 43(6), 51–54, 2003.

[18] Jacob, A. M., *Benefits and barriers to the hybridization of schools. J. Educ. Policy Plan. Adm.* 1(1), 61–82, 2011.

[19] Vaughan, N., *Perspectives on blended learning in higher education. Int. J. E-Learn.* 6(1), 81–94, 2007.

[20] Davis, F. D., *Perceived usefulness, perceived ease of use, and user acceptance of information technologies. MIS Q.* 13(3), 319–340, 1989.

[21] Lederer, A. L., Maupin, D.J., Sena, M. P., Zhuang, Y. *The technology acceptance model the World Wide Web. Decis. Supp. Syst.* 29(3), 269–282, 2000.

[22] Kim, D., Chang, H., *Key functional characteristics in designing and operating health information websites for user satisfaction: An application of the extended technology acceptance model. Int. J. Med. Inf.* 76, 790–800, 2007.

[23] Polancic, G., Hericko, M., Rozman, I., *An empirical examination of application frameworks success based on technology acceptance model. J. Syst. Softw.* 83, 574–584, 2010.

[24] Belenky, M. F., Clinchy, B. M., Goldberger, N. R., Tarule, J. M., *Women's Ways of Knowing: The Development of Self, Voice, and Mind. Basic Books, New York, NY*, 1997.

[25] Ong, C., Lai, H., *Gender differences in perceptions and relationships among dominants of e-learning acceptance. Comput. Hum. Behav.* 22, 816–829, 2006.

[26] Smart, K., Cappel, J., Students perceptions of online learning: a comparative study. *J. Inf. Technol. Educ.* 5, 201–219, 2006.

[27] QAA, *Quality Assurance Agency for Higher Education (QAA) – UK*, available at: http://www.qaa.ac.uk/.

[28] UNESCO, *Teacher Education Accreditation Council (TEAC)*, available at: http://www.teac.org/ *The joint quality initiative*, available at: http://www.jointquality.nl.

[29] *Higher Education/Quality Assurance and Recognition*, available at: http://www.unesco.org/ *World Bank - Tertiary Education*, available at: http://www.worldbank.org/, http://www.usc.edu.au/explore/vision/strate gy-quality-and-planning/quality-and-standards-framework.

[30] Colvin, C. R., Richard, M. E., *e-Learning and the Science of Instruction: Proven Guidelines for Consumers and Designers of Multimedia Learning*. John Wiley and Sons: Hoboken, NJ, 2003.

[31] Robert, F. M., *Measuring Instructional Results. The Center for Effective Performance,* 2000.

[32] William, J. R., Kazanas, H. C., *Mastering the Instructional Design Process: A Systematic Approach*. Josey-Bass, Inc., Publishers: Hoboken, NJ, 1998.

[33] Available at: https://www.learningsolutionsmag.com/articles/311/quality-assurance-for-e-learning-design.

[34] Banditvilai, C., *Enhancing Students' Language Skills through Blended Learning. Electron. J. E-Learn. 14*, 220–229, 2016.

[35] Friesen, N., *Report: Defining Blended Learning*, 2012.

[36] *Blended Learning: A Disruptive Innovation. Knewton*: New York, NY.

[37] Staker, H., Horn, M. B., *Blended Learning,* 2012.

[38] *Valerie*, S., *Three fears about blended learning*. The Washington Post, 2012.

[39] *Blended course design: A synthesis of best practices. J. Asyn. Learn. Netw. 16*, 7–22, 2012.

[40] Lothridge, K., *et al. Blended learning: efficient, timely, and cost effective. J. For. Sci. 45*, 407–416, 2013.

[41] Oliver, M., Trigwell, K., *Can 'Blended Learning' Be Redeemed? E-Learning* 2(1), 17, 2005.

[42] Moskal, P., *Blended learning: A dangerous idea? Int. High. Educ. 18,* 15–23, 2013.

[43] Blended Learning is Not the Only Way to Personalize Learning, available at: *www.personalizelearning.com, 2012* [Retrieved May 4, 2016].

[44] Dale, B., *Personalized vs. differentiated vs. individualized learning. ISTE,* 2014.

[45] Margie, M., *The hybrid online model: Good practice. Educ. Q. 18–23,* 2003.

[46] *Interactive Learning Centers Announces Name Change to EPIC Learning. The Free Library,* 1999.

[47] Bonk, C. J., Graham, C. R., *The handbook of blended learning environments: Global perspectives, local designs. Jossey-Bass/Pfeiffer,* San Francisco, CA, 5, 2004.

[48] Josh, B., *How Did We Get Here? The History of Blended Learning. The Blended Learning Book: Best Practices, Proven Methodologies, and Lessons Learned.* John Wiley and Sons, Hoboken, NJ, 2004.

[49] Plato Rising, available at: *Atarimagazines.com,* 2013.

[50] *Coach resources,* In *the real world |Coach resources. Khan Academy,* Mountain View, CA, 2013.

[51] Friesen, *Report: Defining Blended Learning,* 2012.

[52] 6 Models of Blended Learning. *DreamBox,* Washington, DC, 2014.

[53] Alison, D., *Different Faces of Blended Learning. District Administration,* 2014.

[54] Kim, A., *Rotational models work for any classroom. Education Elements,* 2014.

[55] The four important models of blended learning teachers should know about. *Educ. Technol. Mob.* Learn, 2014.

[56] Blended learning: how brick-and-mortar schools are taking advantage of online learning options. *Connect. Learn.,* 2014.

[57] 6 Models of Blended Learning, 2015.

[58] *Blended Learning 101. Aspire Public Schools,* Oakland, CA, 2014.

[59] Models of Blended Learning. Idaho Digital Learning, St, Boise, ID, 2014.

[60] PERC, Blended Learning *Defining Models and Examining Conditions to Support Implementation. Philadelphia Education Research Consortium (PERC), Philadelphia, PA,* 2014.

[61] Top 5 Benefits of a Blended Learning Platform, 2015.

[62] Mustafa, S., et al. The effect of blended learning environments on student motivation and student engagement: a *study on social studies course. Educ. Sci.* 2015.

[63] Five benefits of blended learning - DreamBox Learning. *DreamBox Learning*, Washington, DC, 2016.

[64] Garrison, D. R., Kanuka, H., *Blended learning: Uncovering its transformative potential in higher education. Int. High. Educ.* 7, 95–105, 2004.

[65] Alexander, S., *"Flexible Learning in Higher Education"*, in P. Peterson, E. Baker, B. McGaws. *International Encyclopedia of Education (Third ed.). Elsevier*, Oxford, 441–447, 2010.

[66] Alexander, S., McKenzie, J., *An Evaluation of Information Technology Projects for University Learning. Committee for University Teaching and Staff Development and the Department of Employment, Education, Training and Youth Affairs*, Canberra, QC, 1998.

[67] The E-Learning Edge: Improving Access With Ontario Learn, Available at: www.conferenceboard.ca. 2016.

[68] Idit, H. C. *Learning to Make Games for Impact. J. Media Liter.* 59(1), 28–38, 2012.

[69] Jacob, A. M., *Benefits and Barriers to the Hybridization of Schools. J. Educ. Pol. Plan. Adm.* 1(1), 61–82, 2011.

[70] Chen, I., *For Frustrated Gifted Kids, A World of Online Opportunities.* KQED, 2014.

[71] Heinze, A., Procter, C., *Online Communication and Information Technology Education. J. Inf. Technol. Educ.* 5, 2006.

[72] Ginns, P., Ellis, R., Quality in blended learning: Exploring the relationships between on-line and face-to-face teaching and learning, *Elsevier J. Int. High. Educ.* 10, 53–64, 2007.

[73] Available at: https://www.statista.com/statistics/190342/us-householders-in-rural-areas-without-internet-by-educational-attainment/.

[74] Available at: http://www.internetworldstats.com/stats.htm.

[75] Available at: http://scholarshipweb.net/tag/rural-communities-penalised-by-poor-broadband.

[76] Cousins, J. B., Aubry, T., *Centre for Research on Community Services, University of Ottawa, Roles for Government in Evaluation Quality Assurance: Discussion Paper*, University of Ottawa, Ottawa, ON, 2006.

[77] Hiagu, O. P., Information and Communication Technologies: Widening or Bridging the Digital Divide? *J. Linguist. Commun. Stud.*, 2013.

[78] A paper presented by UNICEF at the meeting of The International Working Group on Education Florence, Italy June 2000, Available online https://www.unicef.org/education/files/QualityEducation.PDF

[79] Series Editors Allan Pitman University of Western Ontario, Canada Vandra Masemann University of Toronto, Canada Miguel A. Pereyra University of Granada Quality and Qualities Tensions in Education Reforms Clementina Acedo, Don Adams and Simona Popa (Eds.), Volume 16, ©2012 Sense Publishers

[80] Adams, D. (1993). Defining educational quality. Improving Educational Quality Project Publication #1: Biennial Report. Arlington, VA: Institute for International Research.

2

The Role of Higher Education (Quality Assurance-based System) in Promoting Lifelong Learning

2.1 Introduction

Higher education has been always seen as "economic engines" that ensure the knowledge production through research and innovation. Wild-world common agreement and realization about the importance of quality in higher education have as well come to surface and considered no-compromise footstep regardless of any burdens that the higher education institutions may face. The global realization has made high-quality tertiary education more vital than before in order to dominate employment skills and to sustain a global competitive research – based dedicated to expand knowledge propagation that benefits values of society.

The global realization of quality importance also supported the new framework for higher education to cope with the economic development and shift from elite, fellow, and under control institution to act as a sort of dependent and physique system as referred in Tom (2005). The economic developing impacted higher education framework to be more competitive and more operative institutions with internal transfer before meeting the standards of outward, which is engagingly, a must next step. The recognition of the higher education framework, agendas, and contexts in order to sight seek international acknowledgment has included and assured quality concept and adoption which, in turn, encouraged a speedy proliferation of new educational programs in many nations to test traditional practices for assuring academic standards. The rapid growth of higher education systems and the incentive to cross-borders have posed novel challenges to national systems of external quality assurance, particularly those based upon central control of public institutions. Furthermore, high and race competitions released by globalization have required institutions to become more receptive to rapidly varying and on-going developing systems.

On the other hand, students are the first to be affected with this inflation in positive and negative ways. Positively, students are introduced to a new level of higher education programs that by default will feel more interested in. The chance of promoting learning is guaranteed with globalization, assuming the quality base. Students are also more likely to get more flexibility from internal institutions that cope with changing social demands and establish new academic programs, reconfigure existing programs, and eliminate outdated programs. Still optimistic, and it is mentioned that this transfer is supposed to go through adoptive channels of quality assurance and regulations to control this elasticity. On the negative side, and dramatically, the empirical reality of growing private benefits of new academic programs and new framework of higher education has altered the traditional debate about higher education finance, encouraging the inflation in education fees requiring students and their families to pay a larger share of this new transfer costs.

The new practices of external quality assurance also seek to respond to public concerns that institutions provide educational value for money. Immediate response is a must to control the growth while giving a space for moderating the educational system that fits to purpose and support students and community. Therefore, a national framework is a must to be applied with standards to help higher education standards to survive and expand taken into consideration that the quality of these standards is the key to encourage innovation in academic curricula. Representatives of higher education such as education ministries are investigating innovation forms of academic quality assurance to cope with globalization and meet the requirements of community. The first government experiments with new quality assurance practices occurred predictably in the United States, a primary advocate of figure higher education. In the early of 1980, majority of US states adopted protocols requiring that public universities should develop explicit plans for assessing student teaching (Dill et al., 1996), and the aim was to decline hypothetical standards and bring quality into practice starting from public education. Subsequently, new national quality assurance policies were also introduced in France in 1984, the United Kingdom in 1985, and the Netherlands in 1985 (Van Vught and Westerheijden, 1993). These governments and mainly the French government were primarily interested in reducing its dysfunctional quality assurance bureaucracy, the government in Great

Britain sought to achieve a better linkage of higher education with labor market, while the Netherlands adopted a new quality assurance and control framework and structure in overtone with a state-of-the-art approach to navigation of higher education institutions in general and universities in specific.

The successful and fruitful experiments and achievements in these pioneering countries diffused to other governments in Europe, Asia, and eventually around the globe in short time. The achievements were different as different quality assurance frameworks were applied. Discarding many mini-details, we can say that at the end, general models of quality assurance have been introduced wide world and most are following these days (Dill, 1992):

1. The European Model of central control of quality assurance by state educational ministries.
2. The US Model of decentralized quality assurance combining limited state control with market competition.
3. The British Model in which the state essentially ceded responsibility for quality assurance to self-accrediting universities (Dill, 1992).

In the United Kingdom, actions were taken by Thatcher government in 1981. Quality assurance and control concept were taken into action and started to form academic quality in the public universities. The academic quality was delegated to the academic profession itself, which monitored and assured the standards of university degrees through public instruments such as the external examiner system. Higher education also played important roles and has been active in setting frameworks for universities. Higher education worldwide established and monitored regulations on university admittances, academic activities, program curricula, and end-point inspections. Quality assurance and control systems were as contamination moving from one higher education to another as most of higher educations were more attached to globalization theory and impacts. Also, higher education played the role of being more responsible toward the internal community; therefore, higher education kept the communication channels internally and externally to compromise the financial side. In the United States, for example, members of congress argued many times that providing federal financial assistance directly to students rather than institutions was the most efficient and effective means to both equalize opportunities in higher education and harness market forces for enhancing academic quality. Higher education included the definition of quality assurance toolkits that define and manage academic degree frameworks for universities. The new toolkit introduced policies of new assessments practices such as academic audits and/or subject imposts designed to sustain and recover internal quality assurance practices. The quality assurance presented and forced by higher education also has a supported procedure which is accreditation of programs and institutions. The accreditation is a system to control the framework of institutions to perform accordingly to higher education standards. Higher education standards as well direct quality procedures to execute internal institutions and touch basics of such faculty affairs. So, quality assurance also has procedures for performance, funding and contracts, and periodical monitoring and regulation influencing communal provision of academic information such as exams and reviews.

Higher education is also aware of the importance of transparency, especially with students and their families. It is necessary for students and their families to achieve actual consumer power through informed choice of academic programs. So, higher education as well emphasizes on traditional practices carried out by professional bodies including accreditation of academic programs. These practices are transparent to students and their families to be aware and understand how institutions are submitted and committed to the higher education regulations. To get students into the frame, they may be introduced to the external examination system, for example, which is known to be a great precise for quality in the teaching–learning process. Another quality assurance practice which is a kind of important to students is the university ranking system. Most of universities are obsessed by ranking systems which generally inform students how quality is being running inside this university. Higher education is supporting the internal quality practices which in turn support the external quality that promotes education systems to meet the international standards. Finally, it is worthy to mention that higher education established the idea of quality assurance agencies. Quality assurance agencies are performers in higher education governance as well. The main objective is to asses programs and academic plans of institutions to judge whether these are appropriate and effective, mainly in the developing progress and projects. Quality Assurance agencies 'influence varies from many levels of plans and procedures inspections, inspect if the program passes by program certification and ends to the final volume of system accreditation.

This chapter is written to clarify the role and contribution of higher education based on the quality assurance system designed, managed, and monitored in order to reach a sustainability and excellence of higher education.

2.2 The Role of Higher Education in Sustainable Development

The social role of higher education is to provide a promoting life-long learning with an effective strategy for achieving internal and external quality of educational systems, using quality-oriented procedures. The role of higher education is to sustain a knowledge-based system that can be transferred from one generation to another and maintain smooth transmission via quality of teaching, quality of learning, and quality of teaching–learning outcomes. So, the sustainability of higher education itself as an education system depends on the quality of built knowledge. And henceforward, excellence in higher education is a must and should remain the prime objectives and a neutral of any instruction of higher education wide world, mainly universities. To assure the quality of this knowledge, higher education and its institutions form and guarantee excellence in quality procedures. If this excellence in quality exists, then relevance must logically follow. This at least is to retain the traditional role as a servant of society.

This section constitutes a base of exploring how to expand and transmute the role of higher education in promoting life-long learning. The base of higher education life-long learning can be drawn in the following aspects:

1. Increasing the awareness of importance of quality assurance concept in higher education and how following this will impact society in a positive way.
2. Reforming the teaching–learning process on quality-based melody by recognizing responsibilities of both teachers and learners. This should create innovative prospectuses to capitalize experiences of teachers and students as well.
3. Outlining the mutual impact of higher education to enhance individual's life, experience, and skills while still keeping the cultural context of society.

With a rapid change of marketplace, higher education is also expected to become more receptive to the work and life situations of adult learners, helping them not only to attain skills and acquaintance but also to maintain and improve their position in society and ultimately enhance their quality of life. On the other hand, still higher education is responsible to promote a common intangible understanding of life-long learning among universities, stakeholders, and sponsors for the integrations of life-long learning perspectives inside institutional stagey and operations.

Now to realize quality prospects in education systems, higher education must adopt more lively approaches in terms of meeting the needs of society away from being passive agent caring for standards more than planning how to apply them. Higher education started the emerging of quality into education systems by growing higher education quality agents. These agents can be independent from institutions or can be established as a part of institution processing system. And actually this is the most happening now. The fortuitous of growing into quality assurance agents helped to present quality concepts and procedures and gave the higher education the power and control to maintain its role as a real focal point of knowledge and its applications. These agents actually formed a common dialog and exertion between these institutions and higher education standards. They played the role of an internal interpreter to assure mutual understanding of higher education requirements and bring them into practice with a quality-oriented framework which, as a result, established a sustain base of higher education that can be expanded and transformed and contributed to society developing. Higher education standards at the end along with these quality agents should serve to remind universities that their visions should go along with quality relevance and compatibility and at any pressure of flexibility should not be at all expense of quality.

In fact, the continuity of quality and excellence in procedures, systems, and strategies of universities and higher education is very important and badly needed if they are to make a real contribution to the development of society. The social role of a university proves and underlines the inert relationship of higher education and society as it runs the linkage amid the intellectual and instructive role of universities on the one hand and the development of society on the other hand. This link theory is well known in industrialized societies as "Substantial" link.

The role of higher education and contribution in sustainability development can be internationally expanded. The successful experiments can be conveyed to other education systems in other countries. For example, developed countries along with their ultimate higher education can present perfect examples to developing countries. This kind of quality culture exchange does not only help education to grow into healthy quality assurance systems but also provide alive and dynamic procedures that make difference in these countries in short time. This help can also be a form of training of specialists, of professionals, and of highly qualified manpower to meet the needs of the society which in turn help development process.

Universities as higher education institutions should reflect the standards of higher education. The university usually reflects standards by stating their visions and missions. In general, the vision states that the responsibility of the university as a higher education institution should present the society culture, in an innovative and scientific frame of originality that is expected to support and contribute to the society. And correspondingly, the university's mission stays in the same frame that promotes standards to reach a distinct system that is capable of keeping up with developments and national needs. The distinct system adopts innovation, excellence, and scientific impact within a modern participatory concept. Vision and mission of universities should express in the frame of these values:

1. Equal opportunities for all
2. Innovation and excellence
3. Development and sustainable improvement
4. Quality assurance that gives equality of access and learning
5. Social justice, integrity, and accountability
6. Focusing on knowledge delivery and convenience of services
7. Partnerships with related parties that should benefit education systems
8. Transparency and credibility.

Innovations in higher education are not a simple role to play. A vastly high segment of new innovations has come from a small subset of countries. And to a very important extent that knowledge and technology, even when they are being technologically advanced by businesses, depend on academia coming from universities. These innovations have come actually out of quality-oriented education systems with quality reputations. It is very challenging to comprehend how a society converts into an innovative society. As usual, and common sense, we all agree that there is no simple straight path, not a single answer or one way to guarantee results or achievements. For instance, most of high-income countries have businesses depending on academic researches, growth of the outcomes and new innovation emerges. But these companies first need high skilled scientists, experts, and engineers to be doing that kind of study and follow research methodologies to end up with innovation and technology. But even more, the research which is the origin for new innovations is not being done in companies at all. It is being done in national research laboratory, or exploration centers, or at a very substantial extent, in the universities themselves. The link between universities and business is yet a complex and not direct as a third party plays a crucial role. Such a great link between academia and marketplace is commonly controlled by quality-oriented procedures. This pathway of money flowing, if controlled properly with standards and quality assurance procedures that apply rules and regulations for the social justice and accountability purposes, will be a straightforward investment for new innovations. This integration between higher education and sustainable development is a real challenge for each country that dreams to create a national invention system, which is constant with its capacities, with its requirements, and with its occasions.

Being more optimistic, we can say for some countries like the United States and Japan, for example, challenges of sustainable development are under control. Quality assurance and quality control systems are major engines of problem solving; universities and education systems encounter to maintain a stainable development at the end. Sustainable developments' target after all is to achieve life-long problem solving such as having and applying procedures and policies to adopt new liveliness/health system, for instance. We can imagine that with life-long problem solving, many things can change. Rural areas can be grapeseed into civilizations and re-engineered, re-purposed, and re-shaped for recovering a more robust setting with high economic productivity and less impact on the physical environment. And what has been known now for centuries is that the higher education institutions can play a great role in helping societies to contend with their very complex problems, with quality-oriented procedures indeed.

Then, we conclude that higher education plays a crucial role in sustainable development and any educational system should cope with economic development and provide processes to emphasize that education is absolutely critical for monetary development as the fact as education is a conduit for a high-efficiency and -productivity workforce. The quality with control can upgrade the educational process and the curriculum including accreditation, education, and teaching excellence as well as enhancing the excellence in research for the internal and international presence of any institution commits to the higher education values. Quality assurance procedures are meant to enable us to participate more broadly as citizens and participate in each challenge of problem solving following standards and managing control strategies. Higher education along with designated quality assurance work frame has shown itself repeatedly to be crucial for any society positive growth. The endogenous innovation-based growth is always a new knowledge linked to relevant technology-based that developed progressively.

Yet, quality assurance in higher education is not limited to the created innovations internally. They play an important role in any adaptation of technologies from abroad. The technology from outside comes with an imported package in a way of encapsulation but broadly accessible which can lower standards of internal society and damage principles being hardly to be understood. Quality assurance procedures do lots of catching up with this technology growth. The standards cope with a new transferred technology and adopt quickly. The adoption involves training skilled workers, scientists, and engineers who will be responsible to adapt the imported technology to local conditions.

The new field of compatibility should produce highly skilled workers who are important parties in technology transfer, and in turn, universities of course are the main partners for providing that knowledge which is required for that technology transfer. Universities play the role of providing research and development side. They are also fields for training. Training is a long journey that should start with training the trainers, which are mainly the educators, who are going to be employed throughout the social order. And that shows the quality assurance in general and its role in higher education and in specific in problem solving practices. There is always a tremendous amount of innovations that will be required of new systems, procedures, a novel way of thinking, techniques to govern new principles for our behaviors and our organizations of our social lives, and our political systems, and universities are needed to play a key role in that kind of problem solving.

Progressively, universities have been incorporating sustainable development values and practices into their core activities of teaching and research, institutional management, and operational systems.

2.3 Quality Assurance for Higher Education

Teaching, schooling, tutoring, and all learning forms are captivating vouchers to individual's job and higher income. These dreams are nightmares for higher education institutions that get to be afraid of failure to meet the requirements of society. Quality assurance being a tool for problem solving will always keep universities in the enhancing circle of promoting the learning process and operation of administrative services. When we talk about universities, we re-assure the connection between university and society and promoting learning to a landmark of development and culture.

The intended promotion includes applying and making the most current information and communication technologies accessible for upgrading the quality of student services. Transferring learning into quality assurance melody is not a simple task, or a single layer of procedures. The transfer includes multi-layers of quality-oriented building blocks that integrate along with educational systems relying on well-designed strategic planning and right mission and provision to reach an incorporated quality system that behaves with quality and transfer knowledge with quality. This should be a global belief to be shared among countries to have a comprehensive and intercontinental quality assurance framework for framing knowledge and agree on protocols of how this accumulative knowledge-based system is managed and controlled. Universities as learning bodies can be transformed into effective partners with a high level of quality in their cities first place and then in globe as problem solving sponsors. The United Nations sponsored a sustainable development solution network (SDSN). They aim for a very broad membership of institutions of higher education from around the world to join and exchange knowledge and concepts and to deliberate and confer scarce technological approaches, for example, the clean energy. The knowledge network and quality assurance bring universities around the world at the very critical edge of sustainable development, thinking, notions, and methodical know-how. Students can be trained and graduated with required skills around the world forming their own chapter of SDSN.

A comprehensive review of quality assurance literature done by Kahsay (2012) investigates the lack of concept of quality assurance in higher education. Furthermore, Kahsay (ibid, p. 29) quotes from higher education experts who tried before to identify quality:

1. "Notoriously elusive" (Gibson, 1986; Neave, 1986; Scott, 1994)
2. "Slippery" (Pfeffer and Coot, 1991)
3. "Relative" (Baird, 1998; Harvey and Green, 1993; Middlehurst, 1992; Vroeijenstjn, 1992; Westerheijden, 1990)
4. "Multidimensional" (Campbell and Rozsnyai, 2002)
5. "Dynamic" (Boyle and Bowden, 1997)
6. "A philosophical concept that lacks a general theory in the literature" (Green, 1994, Westerheijden, 1999).

Harvey and Green (1993) proposed five "ways of thinking about quality," rather than definitions. These five keys of new thinking had added a new framework in higher education of quality and quality assurance. These five ways to think about quality can be listed as the following:

1. Quality as exceptional/excellence
2. Quality as perfection or consistency
3. Quality as fitness for purposes
4. Quality as value for money
5. Quality as transformation.

Figure 2.1 illustrates Harvey and Green's five ways of thinking about quality in higher education.

Quality as excellence aims to develop academic standards and represent them into an exceptional framework. Working in this framework is not easy or attainable by all. Quality is attained when a higher outcome can be achieved at the same cost, or in other cases, the cost decreases while the outcome is maintained. This working frame should be more practical than just affording quality as excellence theory, but yet maintaining the outcome is not a standard as it holds quality at many aspects in the learning process.

Many other researchers have thought of quality assurance in higher education as a system concerning two aspects: the context and stakeholders (Watty, 2003). Linking quality in higher education to context means that we expand quality and quality assurance to cover perspectives such as quality of assessments, students' intake, and academic programs. Quality as a context is supposed to attach teaching and learning process to quality of teaching and measuring outcomes of learning. Linking quality to context in higher

Figure 2.1 Categories of "Quality."

Source: The chartered Institution of Internal Auditors.

education framework supports the theory of quality-oriented problem solving as quality is the context to evaluate and take the decision. Quality assurance, for example, enlightens problems of old teaching methodologies and quality of teaching is measured. Quality assurance procedures can make decisions when university services do not meet students' requirements such as outdated programs.

Stakeholder's perspectives are the other way of thinking about quality in higher education. There is an assortment of perceptions concerning what quality of higher education is amid different stakeholders. Stakeholders can be students, academics, employers, and policy makers. If we consider the quality from student's perspectives, quality will be more attached to their study circumstances and student-centered teaching and learning. Teachers are more concerned about quality of teaching and how they increase the interaction with students which in turn evaluate their teaching methods. Also teachers worry about their self-development and link the quality to the chances and opportunities and activity involvement they get to enhance their performance, such as research activity.

Quality has always been a part of academic tradition (Newton, 2006). Quality assurance and control are based on peer reviews and self-evaluation and regulation (Van Damme, 2011). Now, we can say that quality assurance in higher education is not steps or procedures to check that it is a cycle of enhancement and improvement. This is a complex framework that starts to be nationally including the quality assurance agencies and then internationally seeking the global recognition. By all means, the quality system needs to be transparent and affords sufficient information to the public and relevant stakeholders, mainly students and their families.

Table 2.1 compiles common definitions of quality assurance in the higher education framework.

The framework of quality assurance in higher education has many purposes that are planned and designed to identify the objectives and needs. Previous studies and literature listed and categorized quality assurance framework based on original purposes and definitions; others as Martin and Stella (2007) had presented the framework and shaped in into two categories: Accountability and improvement.

1. Accountability: Quality assurance is usually linked to the need to afford information for public and certify expectations and minimum requirements of quality assurance standards and goals for the higher education framework.

Table 2.1 A compilation of quality assurance definitions

Year	Author	Definition or description of quality assurance in the higher education framework
1998	Church	It is not about specifying the standards or specification against which to measure or control quality. Quality assurance is about ensuring that there are mechanisms, procedures, and processes in place to ensure that the desired quality, however defined and measured, is delivered.
1992	Barnett	It implies a determination to develop a culture of quality in an institution of higher education, so that everyone is aware of this own part in sustaining and improving the quality of the institution.
1994	Green	Considered important for it enables a university become a learning organization.
1995	Vroeijenstijn	A systematic, structures and continuous attention to quality in terms of quality maintenance and improvement.
1997	Wilger	A collective process by which a university ensures that the quality of educational process is maintained to the standards it has set itself.
1997	Boyle and Bowden	On-going development and implementation of ethos, policies, and processes that aim to maintain and enhance quality as defined by articulated values and stakeholder needs.
1999	Woodhouse	Policies, attitudes, actions, and procedures necessary to ensure that quality is being maintained and enhanced.
2004	UNESCO	A systematic review of educational programs to ensure that acceptable standards of education scholarship and infrastructure are being maintained.
2004–2007	Harvey	A process of establishing stakeholders' confidence that provision (input, process, and outcomes) fulfills expectations or measures up to threshold minimum requirements
2007	Vlasceanu et al.	An all-embracing term referring to an ongoing, continuous process of evaluating (assessing, monitoring, guaranteeing, maintaining, and improving) the quality of a higher education system, institutions, or programs. As a regulatory mechanism, quality assurance focuses on both accountability and improvement (...). Quality assurance activities depend on the existence of the necessary institutional mechanisms preferably sustained by a solid quality culture. Quality management, quality enhancement, quality control, and quality assessment are means through which quality is ensured.

(Continued)

Table 2.1   Continued

Year	Author	Definition or description of quality assurance in the higher education framework
2011	Harvey	Quality assurance, in higher education, has become a generic term used as shorthand for all forms of external quality monitoring, evaluation, or review.
2012	ASEAN and SEAMEO RIHED	A tool for harmonization
2015	European student's union	Quality is a multidimensional concept that touches not only upon quality assurance procedures but also accessibility, employability, academic freedom, public responsibility for higher education, and mobility (Galan Palomares et al., 2013). QA itself serves multiple purposes, enhancing learning and teaching, building trust among stakeholders through the higher education systems, and increasing harmonization and comparability in the European Higher Education Area.
2015	AUN-QA	A systematic, structures and continuous attention to quality in terms of maintaining and improving quality,
2015	Standards and Guidelines for QA in the EHEA	Quality while not easy to define is mainly a result if the interaction between teachers, students, and the institutional learning environment. Quality assurance should ensure a learning environment in which the content of programs, learning opportunities, and facilities are fit for purposes.

Source: Quality assurance definitions adopted from Kahsay (2012) and expanded by the authors.

2. Improvement of Quality Assurance: Enhancement purposes emphasize the importance of internal audience in higher education institutions. The enhancement process enlightens weak sides and recommends ways for improving quality. It appreciates standards and procedures to ensure the cycle of enhancement.

Europe Higher Education Area (EHEA) supports the idea behind students' involvement in quality assurance procedures. They play important roles in higher education policy and shaping quality assurance principles. The Europe Student Union (ESU) had participated and contributed by identifying additional purposes of quality assurance framework at national level. These additional purposes can be summarized as:

1. Providing transparency and information
2. Forming trust among stakeholders over the HE systems
3. Rising coordination and comparability within the region.

These different definitions and frameworks of quality assurance inside higher education concept drive more potential challenges that some wonder how to balance with all these purposes and how to manage all these different concepts of frameworks. More to add, we find that the most important challenge is the possibility of having an integrated framework that fits all higher education institutions all over the world.

To be able to elaborate more about quality assurance in educational systems, we would like to present quality assurance agencies as active tools and allies in quality dialog. The quality assurance agency (QAA) in the higher education frame is an independent body dedicated to check standards and quality in the higher education framework and accordingly works along with its institutions. QAA checks how universities and colleges maintain their academic standards and quality. In most cases, this is done by an external peer review. The reviewer checks and meets academic standards along with higher education requirements. Depending on the reviewer's report, QAA communicates with the university to bring to their attention of any shortages.

Quality assurance agency is expected to work closely with organizations that have an interest in the reputation and excellence of higher education. Taken into consideration that this kind of cooperation should be a quality-oriented process, the main assignment of QAA is to protect standards and expand quality culture of higher education wherever it is conveyed around the world. Four strategic plans are used by quality assurance agencies in order to achieve the mission of quality:

1. Quality assurance agency should have a plan and a process to address the requirements of students. This strategic plan and process help to identify the "Fit for purpose" quality theme.
2. Quality assurance agency should have plans and processes to defend standards and regulations that should play a role of controlling and maintaining the quality assurance and educational systems.
3. Quality assurance agency should design a plan and a process to drive improvements.
4. Quality assurance agency should plan to improve public consideration and understanding of higher education's regulations and social roles.

Now, universities have internal quality assurance units that work in parallel with the higher education QAAs. This harmony and cooperation in between should benefit the university of expanding and practicing the quality culture internally before the step of being evaluated from the outside quality assurance agencies. The internal quality assurance units are actually an active committee that regulates standards inside the system, issues quality documents, and controls the internal quality inside the system. Members of this committee should be aware of quality assurance procedures and control. It represents a robust academic management structure and policies and approaches that help and enable higher education expectations to be fulfilled.

Although there are some metamorphoses between the methods used by quality assurance agencies, they have some keys in common. All reviewers check that higher education's expectations are met; currently, this is done by benchmarking the provision. Quality assurance agencies should cooperate to come with plans and on-going procedures to develop a teaching excellence framework which will help a lot in assessing quality of teaching in universities. Quality assurance agencies as well should work on a common guidelines and handbooks that fit majority of education systems and present them to be learning resources that convey and raise the awareness of quality culture in education. These guidelines can be designed in cooperation with well-established and professional universities and higher education bodies, for example, the European standards and guidelines maintained by the European Association for Quality Assurance in Higher Education (ENQA). Each review results in a published report containing judgments on whether UK higher education expectations are met. Separate judgments remark on academic standards, academic quality, and the public information provided about courses. Reports include recommendations for improvement, citations of moral practice, and declaration of actions taken by the higher education benefactor to improve since the last review.

Quality assurance agencies also have other methods to communicate and evaluate universities. Self-evaluation document is a document with specific evaluation areas done by the university itself and should be submitted to these quality assurance agencies. The other method is asking universities for student's involvements. So sometimes, quality assurance agencies ask for specific surveys, for example, to be filled directly by students. These surveys measure students' satisfaction in terms of quality of university services. These kinds of surveys also widen the awareness of these agencies of student's requirements and also give a chance for students to practice quality culture.

Quality assurance agencies also investigates "systemic failings" by higher education providers where the agency deals with individual complains. This system is designed as a resource for students who have already asked for internal investigation into their complaints and have not found the outcome of this to be satisfactory. Systemic failings are taken to mean a failure by a university or college in meeting its responsibilities for standards and quality. The concern needs to be supported by evidence. QAA deems a full investigation necessary it publishes its findings in a report.

Quality assurance and control agencies all around the world can work through online materials, through joint activities, through common teaching programs and awareness programs, and through common problem solving efforts supported by UNESCO. Quality Assurance agencies work and assure procedures of experiments transfer to other cultures and societies. For example, quality assurance agencies follow procedures of learning and training in the same way that the moral act created the land grant institutions to be available for practical problem solving in other places in the world to achieve close result as possible. Quality assurance agencies work in a sustainable developing process everywhere in cooperation with governments and agencies on the complex challenges of ecological development. This kind of cooperation forms a network that invigorates the extent and force of problem solving, can indeed succeed in great challenges, and can achieve multiple objectives of monetary development, social presence, conservational sustainability, and good supremacy that all of our civilizations need and long for.

In many developing countries, an expansion of higher education fetches deteriorating to quality aimed to rescue educational systems from the random knowledge and practice transfer. To ensure quality in education while responding to sustained increasing demand, it is a compulsory to raise the quality of various aspects, including teachers, students, facilities, equipment, educational materials and methods, and financing. The quality of teachers is particularly obligatory for raising the quality of higher education; therefore, it is an urgent task to inflate the number of students who complete graduate schools. In addition, when higher education rapidly expands, the gap among different higher education institutions widens. In particular, and in many countries, the lower quality of private universities compared to national public universities has been renowned, and there is a requisite to create a quality assurance coordination system that will guarantee the quality of education and research.

The quality assurance is supported with a framework of standards intended to be the guidelines that will contribute to the development of a university in order to achieve the university's objectives and desired outcomes. The quality assurance framework should take into consideration the following points:

1. The framework should check and assure standards and quality presented by university's provision and whether it meets higher education requirements and norms or not.
2. The framework should present quality improvement plans for on-going improvement of the university.

Quality system is an iterative system that starts with planning for quality emerging, formulates plans and strategies for implementations, reviews these plans, and has always a space for enhancement. Figure 2.2 is a simple presentation for a simple quality system.

Figure 2.2 presents a university quality system which has the following steps: Plan, implement, review (with many stages starting from evaluating and asking for data as feedback), and improve which includes decisions of steps that should be improved. These steps are formed in a cyclical system for on-going quality improvement. The system shows principles of the quality and standards' framework which can be easily embedded and demonstrated in all university activities.

1. Plan: The step is to formulate plans, policies, processes, timelines, and responsibilities for achieving outcomes intended to maintain quality, including performance standards, measures, indicators, targets, and methods. These measures and methods are planned for performance monitoring and reporting.
2. Implement: Implementation of planned arrangements, including regular monitoring and reporting on progress.

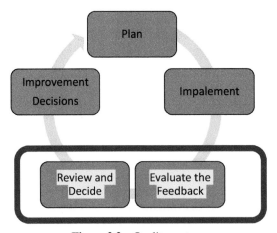

Figure 2.2 Quality system.

3. Review: On-going and summative review based on evidence in order to improve the quality cycle.

4. Improve: Reviewing plans and the way they were implemented should come to generate a new method to improve and incorporate new policies and new implementations taken into consideration what the reviewer forwarded.

References and Suggested Literature

[1] Pavel, A.-P., The importance of quality in higher education in an increasingly knowledge-driven society. *Int. J. Acad. Res. Account. Finan. Manag. Sci.* 2, 120–127, 2012.

[2] Dill, D. D., "Quality assurance in higher education: practices and issues," in *The 3rd International Encyclopedia of Education*, eds B. McGaw, E. Baker, and P. P. Peterson, Elsevier Publications, 2007.

[3] El-Khawas, E., Accreditation of teacher education in the US: An audit approach to subject assessment. Public Policy for Academic Quality Research Program (Chapel Hill: University of North Carolina), 2005.

[4] Lewis, R., External examiner system in the United Kingdom: Fresh challenges to an old system. Public Policy for Academic Quality Research Program (Chapel Hill: University of North Carolina), 2005.

[5] McInnis, C., The Australia qualifications framework. Public Policy for Academic Quality Research Program (Chapel Hill: University of North Carolina), 2005.

[6] Massy, W. F., Education quality audit as applied in Hong Kong. Public Policy for Academic Quality Research Program (Chapel Hill: University of North Carolina), 2005.

[7] The Role of Higher Education in Society: Quality and Pertinence, *2nd UNESCO-Non-Governmental organizations Collective Consultation on Higher Education*, Paris, 1991.

[8] UNESCO Institute for Education (UIE). *Agenda for the Future Adopted by the 5th International Conference on Adult Education*, Hamburg, 1997.

[9] UNESCO Institute for Lifelong Learning. UNESCO Institute for Education (UIE). *The Cape Town Statement on Characteristic Elements of a Lifelong Learning Higher Education Institution* (Hamburg: UNESCO Institute for Lifelong Learning), 2001.

[10] UNESCO. *Communiqué of the World Conference on Higher Education: The New Dynamics of Higher Education and Research for Societal Change and Development* (Paris: UNESCO), 2009.

[11] [Online] Available at: http://www.mohe.gov.jo/en/pages/Vision-Mission. aspx.

[12] *The Role of Higher Education in Sustainable Development*, [Online] available at: https://www.coursera.org/learn/sustainable-development/ lecture/d0fPq/the-role-of-higher-education-in-sustainable-development.

[13] *Higher Education Partnership for Sustainability and Forum for the Future.* Reporting for Sustainability: Guidance for Higher Education Institutions, 2003.

[14] Sierra, L., Zorio, A., García-Benau, M. A., *Sustainable Development and Assurance of Corporate Social Responsibility Reports*, Ibex-35 Companies, 2012

[15] De Beelde, I., Tuybens, S., Enhancing the credibility of reporting on corporate social responsibility in Europe. *Bus. Strat. Dev.*, 2013.

[16] Bennich-Björkman, L., *Has Academic Freedom Survived? – An Interview Study of the Conditions for Researchers in an Era of Paradigmatic Change, in Higher Education Quarterly* (Oxford: Blackwell Publishing Ltd.), 63, 334–361, 2007.

[17] Cavalli, A. (ed.), *Quality Assessment for Higher Education in Europe*, Academia Europaea and Portland Press, London, 2007.

[18] Enders, J., *Higher Education, Internationalisation, and the Nation-State: Recent Developments and Challenges to Governance Theory, in Higher Education*, Springer, Netherlands, 47, 361–382, 2004.

[19] ENQA, *Standards and Guidelines for Quality Assurance in the European Higher Education Area, European Association for Quality Assurance in Higher Education* (Helsinki: European council, Lisbon strategy), 2000.

[20] Fägerlind, I. and Strömqvist, G., (eds), *Reforming Higher Education in the Nordic Countries – Studies of change in Denmark, Finland, Iceland, Norway and Sweden* (Paris: International Institute for Educational Planning), 2004.

[21] Sierra-García, L., Zorio-Grima, A., García-Benau, M. A., Stakeholder engagement, corporate social responsibility and integrated reporting: An exploratory study. *Corp. Soc. Respons. Environ. Manage.*, 2013.

[22] Odriozola, M. D., Baraibar-Diez, E., Is corporate reputation associated with quality of CSR reporting? Evidence from Spain. *Corp. Soc. Respons. Environ. Manage.*, 2017.

[23] Fonseca, A., How credible are mining corporations' sustainability reports? A critical analysis of external assurance under the requirements of the international council on mining and metals. *Corp. Soc. Respons. Environ. Manage.*, 2010.

[24] Zorio, A., García-Benau, M. A., Sierra, L., Sustainability Development and the Quality of Assurance Reports: Empirical Evidence. *Bus. Strat. Dev.* 22, 484–500, 2013.

[25] *A Practical Handbook, Quality Assurance in Higher Education, Central European University Yehuda Elkana Center for Higher Education Budapest, Hungary,* 2016, available online: https://elkana center.ceu.edu/sites/elkanacenter.ceu.edu/files/attachment/basicpage/57/qahandbook.pdf.

[26] Trow, M., Californians redefine academic freedom. *Res. Occasion. Papers Series (ROPS)*, 2005, available online: https://cshe.berkeley.edu/sites/default/files/publications/rop.trow.academicfreedom.3.05.pdf

[27] Dill, D. D., *Quality Assurance in Higher Education: Practices and Issues*, available online: http://www.unc.edu/ppaq/docs/Encylopedia.pdf.

Further reading and websites:

- *Analytic Quality Glossary*: http://www.qualityresearchinternational.com/glossary/
- *Center for Higher Education Development (CHE)*: http://www.che.de
- *Centre for Research into Quality*: http://www.uce.ac.uk/crq
- *Council for Higher Education Accreditation (CHEA) – US*: http://www.chea.org/default.asp
- *European Association for Quality Assurance for Higher Education (ENQA)*: http://www.enqa.eu/
- *Graduate Careers Australia (GCA) – Australian Graduate Surveys*: http://www.graduatecareers.com.au/
- *International Network for Quality Assurance Agencies in Higher Education (INQAAHE)*: http://www.inqaahe.org/
- *National Survey of Student Engagement (NSSE)*: http://nsse.iub.edu/index.cfm
- *Public Policy for Academic Quality Research Program (PPAQ)*: http://www.unc.edu/ppaq/
- *Quality Assurance Agency for Higher Education (QAA) – UK*: http://www.qaa.ac.uk/

- *Teacher Education Accreditation Council (TEAC)*: http://www.teac.org/ *The joint quality initiative*: http://www.jointquality.nl/UNESCO
- *Higher Education/Quality Assurance and Recognition*: http://www. unesco.org/ *World Bank – Tertiary Education*: http://www.worldbank. org/
- http://www.usc.edu.au/explore/vision/strategy-quality-and-planning/qua lity-and-standards-framework.

3

Quality-oriented Strategic Planning
for Higher Education

3.1 Introduction

While expanding, academic institutions try their best to cope with the knowledge and technology transfer. And most of these institutions may underestimate the importance of having a decent planning practice. Engaging with poor planning process may cost the institution a high price that cannot be tolerated. The cost range from disheartened faculty, staff, and students to poor use of resources and most of cases fail to meet the requirement of higher education, which leads to lose accreditation and end up as institution without funding, nor prestige.

In 1960, a new concept of strategic planning started inside the US army. This strategic planning came with significant contributions from RAND researchers to planning and programming (Petruschell, 1968; DonVito, 1969). On the academic side, universities had begun emerging strategic planning into their systems in 1950. The academic experience with strategic planning was limited and premature that focused on instant budget and forecasting essentials involving students and linking outcomes to recruitments, enrollment, and resource growth. Universities by that time were using long-term planning (Kolter and Murphy, 1981) rather than planning for milestones of strategies. The inflammation in higher education caused that most of these plans lost tracks and guidance to control the expansion. This lost and inflammation promoted US public and non-profit organizations including those in higher education to adopt the strategic planning models used in business world (Bryson, 1995; Dooris, 2003).

In the late of 1990s, blue ribbon panels and some educationally related organizations defined new standards and indicators of achievement to estimate output measures in higher education institutions. As expected, these standards spread with wide acceptance and recognition from different institutions as a number of state and federal reports were developed based on these measurements. Higher education institutions were impacted as well and influenced by market forces, so they were also urging to emerge strategic planning processes derived from a business sector into an education system (see the review in Taylor et al., 2008). As time passed, strategic planning in educational systems had abused by cycles and theories of supervision which in turn negatively impacted the academic organization, staff, and faculty. For example, staff and faculty who spent hours to design these strategic plans, and then were suspended and never implemented because of internal circumstances or refusal from stakeholders, would for sure harm them that they would refuse in turn to contribute in further development steps.

3.2 Definition of Strategic Planning

3.2.1 Definition

Strategy as a word has various meanings and definitions. But a common definition involves setting goals and forming actions to achieve these goals. Actions can take the form of decision making to overcome an obstacle or allocate resources that help to accomplish the mission and reach the goals. In the same context, we can define the strategic planning to be a planned and designed process that states a strategy that fits for specific purposes. The strategy involves actions to guide, direct, and make decisions on allocating resources to accomplish this strategy.

For academic institutions, strategic planning is a systematic management activity that sets strategies to define academic institutions' priorities to survive and manage to perform perfectly. Within the academic theme, the actions should be directed toward managing internal activity and reflect quality in performance. For instance, strategic planning can establish agreements around intended outcomes/results of the current educational system. Processes are designed to assess and evaluate system's response and interaction with environmental changes. Academic institutions seek effective strategic planning which not only articulates where the system is going and the actions are needed to make progress, but also identifies if it is successful and sustainable. Strategic planning is *"The process of self-examination, the confrontation of difficult choices, and the establishment of priorities" (Pfeiffer et al., Understanding Applied Strategic Planning: A Manager's Guide). Strategic planning involves "charting a course that you believe is*

wise, then adjust that course as you gain more information and experience"
(Wilder Foundation, Strategy Planning Workbook).

Academic planning can be strategic, tactical, or operational. The following should help in distinguishing between these terminologies.

1. Strategic planning: It comes with a form of long-term planning that compromises all organization's management areas taken into consideration the context of these plans to be relative and relevance to the system. The plans present long-lasting concerns that certify the institution's long-term efficiency and endurance.
2. Operational planning: It comes with a short-term form to help institutions to achieve objectives and carry out actions and activities designed by strategic planning.
3. Tactical planning: It comes with a form of restricted directions, package, or specific programmatic area with a medium term of time wise. Tactical planning is more precise and usually used to tie between strategic planning and operational planning.

In recent years, strategic planning has no longer designed as long-term strategies. They have converted to take the form of premediated planning as many academic institutions find this more appropriate to actions and activities inside the education system. According to this, academic strategies and planning both long-range planning and strategic planning and the difference can be as follows:

1. Long-range planning: It refers to an assigned academic committee's design procedures to oversight the vision and determine what the institution requests to look like at the end of the specified period of time, usually 5 years, and then use that revelation to establish multi-year goals and objectives which form and describe what the institution wishes to undertake and improve. Improvement can be expanded to programs, tasks, and timelines for attaining them.
2. Strategic planning: An assigned academic committee in the education system regulates what it is intended to be in future and how the system can reach institution's goals. The process is designed to develop a vision for the institution and determine the necessary priorities, procedures, and operations (strategies) to achieve that specific vision. The process should involve an evaluation system to guarantee that the system can achieve the designed mission. So, the strategic planning takes the form of cycles of planning, designing, and evaluating for quality assurance purposes that give more space for improving and enhancing.

The main difference between both forms of strategies is that strategic planning is more concerned to place emphasis on strategies itself and how the academic system can achieve its mission, while long-range planning places concerns on determining the vision. It predicts future conditions and realities, and plans how the system can perform effectively within these conditions. Because long-range planning has multi-year projections, it should take the general form more than being specific as short-time strategies. The long term generates comprehensive annual work plans that detail annual objectives, tasks, methods, responsibilities, and timelines. Keep the attention that both long-range planning and strategic planning should have and include measurable systems to evaluate and enhance, and more to meet realistic circumstances and form goals that can be attainable by the relevant education system.

It is difficult to have long-range planning strategies, when the academic institution is yet a young in its domain. The same situation persists when leaders are new and without relevant experience to plan. Therefore, most universities when they are still new establishment take the option of short-range planning and usually it is done for 1 year. Leaders of this new institution need to accommodate experience and practice quality-oriented strategies on how to run the system and manage plans that help the institution to survive. They also need time to have good intellect of community and external environment. Short-term planning also gives a chance for new institutions to grasp harmony, learning, and outcomes to help for longer term planning. This kind of careful strategic planning gets the university to be engaged more with environment to achieve beneficial side and to adapt to the rapidly shifting environment. We believe that institutions that have quality-oriented strategic planning are better understanding and better promoting the life-long learning process. Simply, *"Institutions of higher education that do not rethink their roles, responsibilities, and structures ... can expect a very difficult time in the next decade and the next generation. Some will not survive. Most will be expected to do much more with far less"* (Glassman and Rossy, n.d.). And what we are trying to adopt in our book is quality-oriented strategic planning as a main factor on the way to achieve excellence and sustainability in higher education.

A recent research entitled *"Strategic planning sustainability in Canadian Higher education"* (Bieler and McKenzie, 2017) reviewed the representation in Canadian higher education institutions by linking its depiction in the strategic plans. The study analyzed strategic plans of 50 higher education institutions in order to determine the extent of sustainability included as a significant policy priority in these institutions' strategic plans. The study mentioned 41 strategic plans embedded sustainability in their discussions, and consequently the study was able to identify three characteristic types of responses (Bieler and McKenzie, 2017):

1. Accommodative responses included sustainability as one priority and addressed one or two sustainability domains.
2. Reformative responses involved some alignments of policy priorities with sustainability values in a few domains.
3. Progressive responses made connections across four or five domains and offered detailed discussions of sustainability.

Figure 3.1 is a chart of the responses.

The study presented significant implications for policy makers and sustainability actors in Canadian higher education, as well as internationally. The study discussed and enlightened crucial concerns as follows:

1. Disconnection between strategic planning and other kinds of sustainability policy initiatives within 35% of institutions.
2. The weak language and lack of specific suitability goals within many accommodative and reformative plans.
3. The review pointed to the need to work toward integrative and holistic concrete policy targets at the strategic planning level.
4. The review pointed to the need for further research on strategic planning policy and associated it with sustainability in higher education referring that this would develop a better understanding and acceptance of sustainability transformation in higher education.

In general, practicing excellence strategies in education systems will lead to a sustainability model in higher education. These excellent strategies evaluate the impact on the level of students, and can be at level of each student, employee, and institution. A recent study entitled "The Role of the Practice of Excellence Strategies in Education to Achieve Sustainable Competitive Advantage to Institutions of Higher Education-Faculty of Engineering and Information Technology at Al-Azhar University in Gaza" (AlShobaki and Naser, 2017) helped to identify necessary enhancements that should be applied at the institution level in terms of institutional performance, employee, and students and it contributed to raise the awareness of sustainable competitive advantages that promote excellence culture in the higher education framework. The research work focused on the development and improvement in all aspects of administrative and academic process at Al-Azhar University. At the academic level, efforts were designated toward

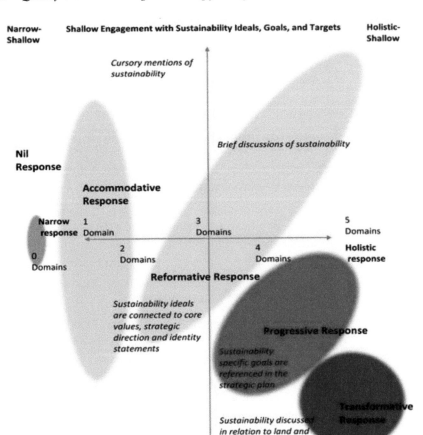

Figure 3.1   Types of possible responses to sustainability in strategic plans (adapted from Sterling, 2013).

Note: "The horizontal axis indicates the breadth of engagement and the vertical axis indicates the depth of engagement. The text in italics is used to indicate some of the characteristics of different depths of engagement with sustainability in the strategic plans. The text in bold indicates the different types of responses. The background colors suggest the range of different responses in relation to depth and breadth." (Bieler and McKenzie, 2017)

assessments to evaluate academic faculty performance, academic programs, and the performance of students which actually represents the learning outcomes. To achieve the research goal, university had participated and contributed by organizing and developing a wide range of activities for faculty and administrators. The research recommended the following:

1. The need to implement more programs of excellence in education systems.
2. Urgent call for integrating the competitive sustainability advantages and excellence strategies to reflect institution's mission and objectives.
3. The call to adopt strategies of excellence in education policy and emerge them with development procedures.
4. The need for staff's contribution in excellence programs.

We conclude that the support at the institutional level for excellence strategies generates appropriate atmosphere for innovation and growth, which in turn makes the difference (SKELTON, 2005).

3.3 Components of Strategic Planning

Higher education institutions often face challenges with strategic planning itself as it is a wide scope and yet untamed to perfectly suiting the academic theory. They have made many attempts to convert strategic planning from the form of business-based plans to the culture quality and excellence in higher education. The conversion had many failure scenarios such as designing a process to motivate assessments-based change without taking into consideration that appropriate timelines to be assigned which turned at the end to a failure change. The change should be supported with information and quality assurance procedures and orientations. With time and practice, universities started to emerge strategic planning and deploy it to articulate institutional mission and vision, to help prioritize resources and promote organizational focus. This was a great transfer and emerge if plans are implemented properly. Plans need to be supported with actions and procedures as spending hours in planning does not guarantee proper implementation. It happened in academic systems that time was needed to activate these plans and recommendations and put strategic planning methods on work, which disappointed staff and faculty who were assigned to plan.

Quality-oriented strategic planning made the difference against challenges and provided practical tools for universities to practice planning competitively with others in domain and in the same/different environment. Californian's university was one of the pioneers who believed in the positive impact of adopting quality-oriented strategic planning into their institutional vision. They re-designed an effective strategic plan for developing human resources of the university. Benefits of emerging strategic planning into academic system are countless. We will list some:

1. Quality-oriented strategic planning creates a framework to guide and direct which pathway to achieve university's vision.
2. Quality-oriented strategic planning provides a framework to practice competitive advantages.
3. Quality-oriented strategic planning allows stakeholders to contribute and work together toward accomplishing goals.
4. Quality-oriented strategic planning allows common dialog between the participants improving the quality system.
5. Quality-oriented strategic planning aims to align the university along with its environment.
6. Quality-oriented strategic planning allows the university to set priorities.

Modern strategic planning has multiple components and each component serves a specific purpose. Strategic components are in fact planning tools used separately or in groups, but their development is linear progression. Strategic planning actions and processes need to ensure that these individual components are aligned with each other with harmony and synchronization. The processes are managed and monitored by quality assurance procedure and the aim here is to provide a mechanism for evaluating progress toward the vision of the associated educational system. Quality procedures are also to maintain the value statement to describe a manner in which the institution will work to achieve its goal. Figure 3.2 presents a simple imagination of strategic planning components.

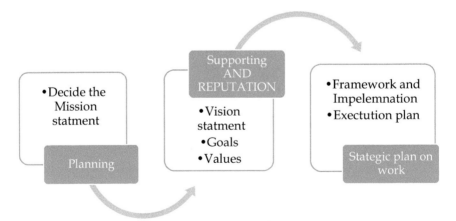

Figure 3.2   Components of a strategic plan.

Please pay attention that strategic planning cannot be planned at one layer. Clear alignment and response of the various stages of the strategic planning system to the environment will support the efficacious implementation of a university strategic plan with the greatest chance for success. To start a strategic planning, the university should response to the following questions:

- Where are we?
- Where do we want to go?
- How can we get there?
- How do we ensure we get there?

These comprehensive questions should be able to help starting a strategic thinking of how to plan and how to perform. Agreeing on values, vision and mission are usually best proficient as a part of planning departure. The mission statement is the groundwork for strategic planning because everything contained in the strategic planning must be aligned with the mission. The vision statement is the expression of the institution's aspiration and is based on the analysis of the institution's environment. The value statements label the manner in which the institution will work to attain its goals. To elaborate more, we will take each separately and detail in the framework of higher education.

1. Mission Statement

Mission is the foundation of any strategic planning and other components should be effectively aligned to the mission statement. A well-written mission statement needs to identify specific purposes of the university and its field of activity in terms of outcomes and differentiate one university from others (Ulgen and Mirze, 2004). The importance of the mission statement goes beyond just a well-written and -defined statement. It is meant to be a communication element that presents the basic purposes, borders of activities of the university. The mission statement should include principles and values to help the university to reach its aim and express strategic objectives (Tutar, 2004). According to Dincer (2004), mission is a long-term goal that concerns about quality, not quantity.

Bart and Baetz (1998, p. 827) examined the relationship between the mission statement and organizational performance. The study came with a conclusion that the well-defined mission statement will increase the performance of the relevant institution for many reasons as follows:

1. A well-defined mission statement provides better staff motivation to achieve the institution's purposes.

2. A well-defined mission statement allows better management practices by motivating employees through institution's goals.

These two major benefits can be realized if the institution's philosophy and values are accepted by the employees without questioning. And to achieve this far, we need to mention that the well-defined mission statement should be clear, concise, and intense. It should define the purpose of the institution, identify the people and organizations served, state the field of activity of the university, mention the needs served by the institution, and express the legal duties of the institution (DPT, 2006; Ercentin, 2000).

Historically, the mission statement was log with a detailed description of the institution's goals. But now, universities seek a well-defined mission statement that reflects the goals of the university without losing the reader in long description and detail. The following examples were copied from websites of universities to give an example of influencing mission statements:

- From Penn State College of Education [10]:

  "The mission of the College of Education is to help prepare out-standing educators, scholars, and researchers, and to advance the profession of education, as broadly defined, through research on the science and art of teaching and learning, the application of clinical processes, the effective uses of technology, and the analysis and development of leadership and educational policy."

- From Harvard, John A. Paulson, School of Engineering and Applied Sciences:

  "The SEAS HR office strives to develop and deliver innovative human resource programs and services designed to support the mission of SEAS and Harvard University. Our core services and competencies include recruitment and staffing, diversity, employee and labor relations, compensation, performance management, employee development, HR information management, and regulatory compliance. Our goal is to be the most effective, accountable, and engaged HR team possible."

2. Vision Statement

The vision statement is another important component of a strategic planning for universities and academic institutions. The vision statement refers to the long-term objectives of an institution and it defines what the institution wants to become in the future and which position it desires to acquire (Efil, 2004) and is the expression of a dream concerning a future desired state. The vision statement shapes future practices (Zel, 1997) and helps to predict future events and be prepared for changes and innovations. The academic objectives might be presented as education, research, technology invention, or cooperation for all mentioned.

A well-defined vision statement should emphasize a unique characteristic of the university that differentiates it from others, and take all future activities planned for internal and external environments of the institution into consideration.

3. Values

Values have been uninvolved in the mission to their own values' declaration component. The university's values elucidate what the institution stands for and the way in which it intends to conduct its activities. In some cases, these values are so significant that the institution has agendas and assessment measures to sustenance and sustain them as key elements. But nevertheless of their priority, within the context of planning and evaluation, the value statement should be declared.

The value component is significant that helps to understand the cultural bases of the stakeholders to work harder or demand more. Distinctive competence shows the basic difference of a university from others which can be related to activities, operations, culture, purposes, interests, or attitudes.

Quality orientation is the core characteristic of strategic planning and approaches that can be expected and included in all strategic planning components. The quality orientation helps to develop an action plan that addresses goals and specifies objectives and work plans on an annual base. Once the elements of the strategic plan have been developed, there will be enough time to ensure specific work plans to be implemented. Strategic planning assures that strategies must reflect current conditions within the educational system and its environment, and thus, it is not practical to develop detailed annual objectives except for the first year, or first 2 years by the maximum. However, annual action plans are needed and should be based on a time measurable system to evaluate and monitor tasks and activities within the time frame.

Often, the higher education system creates a quality assurance committee (QAC) to monitor the implementation of the institution's policies, strategies, and resources for the management of the quality and monitor the core functions of teaching and learning, research, and social contribution. The responsibilities of this committee are expanded to ensure the effective quality assurance policies and quality assurance systems and promote a culture of continuous improvement within the university. There is also a board-staff committee to supervise the entire planning process. The strategic planning process happens at two layers: the first layer is the planning session and most likely to be done by staff and the second layer is the board session which is responsible to review and approve action plans. The key planning sessions perform the best when staff is knowledgeable about the institution and about the community-based organization. Staff is expected to be skilled in group processes and experienced in strategic planning. Some institutions afford management assistance grants that can support the entire process. The board should approve the action plan and staff can perform then on developing side of the defined plans. To summarize strategic planning and QAC, we can say:

1. Assis in the preparation of the institution's strategic plans.
2. Assis in modification and improvement process of strategic plans as required.
3. Design and plan for a mechanism to collect information, data, statistics, and feedback.
4. Make recommendations and suggest adopting new strategic plans to support the existing ones.
5. Create a strategic plan monitoring committee to monitor the implementation of the institution's strategic plans.
6. Consider strategic improvement suggestions from quality peer review reports.
7. Create an assessment committee to monitor standards and assessment processes within the academic system.
8. Plan well for accreditation process by assigning expert faculty and staff, get required consultations from agencies, and prepare the self-assessment documents.
9. Plan well for globalization step by assigning experts and coordinate with higher education council to understand the process and evaluate the current readiness of the system for internationalization.
10. Budget and funding concerns.
11. Allocate learning resources and plan for sharing and maximum usage.
12. Generate reports about the quality level within the institution.

Please note that detailed discussions will be in the following chapter to clear all these points.

One of the purposes of planning processes is to ensure that individual components are aligned with other components and in supportive mode of each. And once the strategic planning motivations and goals are clear, the strategic plan's board goal should guide the process of planning. The guidance can be top-down guidance which emphasizes on vision, mission, and goals. The top-down method runs a linear flow, breaking down higher goals and objectives to sectional objectives. Bottom-up planning is also a practical guidance and the method synchronizes actions toward more abstract mutual goals but do not stress breaking down objectives, especially from one level to the next. The strategic planning processes are expected to secure the commitments of the majority of organization's members as the whole process depends on the participation and contribution of those members. Many procedures and communicating with staff and members will help to spread the quality culture and raise the awareness of strategic planning processes. A plan can be designed to achieve this level of communication, and processes can also be designed to assure the following:

1. Understanding the critical aspects of a current reality in order to transform it in future.
2. Creating a shared vision and define processes of how to achieve it. The process should be transparent to all members.
3. Selecting the appropriate strategic to achieve stated goals.
4. Communicating with staff, faculty, members, and students about strategic planning and processes the system applying which raise the spirit of contribution and involvement.

Asking all staff and stakeholders for more involvements and contribution to the quality cycle does not mean that that they are assigned to be a planning committee member. Only official members get a charge letter to assign them. A sample of this charge letter is shown in Figure 3.3. Charges and duties can differ from one university to another, but we can suggest few important tasks on close for planning processes:

1. Organize orientation sessions and induction for new students and staff.
2. Re-visit the schedules of the university system (administrative and academic ones) and assure that tasks are done by due dates and reports are submitted.
3. Coordinate with system's components in order to generate a quality of work, report, specification, and documents per guidelines and pre-determined standards.

Charge to the Strategic Planning Committee

Thank you for agreeing to serve on the Strategic Planning Committee. Your participation is not only critical to the strength of the College, it is also very much appreciated. The planning process is our way of prioritizing the activities and resources that support our mission. The Strategic Planning Committee is charged to support and monitor the planning process in the following areas:

I.   The Committee

The Strategic Planning Committee is a standing committee established to develop and monitor the strategic planning process for this institution. The composition of the membership includes:

- 3 Senior Staff
- 2 Faculty Representatives
- 1 Student Representative (SGA President)
- 1 Staff Representative
- 1 Alumni Representative
- 2 Staff Support

Terms of service for administrative committee members will be continuous. Terms of service for non-administrative committee members will be set for a predetermined length of time.

Leadership

- The President will serve as Chair of the Committee.
- The Provost will serve as Vice Chair.
- The Director of Institutional Research will manage the completed strategic plan and support assessment of the implementation plan.

II. Strategic Planning Process

The strategic planning process will include the following:

- Development and oversight of all appropriate planning documents (Vision Statement, Strategic Plan, Implementation Plan, etc.)
- A 5-year cycle for implementation of The Plan
- A regularized annual cycle of implementation and assessment
- Institution-level and department-level components

III. Roles and Responsibilities

Committee members will be responsible for the following:

- Understanding the components of an institutional strategic plan and developing those that are necessary (i.e., Vision, Mission, Values Statement, etc.).
- Developing and supporting the objectives and goals of the institutional strategic plan.
- Engaging identified stakeholder groups in the development of the objectives and goals for the institutional strategic plan, and providing feedback to those groups on a continuing basis.
- Overseeing review of annual plans for progress.
- With the support of the Director of Institutional Research, identifying or developing key indicators and assessment measures to document implementation of the Strategic Plan objectives and goals and reviewing those indicators and measures on an annual basis.
- Actively participating in committee activities and discussions.

IV. Other Responsibilities

In addition to the roles and responsibilities outlined above, Committee members will also:

- Promote and advocate for implementation of the institution's Strategic Plan to all internal and external stakeholders.
- Actively engage in disseminating information about the planning process, the Strategic Plan, and its implementation.
- Be aware of strategic issues in the internal or external environment related to the institutional planning process and ensure that the Committee is informed.

Figure 3.3   Sample of a charge letter.

"The charge letter text contains the basic elements of a written charge to any planning committee. The letter is usually modified to fit the culture and unique needs of each campus; however, these basic elements are always included" (Hinton, 2012).

Source: https://oira.cortland.edu/webpage/planningandassessmentresources/planning resources/SCPGuideonPlanning.pdf

4. Focus on processes of developments such as seminars for students and faculty to keep them up-to-date.

5. Focus on processes of improvement cycles and practices such as suggesting plans and academic advisory to deal with students who are struggling through their studies to understand their problems.

6. Early evaluation systems will help a lot to identify obstacles and weakness.

7. Periodically check students' work, assignment, and project to be in the frame of their performance.

8. Encourage cooperation between tutors and students for the teaching–learning enhancement process by conducting more workshops and involve them in quality procedures.

9. Raise the awareness of survey's importance to get all involved and contributed in locating obstacles.

10. Pay more attention for academic affairs' committees to help both tutors and students.

11. Ask expert faculty for remarkable contribution by training others or acting as leaders.

The letter shown in Figure 3.3 holds specific position names such as coordinators, leaders, and senior staff (referring to experts). Creating a committee should be with specific procedure to show the time, expectations out of this committee, responsibilities, and duties. The more specific duties and responsibilities are, the more quality outcomes are expected.

Effectiveness can be extended to validate the vision and the relative priorities of the strategic plan with members of the institution, and this can be easily accomplished in many ways. The most effective way is to develop a culture of planning and quality-oriented strategic thinking inside the institution's system. And specify expectations out of the planning committee which usually opens a group discussion to collect important quality-oriented information and extend contribution.

The university system should assign a process to request an annual plan form each department. The submitted plans should detail any concerns or issues faced by the department. These issues are problems that the department is facing or expected problems in the near future. Incorporating these issues into the strategic planning process will coordinate efforts of planning committee and real needs of the education system.

There are many different models of strategic planning. Some can be summarized as:

1. Conventional strategic planning model and this actually is the most common one although it does not fit for all universities as they have different strategies and structures. The conventional strategic planning is perfect for institutions that have sufficient resources to pursue the vision and mission of the institution. The model includes many phases such as: developing institution mission, vision, and values. Develop the action plans accordingly specifying responsibilities and duties and outline the outcomes. The model is keen for identifying associated plans such as staffing, funding, graduates' plans, and so on.

2. Issue-based strategic planning model which works to achieve the best for universities with limited resources. The model includes: identifying and addressing important issues, concerns, and obstacles facing the institution and preventing it to attain the objectives. The model then designs action processes to work accordingly while documenting the procedures for the next cycle of enhancements. The next cycle evaluates the result and it may end up to replace the old action plans, develop them or keep them as they give perfect results, the model then specifies other issues and concerns, and so on.

3. Organic strategic planning model which is perfect for long-term planning strategies and includes phases such as forming a committee to articulate long-term objectives and suggest how to work toward them.

4. Real-time strategic planning model which suits universities that are under changing circumstances and cannot plan for long-term objectives. So they plan for strategies per situation (real time) and it has phases as well such as: articulating mission in general, and may defer the vision and objectives. Ask for intensive consultation and experts to scan the external environment addressing the threats the university may face and try to plan to overcome in the real time.

5. Alignment model of strategic planning which is mainly to ensure strong alignment of university's internal operation with achieving an overall goal and objective. It is meant to increase the productivity and profitability and successfully integrate new functions or systems such as transforming the whole system to a computer-based system. The model comes with phases as well such as: deciding on the overall goal for the entire system, analyzing internal operations, and aligning them to achieve this goal. Methods that may use include global standards such

as ISO and total quality management systems (which will be discussed in detail later).

6. Inspirational strategic planning model which is perfect when planners have short time for planning and it has also phases such as: creating an inspirational vision for the university and focusing on the vision statement, mainly to attract stakeholders. This model can be highly energizing, and it may produce a plan which is unrealistic for some degree as no enough time was given for action plans and procedures.

Strategic planning techniques need methods of analyzing data associated to the planning and associated procedures. SWOT analysis system can be used for this purpose. SWOT is an abbreviation that stands for strengths, weaknesses, opportunities, and threats. It structures the planning method used in strategic planning to evaluate the above-mentioned four factors in the institution. SWOT involves specifying the objectives of the institution identifying internal and external factors to achieve the institution's objectives. Using other words, SWOT is a perfect strategic planning tool for quality assurance purposes as it identifies and locates quality deficiency within the system, and can be embedded inside the quality enhancement cycles to attain quality. It is closer to be "Fit to purpose" theme as it helps to coordinate factors and compromise them to work toward attaining the main objectives.

A SWOT analysis can be used for the following:

1. Identifying problems and acting like a problem solving process
2. Identifying limitation preventing the institution to attain its objectives
3. Helping with diction making to suggest the best and most effective directions and process
4. Contributing to the quality enhancing cycles
5. Revising plans and encompassing action processes.

Deploying SWOT to promote the quality system has two methods: either to match strengths to opportunities or to convert weakness and threats into strengths or opportunities, such as funding threats can be converted trying to opportunities to find a global provider, for example, or changing the internal market into external market, and so on. Also SWOT can corporate with other analysis strategies such as PEST. PEST analysis stands for political, economic, socio-cultural, and technological, and it is used to describe a framework of macro-environmental factors used in the environmental scanning component of strategic management. Analysis of environmental factors is important for SWOT and can be summarized as the following:

1. Stating objectives.
2. Environmental scanning: can happen internally: promoting the SWOT analysis to include assessments of the present situation.
3. Analyzing existing strategies to identify the four factors: strengths, weakness, opportunities, and threats.
4. Developing strategies according to the analysis findings.
5. Establishing and determining success factors to promote them.
6. Identifying and preparing operational resources for strategic implementations.
7. Monitoring and evaluating results.

In this section, we would like to customize that the strategic planning models and strategies can be the next simple strategic planning approach and can be summarized below assuming a cooperative effort between the board and staff of the strategic planning and QAC. Some of the work can be done in committee, but still cooperating and understanding from all members are crucial. Please note that steps from 1 to 3 may occur before strategic planning retreats. Steps 4–7 are more likely to happen during retreats and steps 8–10 are more likely to occur after the retreat.

Step 1: Agree on strategic planning and process

During the session

a. The committee provides understanding of strategic planning and how it will be executed.
b. They discuss its potential value and provide the common vision.
c. They consider the costs of performing the strategic planning. They discuss about other resources and what might to be given up in order to develop a plan, if the institution is in crisis, for example
d. They consider whether the organization is ready for long-range plan or not yet.
e. They decide whether the strategic planning seems appropriate and plan the action processes to achieve the strategy goals.
f. They should agree upon processes and establish responsibilities for the various steps in these processes

Step 2: Carry out an environmental scan

It helps to provide an understanding of how the organization relates to its external environment. The scan usually includes external components in order to identify and assess opportunities and threats in the external environment.

And it also includes the internal components trying to assess strengths and weaknesses inside the education system. This process is known as "SWOT" strengths, weaknesses, opportunities, and threats. Figure 3.4 summarizes the idea of SWOT.

The uses of a SWOT analysis is to organize information and provide insight into barriers that may present while engaging in social change process. A SWOT analysis's benefits can be summarized as:

1. Enhance "credibility of interpretation." This will help to understand the process and its results.
2. Explore new solutions and apply them to problems.
3. Identify barriers that limit goals and objectives.
4. Revise plan to best navigation of the system.
5. Decide directions with most effectiveness.
6. Reveal possibilities for change.

The external component of the environmental scan should include a review of the target or service community and the broader environment in which the organization operates to identify the opportunities and threats facing the academic institution. The internal component of the environmental scan

SWOT - QUESTIONS

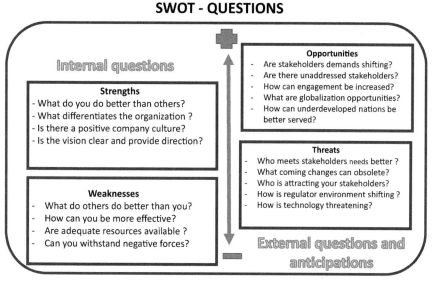

Figure 3.4   SWOT (strengths, weakness, opportunities, and threats).

includes an assessment of strengths and weaknesses. This may include a number of components or approaches.

Step 3: Identify key issues to be addressed as part of strategic planning

Addressing key issues helps the institution to set priorities in terms of time or importance.

Step 4: Defining and reviewing institution's values, vision, and mission

Stating, defining, and reviewing the three important components of a strategic plan which are: values, mission, and vision are crucial processes in strategic plans. The strategic plan should be able to clearly answer questions about the institution: why the institution exists, what goals or outcomes it seeks to active, and what it stands for.

Step 5: Develop a shared vision for the institution

The vision of the institution can be developed after the discussion of community vision, and how it is predicted to come. This is with the assumption that shared institution's vision may be dependent on shared community vision. However, it is important to agree on where the institution wants to be in 5 years, for example. The agreed vision should describe the institution's broadly in terms of its mix programs, reputation, and status inside and outside its primary community.

Step 6: Develop a series of goals that describe the institution in specified number of years

It is a short step from visions to goals. It is very beneficial to transform the vision into a series of goals for the institution and communicate them to the society.

Step 7: Agree upon key strategies to reach goals

This should also include and address key issues identified through the environmental scan.

Step 8: Develop an action plan that addresses goals

Once the long-term elements of a strategic plan have been developed, it is time to ensure a specific work plan to begin implementation. This part should specify objectives and work plans on an annual basis.

Step 9: Finalize a written strategy plan that summarizes results

This includes the outputs of each major step.

Step 10: Create procedures for monitoring and modifying strategies based on changes in the external environment

This is the final step, but actually is a start for the next cycle of assuring and monitoring quality. It is an iterative procedure that aims to enhance and develop the strategic planning and bring them on work.

The above cycle of strategic planning method to determine plans and processes along with evaluation system to enhance and develop these actions represents an integration between quality assurance and strategic planning. In other words, we can say that the above cycle is a cycle of quality-oriented strategic planning method. The cycle is supported with evaluation while implementing which is critical to the success of implementation time. Assessment and evaluation that occur at the beginning are to review institution's achievements of previous plans and modify the goals and steps for the next period of performance. The mid-review provides an opportunity to ensure goal completion.

Universities may hire a planning consultant to help in the strategic planning process which is beneficial up to some level. Many reasons behind this hiring, but mainly, that the university has decided to initiate an effective strategic plan either because it has been compelled to do so by accrediting commission governing board or because of own violation to any of academic standards. But I would go with another reason, which that most of academic institutions decide to initiate a strategic plan and become keen to run strategic processes without any knowledge of how to achieve an effective plan. So, they will be in need to hire an external consultant to help them. A few number of academic institutions can boast of own staff with enough comprehensive experience to lead and support strategic plan without asking for external help. The planning consultant should work with a quality-oriented strategic planning that copes with the institution's vision, mission, and values.

3.4 Integration of Strategic Planning and Quality Assurance

Higher education is an institution of trust and quality. Internal and external competitions among higher education institutions are based on high quality of performance and aligning with quality standards stated by higher education. A part of the quality race in education is planning strategies to help the academic institution to represent itself to the community and achieve pre-determined goals. We have described these strategies with the term of "Strategic planning," as well we have also raised awareness of the important role of strategic planning in educational systems. These strategies cannot hold by themselves if they are not supported with the internal quality-oriented evaluation system that monitors and manages the processes. All these quality assurance steps and internal systems are meant for quality purposes and enhancement cycles.

Institution culture is a key factor that we cannot deny its impact on strategic planning and quality systems. In fact, we can tell that the institution culture may change the frame of strategic planning to turn it into completely a failure or a success. If we pass the same strategic plan to a number of academic institutions, each institution will import other main factors into the procedure and get impacted with culture that ends with a completely different interpretation of this plan and produces different actions and processes. Different implementations of the same strategic plan can be explained by the different factors which impacted the process: institution's environment, internal structure, and competence (administrators, staff, faculty, and students), and mainly the internal culture of quality, planning, and commitment.

In this section, we will try to re-take the strategic planning into a quality-oriented mapping system. And to elaborate, first, we need to introduce the concept of quality mapping system. Quality map is a graphical design and illustration of the quality assurance system. The representation of the quality map is designed to describe the quality framework of higher education institutions taking into consideration: strategic planning, institution's environment and internal processes, and structure of the corresponding academic institution. We have explained what the environmental scan is and how it contributes to the quality system. The scan helps the quality assurance system to raise the awareness of external and internal components and deploy them to serve the quality system. This is actually not an individual step; it is a part of iterative and continuous improvement and enhancing system integrated with the strategic planning action processes to serve customer-oriented goals and achieve constructive external impact on the environment. This is a quality mapping aligning quality to strategic plans in order to reflect the main goals of the academic institution, being transparent in terms of quality internal system, and consequently meet the high-quality standards of higher education. The constructive and positive impacts on the environment always help the institutions to gain reputation and prestige. It is worthy to mention here that the quality map provides a common framework to audit the quality assurance system to prevent audits to be based on backgrounds and subjective practices of the evaluators.

The quality mapping concept helps the academic system to communicate the quality culture to the employees, external evaluator, and other stakeholders. Also it opens more opportunities for quality involvement and long-term commitment as the continuous improvement is principle to achieve high quality at the end. These steps can be simply described by the traditional quality cycle introduced by Deming (1986). The design has two phases: "plan" and "check" phase and "do" and "act" phase. Two phases are integrated into the quality cycle to keep the quality and improvement as main targets to manage the cycle and attain the goals. The internal communication among these steps is a must and done in a constructive way. Figure 3.5 presents the quality and quality map cycles and helps to imagine these phases and clarify the way of internal communication.

The sequential phases of: Plan, do, check, and act as described by the father of quality management (Deming, 1986) set the quality cycle in a four-step problem solving process based on iterations that involve all related stakeholders to perfect operation and output level, which reflect the transparency of the quality system. This cycle also reflects a sustainable developing process if executed properly.

Taking a closer look into the higher education quality cycle, we should mention that the academic frame is excepted to widen the quality processes to

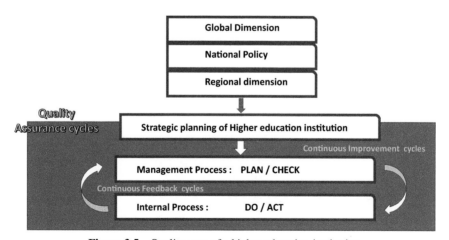

Figure 3.5 Quality map of a higher education institution.

cover academic issues such as research and staff development and processes to support academic services and education policy. These indicators are very important when an education system initiates a strategic plan. Figure 3.6 re-represents the four quality steps assuring the above indicators.

Institutional research and development serves the process of support institution's services. And the support services' processes continuously develop the education policy and process. The whole cycle is supported with a continuous improvement and development steps: plan, do, check, and act. The purpose of the "plan" step is to define objectives and processes necessary to deliver results in frame of the pre-determined specifications. It also includes the strategic plans action plans and process descriptions and operational rules. The input of this phase can be the needs of employers and students, curricula development needs, workload plans for teachers, assigned budget, and human resources. Analyzing the input will define the way of planning and needs should be taken into consideration while planning.

The purpose of the "do" phase is to put the plan on work and implement them properly. The "do" phase depends on teachers, general management, the production of support services, the recruitment of staff, research and development supervision, and project management.

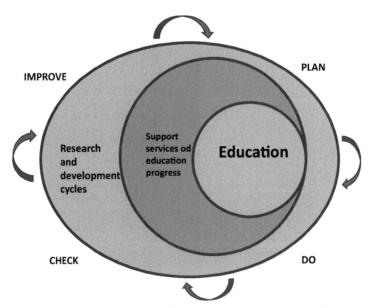

Figure 3.6   Quality cycle of continuous improvement in higher education.

The purpose of the "check" phase is to check and evaluate the processes and report the outcomes. This phase helps to transform the quality system and cycle into actions close to teaching and learning by evaluating the outcomes and recommending improvements. By default, this phase should include a feedback mechanism to collect evaluation and feedback from advisory boards, employers, students, quality assurance agencies, and also ministry of education. Seminars can also be arranged with faculty for annual target discussion to evaluate the achievements of strategic objectives and processes. If the results are unsatisfactory and indicators showed that goals and objectives have not been attained, systematic reflection on what approach and deployment led to these results may suggest useful improvement measures (Woodhouse, 2003).

The purpose of the "act" phase is to apply actions to the outcomes. This step includes deep reviewing all steps of quality cycle and modifies the process, preparing to the next implementation. Modification and improvements at this stage depend on the feedback from the "check" phase. For example, if an assessment shows that learners are not learning as planned and expected, extra tutoring effort can be suggested and embedded into action plans, which is actually an input for the "plan" phase in the following quality cycle.

The quality map takes the role of environment in strategic planning as crucial. It is very beneficial to take into account of education policy and environment needs and align them with internal resources of the academic institution. It brings the importance of aligning resources to the strategic plans, which is a key component of successful strategic planning but unfortunately ignored or deferred at the best till the end of strategic planning. When staff contributing in strategic planning process becomes overly excited and ambitious and push their goals and action plans beyond the university's resources seeking chances to cross-disciplinary and plan for innovations, then it gets stacked for budget limitations. As Goldstein (2012) mentioned in his comprehensive guide to higher education budgeting, *"a budget will not achieve the institution goals unless it is undoubtedly interconnected to the institution's strategic plan"* (Goldstein, 2012). To be able to achieve this, universities should adopt performance-based budgeting practices to train members how to align resources to strategies. The university should also prioritize funding requests and allow a quality cycle to include them while planning phase in order to develop clear objectives based on clear and sufficient information about resources. The university as well can expand the procedure and ask departments to validate their funding requests according to the approved strategic plans.

3.5 Strategic Planning and Assessments

Approved strategic plans are expected to be flexible and updated consciously. The updating should be with assessments, evaluations, and reviews. If the evaluations are positive, the strategic plans can continue on work for a longer term. Reports on the achievements of strategic plans should be documented for the next planning process. Assessments usually produce a final accounting of achievement for the life of the strategic plans. Achievements represent the institution's ability to be flexible and opened for new opportunities that help to maintain goals and objectives. The assessment of a strategic plan should not take Figure 3.7, which shows the cyclical process of reviewing the effectiveness of a planning process.

Planning-related assessments are in fact urging calls for accountability and improvement process. These planning-related assessments at the institutional level occur in two forms:

1 Institutional effectiveness

Institutional effectiveness determines whether the university is meeting the stated mission, and at which level, this achievement is happening. The assessment should be based on impact orientation of the institution culture and recognition by its environment and higher education. The institutional

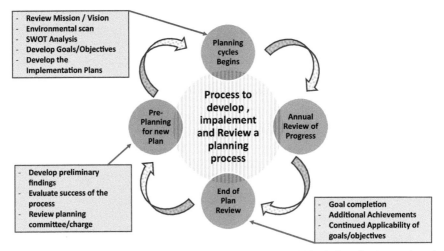

Figure 3.7   "The cyclical process and when the process should include reflection on how it worked and what changes might make it better" according to (Hinton, 2012).

Source: https://oira.cortland.edu/webpage/planningandassessmentresources/planningresources/SCP GuideonPlanning.pdf

effectiveness is measured and evaluated by how much knowledge is generated by the institution and the impact of its activities and practices on the internal system and externally on environment.

Assessing institutional effectiveness helps to sustain a university effective performance and community in diverse, comprehensive, and equitable. It also promotes a system-wide culture of organizational integrity, effectiveness, and openness that enables pursuit of institution's academic ambitions. Another benefit of institutional effectiveness assessment is to steward diversity resources invested by students and the public and private stakeholders in an accountable way to ensure the economic feasibility of the institution.

2 Learning Outcomes

Learning outcome is the perfect measuring tool for successful strategic plans. Measuring the learning outcomes means evaluating the students' performance in the teaching–learning process. It is never about evaluating the teaching and educating process in terms of students passing the course or not.

Learning outcome is considered as a key component of any institutional planning process that must be emerged to the strategic plans. The outcomes/results provide process improvement data and information to develop the next quality cycle.

To end this chapter, we would like to emphasize that strategic planning plays a significant role in any education system. And by now, we can conclude that quality-oriented strategic planning is a main factor to achieve excellence in higher education. If the strategic planning in higher education institutions is planned and designed carefully, it opens doors for collaboration, partnerships, and business investments. As we also emphasize that planning a strategy is not the main objective if the selected strategy is not implemented properly. Reaching a level of well-implemented strategic planning process will be by default a self-sustaining process.

References and Suggested Literature

[1] Hinton, K. E., *A Practical Guide to Strategic Planning in Higher Education, Society for College and University Planning*, available at: www.scup.org, 2012.

[2] McKay, E. G., *Based on materials originally prepared for use with SHATIL, the technical assistance project of the New Israel Fund.* Modified for the National Council of La Raza, and further modified for MOSAICA, May 1994 and July 2001.

[3] Goldman, C. A., Salem, H., *Getting the most out University Static Planning, Essential Guidance for Success and Obstacles Avoid*, available online, 2015.

[4] Goldstein, L., *A Guide to College and University Budgeting: Foundations for Institutional Effectiveness*, 4th ed., National Association of College and University Business Officers, 2012.

[5] Dooris, M. J., Two Decades of Strategic Planning. *Plan. Higher Educ.* 31, 26–32, 2003.

[6] Pavel, A.-P., "The Importance of Quality in Higher Education in an Increasingly Knowledge-Driven Society," *International Journal of Academic Research in Accounting, Finance and Management Sciences*, 2, 120–127, 2012.

[7] EUA Publications, *Examining Quality Culture Part II: Processes and Tools – Participation, Ownership and Bureaucracy*, Brussels, available at: www.eua.be, 2011.

[8] Glassman, A. M., Rossy, G., and Winfield, J., (n.d.) *Toward an Understanding of University-Based Strategic Planning*, California State University, Northridge (Unpublished Manuscript).

[9] Glassman, A. M., *Personal communication*, 1999.

[10] Breneman, D., *Presentation at CSUN: California Higher Education: A State of Emergency?* 1995.

[11] Carron, G., "Strategic planning: Concept and rationale Education Sector Planning Working Papers," *International Institute for Educational Planning*, UNESCO, 2010.

[12] Genç, K. Y., The relation between the quality of the mission statements and the performances of the state universities in Turkey. *Procedia - Soc. Behav. Sci.* 58, 19–28, 2012, ELSEVIER, available at: www.sciencedirect.com

[13] Kettunen, J., "Integration of strategic management and quality assurance," in *Proceedings of the 11th Annual Convention of the Strategic Management Forum*, 2008.

[14] Sidhu, J., Mission statements: is it time to shelve them? *Eur. Manage. J.* 21, 439–446, 2003.

[15] Smith, L. J. E., A vision for the Canadian tourism industry. *Tour. Manage.* 24, 123–133, 2003.

[16] Bark's, B.R., Glassman, M., Mcafee, B., Mission statement quality and financial performance. *Eur. Manage. J.* 24, 86–94, 2006.

[17] Available at: http://ed.psu.edu/

[18] Available at: http://www.seas.harvard.edu/

[19] Özdem, G., *An Analysis of the Mission and Vision Statements on the Strategic Plans of Higher Education Institutions*, Kuram ve Uygulamada Eğitim Bilimleri Educational Sciences: Theory and Practice – 11(4) • Autumn • 1887–1894 ©2011 Eğitim Danışmanlısğı ve Araştırmaları İletişim Hizmetleri Tic. Ltd. Şti.

[20] Kettunen, J. Kantola, M., The Implementation of the Bologna Process. *Tert. Educ. Manage.* 12, 257–267, 2006.

[21] Kettunen, J., and Kantola, M., Strategic planning and quality management in the Bologna Process. *Perspect. Policy Pract. High. Educ.* 11, 67–73, 2007.

[22] Laredo, P., Revisiting the third mission of universities: Toward a renewed categorization of university activities? *High. Educ. Pol.* 20, 441–456, 2007.

[23] Beckford, J. (2002). *Quality*. London: Routledge.

[24] Bieler, A., McKenzie, M., *Strategic Planning for Sustainability in Canadian Higher Education (Sustainability 2017)*, available at: http://www.mdpi.com/2071-1050/9/2/161/htm

[25] Al Shobaki, M. J., Abu Naser, S. S., The Role of the Practice of Excellence Strategies in Education to Achieve Sustainable Competitive Advantage to Institutions of Higher Education-Faculty of Engineering and Information Technology at Al-Azhar University in Gaza a Model. *Int. J. Digital Publ. Technol.* 1, 135–157, 2017.

[26] DonVito, P. A. (1969). THE ESSENTIALS OF A PLANNING-PROCRAMMING-BUDGETING SYSTEM. Available online, http://www.dtic.mil/dtic/tr/fulltext/u2/690394.pdf

[27] Charles A. Goldman and Hanine Salem, Getting the Most Out of University Strategic Planning: Essential Guidance for Success and Obstacles to Avoid. Available online, https://www.rand.org/content/dam/rand/pubs/perspectives/PE100/PE157/RAND_PE157.pdf

4

Quality Assurance Management and Control

4.1 Introduction

When we talk about the relationship between quality and higher education, we can say that "quality" refers to standards of procedures and measurements for comparison purposes, against a similar kind or the degree of excellence. And when we seek excellence in higher education, we seek distinctive attributes or characteristics possessed by the desired level of educational systems. According to the ISO 8402-1986 standard, quality is defined as "The totality of features and characteristics of a product or service that bears its ability to stratify stated or implied needs." Most of researcher in "quality" field believe that it is a relative tenure that takes different meanings to different people which takes us to say as some defined "quality is fitness for use or purpose" while others mentioned quality could be a "conformance to standards." In business world, quality standards are expected to cope with customer's needs and should keep on performing its functions as required from stakeholders as agreed upon standards (Murad and Rajesh, 2010). Academic researches and efforts have been on-going to provide a synthesis of literature on how to define quality and how to emerge this practice into the higher education framework. From here, the term of "quality assurance (QA)" has risen and deployed to signify the level of academic achievement accomplished by higher education institutions. The hypothetical standards should be monitored and controlled on progressively base. QA is companied with dedicated procedures and action processes to outline quality and assure the practice. This is very challenging to any educational institution and the most challenging in being aware of the concept of quality itself and how to communicate the quality culture and as well how to put into default practice, taken into consideration that quality cannot be precisely defined or quantified (American Society for Quality, n.d; Bobby, 2014; Martin and Stella, 2007; Mishra, 2007; Westerheijden et al., 2007). To overcome such an obstacle, higher education deliberated experiments and strategies to deliver a conceptual model of quality-oriented educational structure that can be adopted and applied by most of the academic institutions.

The conceptual model is actually a bilingual structure that means it can communicate and adapt to the internal changes inside the academic institution while assuring "quality." The model is flexible and open to examine considerations for defining QA and provide recommendations to serve fetching a gran configuration to the existing practice inside the academic institution. This means that quality should not be a burden or with high cost to any institution seeking excellence in higher education theme. There are no doubts that more studies are needed indeed to control the feasibility of developing a universal definition of quality that would smear to different institutions regardless of geographic locations. The universal definition subjects the terminologies of quality to culture changes and nation developing progress that can lead to better comprehend the inspiration and influence of culture and identify more reasons as "why quality is difficult to manage in higher education?"

Sahney et al. (2004) indicated that different resources are attached to the term "education" and considered as important influencing input such as: students, teachers, and physical and financial resources and are expected to go under processes as teaching, learning, research, administration, and knowledge transformation. Mentioning the "input" term in education refers to the fact that education systems are viewed as "a system or a network of interdependent components that work together to try to accomplish the aim of the system" (Deming, 1993, p. 98), which is a "product" at the end. Then the academic consists of inputs, transformation processes, and outputs. Graduating with high quality level of learning knowledge accumulating and training is a "quality-oriented outcome" or "quality-based product" that all higher education institutions aim/look for.

Before in Chapter 2 we have stated that higher education plays critical roles to support the society and provide effective tools for self-governing civil society. Quality is essential in higher education systems to provide the social norms of communication and interfacing such as logical thinking and reasoning to countersign the control of its components with full of harmony without any conflicts or biases. Quality is also a crucial benchmark in the enhancing and developing cycles that provide prototypical models to measure the effectiveness of higher education systems, continuity of strategies, and reliability of action processes accomplishing the mission of this system. And nowadays, heading toward globalization and internationalization demands accumulatively for quality adoption in academic systems and seeks evaluation procedures that work and monitor in parallel with higher education to assure quality practices. Then logically, "quality" with "procedures" turns to be the QA procedure and process concept that obviously measures the framework seeking workable guidelines, signs of good practice, and tools that will facilitate procedure to achieve the mission of that system. Next, QA will be taken in detail to help understanding how institutions can adopt it.

4.2 Quality Assurance in Higher Education

4.2.1 Concept of Quality Assurance

Quality assurance appeared as a business policy in western world throughout the 1050s and early of 1960s. The notion of "quality assurance" is intangible as it articulates relative alterations between one component and another. The concept of QA is an accredited factor and a measurement of excellence in any successful business. When the QA concept was emerged into higher education, it innovated new terminologies and methodologies to suit the education process. It presented a great number of different perceptions and suggestions to support the education and fit to the purpose being employed for. After decades of adopting the concept, the usage of QA is still in confusion and it has been proven to be a deceive factor in determining the success or failure of this emerge in higher education.

Trying to define "quality" is a privilege to state the QA concept and as a result of having more than one definition of quality, we have significant challenges to outline the QA concept. In common, if we want to assure something, we need to prepare a list of pre-agreed and pre-determined procedures that all submit and apply it. Therefore, we can re-define the QA again as a set of processes, policies, or actions performed extremely by QA agencies and accrediting bodies or by the internal procedures of the academic institution (Borahan and Ziarati, 2002; Commonwealth of Learning, 2009; Opre and Opre, 2006; Peterson, 1999; Quality Assurance Agency for Higher Education, 2012; Vlăsceanu et al., 2007). The new definition impacts accountability and assures a continuous improvement which cannot be stepped down as there are increasing demands and emphasis on quality enhancing cycles (Campbell and Rozsnyai, 2002; Nicholson, 2011; Singh, 2010; Srikanthan and Dalrymple, 2004). Some others believe that identifying specific aspects of quality that will be assured in order to guarantee fulfilling own purposes and meet requirements of higher education will contribute to decide quality level (Martin and Stella, 2007). These specifications can be re-defined as indicators as we will explain in the following.

Application of QA in higher education may take the side of specifying indicators for identifying quality. Many recent publications have presented this strategy, such as the quality matters rubrics does not include a comprehensive definition of quality, but it signposts specific indicators of quality (e.g., "A variety of instructional materials are used in the course"; Quality Matters, 2014). For the higher education system, four indicators have been identified: administrative, student support, instructional, and student performance. The first three (administrative, student support, and instructional performance) refer to the inputs, while the last is the desired output such as improvements in learning and reflects good performance which in turn assures quality (Tam, 2014). Table 4.1 shows these categories.

Quality assurance strategy of deploying indicators to measure quality can lay out a conceptual model to imagine how to define and assure quality by specifying its conceptual requirements. Figure 4.1 reflects the prominence of eliciting stakeholders' perspectives to define quality and the indicators used to measure quality (Bobby, 2014; Cullen et al., 2003).

Regional framework can have influence on defining QA and limit its practices in higher education. In some regions, QA and accreditation are used synonymously, while in other regions, the terms are not even related (Organization for Economic Cooperation and Development, 2005). For instance, Vlăsceanu et al. (2007) stated that QA can be "an ongoing, continuous process of evaluating (assessing, monitoring, guaranteeing, maintaining, and improving) the quality of a higher education system" (p. 48), while the accreditation as "the process by which a (non-)governmental or private body evaluates the quality of a higher education institution in order to formally recognize it as having met certain predetermined minimal criteria or standards" (p. 37). As an international example, the American Council on Education (2015) stated that accreditation is a pre-requisite step for quality in higher education institutions, but they on the other hand raised the concern of accreditation being insufficient for demonstrating the overall institutional quality. The illustration presented the impact of region, and therefore, there are calls to consider the regional context when we try to outline the QA.

Table 4.1 Categories of quality indicators

Categories	Definitions
Administrative indicators	A set of quality indicators that pertain to the administrative functions of an institution, including developing a relevant mission and vision, establishing institutional legitimacy, achieving internal/external standards and goals, and procuring resources for optimal institutional functioning (Cheng and Tam, 1997; Commonwealth of Learning, 2009; Hill et al., 2003; Iacovidou et al., 2009; Mishra, 2007; Online Learning Consortium, 2014; Owlia and Aspinwall, 1996; Zineldin et al., 2011)
Student support indicators	A set of quality indicators that pertain to the availability and responsiveness of student support services (e.g., the degree to which student complaints are adequately addressed; Garvin, 1987; Hill et al., 2003; Iacovidou et al., 2009; International Organization for Standardization, n.d.; Lagrosen et al., 2004; Mishra, 2007; National Institute of Standards and Technology, 2015; Oldfield and Baron, 2000; Online Learning Consortium, 2014; Owlia and Aspinwall, 1996; Quality Matters, 2014; Wong, 2012; Zineldin et al., 2011)
Instructional indicators	A set of quality indicators that pertain to the relevancy of educational content and the competence of instructors (e.g., programs and courses that prepare students for employment; Biggs, 2001; Commonwealth of Learning, 2009; Harvey and Green, 1993; Hill et al., 2003; Iacovidou et al., 2009; Online Learning Consortium, 2014; Quality Matters, 2014; Tam, 2014; Wong, 2012)
Student performance indicators	A set of quality indicators that pertain to student engagement with curriculum, faculty, and staff, and increases in knowledge, skills, and abilities that lead to gainful employment (e.g., increased critical thinking skills; Bogue, 1998; Cheng and Tam, 1997; Harvey and Green, 1993; Harvey and Knight, 1996; Haworth and Conrad, 1997; Iacovidou et al., 2009; Scott, 2008)

To be able to complete about illustrating quality and QA emergence in education, we need to agree first that "quality" and "education" are now two sides of one coin which means that quality has to be a building block of education pyramid aiming the excellence at the top. And as the education itself, especially the basic one, is not adequate or complete, it should be supported achieving the lifelong learning to become one of the props of the development system. Quality as a concept is not enough to support the education unless we promote the tools it needs to get to the purpose, and these tools should have interactivity to be controlled and managed.

Figure 4.1   Conceptual model of quality-oriented definition for higher education institutions.

Quality management is a concept related to quality and QA that aims at attaining quality goals through planning, monitoring, assuring, and improving quality components at all levels as the quality is not limited to one level. Quality management, can also refers to the integral of more than one system starting form strategic planning stage, applying action processes and management procedures adopted to confirm achievement of pre-arranged quality or sustain the enhancements were achieved so far and give the required confidence to key stakeholders about the "control" level, this education system has managed to attain. Quality control is also a key concept when we try to detail the quality and QA concepts. Quality control is seen as a system of routines and activities that should take the technical form designated to control inventory quality while going through the development cycles by providing consistent checks to ensued: data integrity, correctness, and completeness. This control system also assigned to document and archive all steps being executed, identifying errors and omissions. In general, quality control should include technical methods as accuracy checks on data acquisition and estimating uncertainty that help to build the system on facts and real interactions. We reach to the fact as well that QA and quality control are also two sides of one coin. QA plans should include activities to assure the implementation of quality control procedures and support its effectiveness.

Quality management and control systems must be consciously assuring quality in order to stratify quality demands. And referring to our claim that "quality" and "education" are two sides of one coin, we can reach to a fact and agree that quality management and control are important and should be considered as a fundamental part of the whole education system's management and administration. The fundamental base role is to manage and control the implementation of guidelines and plans derived from the established quality system and assure carrying out activities within the whole educational system. The fundamental base also helps to understand the system's structure, procedures, processes, and other necessary resources required to the application of quality. The above conclusion and fact should also be seen as an advantage supporting the education system as one unit not only to attain quality objectives but also to develop quality coping with new requirements by stakeholders and/or globalization goal.

Responding to the need of higher education to become an international service involves lots of concerns over the world about global quality, global standards of evaluation systems, and universal recognition as well. Measuring and managing quality in higher education has been proven to be a challenging task because of the fact that two main reasons have been identified: Diversity of "quality" itself and having different meanings for different stakeholders and the second reason is that higher education is necessitated to demonstrate accountable engagement in their professional practices and accountability in outcomes they achieve with the resources being used (Jackson, 1998, p. 46). Elton (1992) presented the quality as "accountability," audit, and assessments which involve lots of concerns about the quality control and people who control the quality. Harvey (2005, p. 264) stated further that accountability underpins QA processes but on the other side lowers the standards of efficiency and effectiveness. Then, it is not only about quality but also who is managing and controlling quality. This adds more burden on higher education institutions if those people constraints the financial plans while there still pressure to improve the quality standing in newspaper's higher education institutions' performance tables to support recruitment (Tambi et al., 2008).

Despite of the huge number of theories and studies about applying quality control and management in higher education, there is still a real need for conceptual models and recommendations of quality management dimensions in higher education. These dimensions are expected to be combined together to achieve QA and focus on the maintenance and improvement of all functions within in the educational system with the aim of meeting stakeholders' requirements (Flynn et al., 1994; Kaynak, 2003). Before pursuing with the quality emergence in higher education, we are expected to be able to differentiate between quality management and quality control for higher education systems. The conceptual model of quality components and terminologies must define "quality" as the degree to which the system meets the requirements and being able to attain the goal (PMBOK, 2009), which actually makes it most of the times to be sensitive to manage and control the quality in order to achieve the desired outcomes and encourages more to look briefly at the distinction between quality management and quality control.

Quality management is the practice of sketching and outlining plans. These plans can take the form of strategic planning and designated to determine the standards the system needs to apply. Also, QM involves the decision of who would be involved in managing quality and their specific duties which mainly is to decide if the system is conforming to the quality specifications laid out in the quality management plan and laying out the metrics that are used to measure quality.

Quality control is the set of processes that measure the metrics of quality by assessing the results against standards. Quality control processes monitor and control functions whereas quality management is done during the initiating and planning phase (PMBOK, 2009). Being the other side of the same coin, quality control should support and polish the quality management cycle.

In conclusion, quality management takes into account the lower level details of how the output of the project is to be tracked and measured. Quality control is the process of ensuring that the quality metrics are met. Using other words, quality management is the process of planning and managing to assure quality while quality control is the process of measuring the level of output against quality standards that were decided upon in the quality management plan.

Quality management and quality control are actually too immense to be covered in a few lines, but we attempt here to provide concise and articulate definitions of these important terms. Nowadays, there is intentions to use terms of quality management and quality control interchangeably due to the perception that quality control encompasses the planning aspects as well. The idea has been common and in use for universities that don't have isolated quality department inside the educational system. Other universities run quality, management, and quality control separately and assign them to different people.

4.2.2 Mechanism of Quality Assurance in Higher Education Systems

Quality assurance practices in higher education institutions have a conceptual model and a well-designated mechanism that takes a form of cycles. The internal steps can be customized as follows:

1. Awareness (stipulating) of higher education standards: Being aware of higher education standards helps universities to cope with these specifications and take them as conditions while planning.
2. Quality management and planning stage: This stage is incorporated with the strategic planning concept in higher education systems to integrate quality requirements and higher education specification to outline the framework of the educational system.
3. Quality control that outlines how the execution of the higher education standards will be taken place, namely, the activity of fulfilling a standard or criterion.
4. Evaluation of the execution of the higher education standard which is the activity of comparing the output of the activity of fulfilling a standard or criterion with the higher education standards.
5. Control of outcomes in order to match the higher education standard, specifying the activity of analyzing causes that arise in the failure to achieve the standards. This step includes lots of documentation, reporting, and archiving.
6. Enhancement to match of the higher education standard, specifying the repair activity that is being designed to repair and fix errors when ever found. Then back to the first step.

It is a fact that QA is not only one cycle as mentioned above, it is a system of similar cycles designed for each level of quality in the educational system to work and detect quality deficiency when ever found. QA procedures and controlling systems refer to repeated references to various components of quality in an education system that can be taken to form a useful analytical framework with interconnected components such as: effectiveness, efficiency, equality, relevance, and sustainability. It is worthy to mention here that effectiveness is not the internal effectiveness of the system which refers to the level of the educational system meeting and achieving the desired objectives, but the external effectiveness is also important to take into consideration. External effectiveness refers to the total score of the education process that meets the needs of individuals and society as a whole. Designing a network of such cycles to work and get them integrated inside the educational system will be problem solving procedures to support the educational system attaining the quality as specified by higher education.

Higher education is responsible about defining the standards and forward guidelines to each academic institution in order to communicate quality culture in a proper way. Standards of a general higher education institution may involve the following:

1. Vision, mission, objectives, and strategies.
2. Governance, leadership, system management, and QA.
3. Student's services and graduates.
4. Human resources, faculty members, and supporting staff.
5. Curriculum learning (outcome) and general quality culture.
6. Finance and physical resources including technology and information system.
7. Research, social responsibility, and open cooperation.

4.2.3 Quality Enhancement Cells

Academic QA systems are still under discussion and no solid agreement can be mentioned till now. Yet, most of the academic institutions that seek the accreditation from the higher education activated quality-oriented cells internally, and being dedicated to enhance the QA procedures and enable it to get into practice with a reasonable flexibility to change accordingly with new demands. The quality enhancement cells (QECs) present an active factor managing and maintaining all QA policies and practices, and assuring them to play the roles not only in developing quality cycles but also in ensuring academic practices through academic quality standards designed and maintained by the QA system. The next will be a brief for how to establish such cells and what level of academic standards they look for to form.

4.2.3.1 Academic Standards

1. Academic quality standards

The academic standards are important dimensions to decide whether the procedure of the QA system is performing effectively in an academic institution or not as they are attached to the teaching–learning processes. The quality-oriented academic standards are the basic requirements of higher education. The following outlines can help to form these standards:

- Intended learning outcomes (ILOs): The ILOs refer to the required knowledge, understanding, and skills to form the system integration with the mission statement. It also echoes the use of external reference standards at the proper level.

- Curricula: It facilitates the recognition of the ILOs. The quality of curricula refers to the significant role in defining the quality of teaching and learning outcomes.
- Student assessment: It is a set of academic-related processes such as examinations and other activities conducted by the institution to measure the performance and effectiveness of the ILOs of a course/program.
- Student achievement: The QA system of universities is expected to effectively assure the quality of student's performance. This will help in evaluating and enhancing the quality procedures if weakness encountered.

2. Quality of learning opportunities

Academic quality standards cannot hold it without supporting it with ongoing development cycles. The enhancing counts on also developing a common understanding on quality of learning consequences. These learning opportunities use indicators for evaluation purposes:

- Teaching and learning indicator: Indicator reflects the most effective teaching and learning strategies being innovated by the educational system and the method being used to identify them.
- Student support: Student support indicator refers to the general provision and support from the education system to afford services to students and dealing with possible academic problems seeking at the end, to attain students' satisfaction.
- Learning resources: Learning resources are important to make the teaching–learning process a success. They help to:
    - The facilities at the institution for learning are appropriate, suitable, and used effectually.
    - The system's members are dedicated to adequate and meet the provisions of academic standards.
    - Competence spirit of utilizing these learning resources for innovation, creativity, and research purposes.

3. Research

The QECs hold the responsibility of establishing a research foundation and creative academic activities related to the teaching–learning enhancing process. Research is considered an indicator for quality enhancement cycles inside the educational institution demonstrating the effectiveness of quality-oriented academic plans, diversity of topographies, and the level of interconnections inside the institution in terms of academics.

4. Community participation

The community contribution in the whole development of teaching–learning process is important to accomplish the preferred level of QA. Therefore, more efforts are needed by the QECs to introduce the theory where it does not exist formerly and to style it more effective where it exists in underlying mode and is difficult to be practiced.

QA guarantees that the higher education institution, driven by its mission, makes a significant contribution to the community it belongs and this can be evaluated by: The contribution it makes, the range of activities, relevance to the institution's mission and plan Examples of effective practice

5. Quality management and enhancement

While executing the quality management and enhancement processes at universities, the management level should focus on the following areas:

- Governance and leadership: They are expected to be appropriate to execute existing academic activities and interact with development and change demands.
- Academic leadership: It is expected to provide sustainable basis for academic activities to mature in a productive learning environment.
- Self-evaluation: Tasks such as self-evaluation, reporting, and improvement strategies should be transparent to be able to identify quality gaps and take them into consideration the next cycle of enhancement.
- Management of stakeholders' feedback: The strategies should afford a suitable mechanism for receiving, processing, and reacting to the assessments and feedback coming from an assortment of stakeholders. The feedback utilization is meant to promote the effectiveness of the QA system.

4.2.4 Quality Assurance and Management: Extending the Framework

Managing quality system in higher education can come out of standards as mentioned before. Taking European experience, the quality system is inseparable from international and European norms (standards) of quality. This standard is actually a solemnization of rudimentary principles of quality management. A cumulative number of factors, not only the business side, are determined to adapt its own quality system with requirements, mainly of the International Organization for Standardization (ISO) 9000 which is a family of quality management system standards. These standards are designed to make it possible for an academic institution to ensure meeting the requirements of stakeholders while keeping the statutory and regularity needs related to higher education. Any higher education institution that desires to be accredited to the credential must go through several stages and requirements of ISO 9000: the development of a quality scheme that implements the requirements of ISO 9000:2000; the selection of an accredited certification body; pre-auditing of the quality system by the certification body; the ultimate audit of the quality system after which the certificate is issued; and a series of smaller audits at least once a year. ISO 9000 is related with the fundamental of quality management systems including the seven principles of quality management which are: customer service, leadership, engagement of people, improvement, evidence-based decision making, and relationship management (ISO 9000:2015), while ISO 9001 deals with the requirements of the educational system wishing to meet the standards that must be fulfilled (ISO 9000:2015). ISO 9001 contents are as the following: Scope, normative references, terms and definitions, context of the university, leadership, planning, support, operation, performance evaluation, and improvement. Most of these attributes were discussed before with the theme of strategic planning: Plan, do, check, and act cycles. Third-party certification bodies are required to provide independent confirmation that the university meets the requirements of ISO 9001. In the process of developing and implementing QA and quality control plans, it will be beneficial to mention guidelines and standards specified by the ISO in Table 4.2 including the ISO 9000 series.

ISO AS A DATA QUALITY MANAGEMENT SYSTEM

"The International Organization for Standardization (ISO) series programs provides standards for data documentation and audits as part of a quality management system. Though the ISO series is not designed explicitly for emissions data development, many of the principles may be applied to ensure the production of a quality inventory. Inventory agencies may find these documents useful source material for developing QA/QC plans for greenhouse gas inventories. Some countries (e.g., the United Kingdom and the Netherlands) have already applied some elements of the ISO

Table 4.2   ISO standards

ISO 9004-1	General quality guidelines to implement a quality system
ISO 9004-4	Guidelines for implementing continuous quality improvement within the organization, using tools and techniques based on data collection and analysis
ISO 10005	Guidance on how to prepare quality plans for the control of specific projects
ISO 10011-1	Guidelines for auditing a quality system
ISO 10011-2	Guidance on the qualification criteria for quality system auditors
ISO 10011-3	Guidelines for managing quality system audit programs
ISO 10012	Guidelines on calibration systems and statistical controls to ensure that measurements are made with the intended accuracy
ISO 10013	Guidelines for developing quality manuals to meet specific needs

"*Source*: http://www.iso.ch/" – (We do advise to read more about the ISO standards)

standards for their inventory development process and data management."(IPCC Good Practice Guidance and Uncertainty Management in National Greenhouse Gas Inventories, Chapter 8, p. 7).

It is worthy to mention that there are many recent research and studies about adopting ISO 9000 and its series in higher education. A study entitled "BUSINESS PROCESS AND QUALITY MANAGEMENT IN HIGHER EDUCATION INSTITUTIONS" (Surbakt et al., 2014) contributed with a development business process and quality management framework in higher education institutions. The proposed framework is based on Deming cycle: Plan, do, check, and action and business management common body of knowledge (BPMCBOK). The authors propped to use IWA2 guidelines (quality management systems guidelines for the application of ISO 9001:2000 in education) for self-assessment purposes. Different researchers looked to identify the reason for failing to obtain IS0-9001 of a huge number of universities. One of these studies is entitled "A Suggested Proposal to Implementation Quality Management System ISO-9001 in Egyptian Universities" (Alalfy and Abo-Hegazy, 2015), which aimed at identifying reasons of most of the Egyptian universities failing to achieve the ISO-9001. Authors also suggested a proposal for the application of quality management system ISO-9001 in Egyptian universities to avoid mistakes and problems that have taken place in the past.

As with ISO standards, we have noticed that emerging and implementing of QA cycles requires resources, expertise, and time. The experience is expected to control and manage many tasks regarding the quality as:

1. Resources allocated to quality management level and quality control level using different sources:
   Categories and compilation process to avoid conflicts.
2. Time allocated to conduct the regular checks and review of errors.
3. Availability and access to information and data.
4. Procedures to ensure confidentiality of inventory and information when acquired.
5. Requirements for documenting and archiving.
6. Frequency of quality managing and controlling on different parts of the system.
7. The level of quality control should be appropriate to each source category.
8. Whether sufficient expertise is available to conduct checks and reviews.

With all these requirements and standards of higher education to recognize and accredit the educational system, universities are confused. Within many universities, there are often quite variations in views about the essential characteristics of QA, quality management, and quality control systems as well as what characteristics of institutional work are regarded as being greatest value and why. Universities have concerns about academic recital at the highest level and how such performance can be recognized. Also, it is sometimes about a measure of disagreement within many of higher education institutions about the constitutions of good teaching, which graduates have the most valued characteristics.

Away from such confusion, research continues in terms of quality and related issues. Many of these studies see quality as a relative concept, meaningful only from the perspective of particular judges at particular points of time measured against some either explicit or contained standards or persistence. This means that QA can be simplified and applied in educational systems with extra complications, if the system considers that QA is a supportive mechanism and a process used to lead to the improvement of quality outcomes increasing stakeholder's confidence and belief of achieved goals and system's outcomes. If the university takes the quality management and control seriously, this automatically impacts the quality of teaching and quality of graduates. Stakeholders are also impacted positively by quality reinforcement, and we mention stakeholders as a term we do emphasize that it is not limited to customers or clients. Stakeholders are defined as all who have interest in what university does and in the quality of their outputs, which means that stakeholders can include: students, graduates, employers, parent, professional bodies, and sometimes the government. Then all are contributing and included in the quality concept and as Frazer sees quality having inter-related components to assure that all are involved. Everyone in the university has the responsibility for understanding, maintaining, enhancing, and managing quality concept and outcomes (Frazer, 1991, pp. 3–4). As a result and at the end, it has also come to mean a guarantee or certification that higher education standards are being met.

The quality assurance system has steps of evaluation as well. When we talk about quality control, we refer to processes within institutions used to

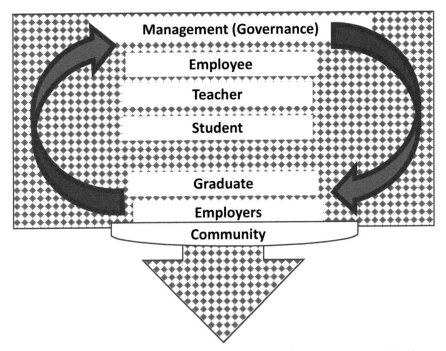

Figure 4.2   Bilateral relations between educational and institution elements and society.

ensure compliance with quality standards or achieve improvements in performance. But when we talk about evaluation, we refer to quality inspectors or quality controllers from independent bodies such as QA agencies, who have been assigned by higher education to monitor the system process and particularly the quality outputs. Quality audit is also a term to be presented here, and it refers to the processes of external scrutiny used to provide guarantees about quality control in any educational system. Quality audit is based on a large element of peer review and on the notion of developing a detailed report that will be forwarded to the university later to assist in improving procedures and achieve enhancing outputs.

Quality assessment is also a procedure and it is conducted externally to determine whether quality activities in a university are coping with the planned arrangement and whether the "outcome" of an educational system is implemented effectively in order to achieve the stated objectives. All the previous terms related to quality in general and QA in specific mean to focus on the educational process and outcomes on one track and the mechanism of achieving quality on other parallel track. These parallel tracks work in harmony presenting the internal preparation of the detailed evaluation document (self-evaluation document) concerning, for example, goals, mission, strategies, and activities for review by an outside review panel who will visit the institution and provide a written report for quality enhancing purposes and accreditation at the end and as expected and desired for.

Quality assurance agencies as mentioned in Chapter 1 are main higher education bodies and we here would like again to enlighten their mission in terms of monitoring quality practices inside the educational systems. QA agencies are expected to first raise the awareness of quality culture and play a role of serving centers to afford all required information about implementation and QA systems. QA agencies should encourage higher education institutions to improve the quality of education, especially by creating an internal system to evaluate it. Evaluation process includes self-assessment and external assessment by an independent body, where the QA agencies play the role of coordinating. At last, and based on the evaluation, QA agencies decide and confirm whether the university meets the higher education standards for accreditation purposes.

Higher education in general contributes a lot to support the academic institutions by numerous means, but at the end and to be able for universities to achieve the satisfactory level of QA strategies and ensure desired outcomes, they are expected to commit to:

1. Create and maintain continuous improvement cycles and approaches for enhancing quality purposes.
2. Include evidence-based and inclusive decision-making processes. This helps the system to take decisions based on facts and feedbacks.
3. Monitor outcomes against its stated goals and performance indicators.
4. Employ the internal and external reviews for improvement cycles.
5. Create a harmony and good environment for the value of honest, teamwork, and passion.
6. Promote a culture of ownership, participation, and responsiveness where all staff understand and feel responsible for quality in their system.
7. Good strategic planning stage will help to embed plan, do, check, and action cycles properly at all levels of quality inside the educational system.

It is clear enough to conclude that the university is committed to create a well-thought-out and scientific method to emerge and practice quality into the correspondent educational system with full quality management, assurance, and control schemes and coordination; this university would find itself adrift in the ocean called organizational development and hence would be unable to meet the myriad challenges.

4.3 Total Quality Management and Higher Education

4.3.1 Definition of Total Quality Management (TQM)

Emerging "quality" into higher education systems in order to sustain the education and pursue the journey toward the excellence in higher education, a philosophy often referred to as total quality assurance (TQM) is required. The notion of the total quality management (TQM) was proposed and promoted by an American, W. Edwards Deming, afterward World War II mainly to improve the production quality of goods and services. The idea was not taken seriously by Americans until the Japanese, who adopted it in 1950 to resurrect their postwar business and industry, used it to dominate world markets by 1980. By then, most US manufacturers had finally accepted that the 19th century assembly line factory model was invalid for the global market.

Total quality management is a perception that educational institutions being able to accomplish through a long period of planning. TQM requires the creation and execution of the annual quality program, which substantially moves toward completing the university's vision. Submission to the TQM's concept is a long way in reviving the higher education system. Some institutions already enjoyed the advantages of TQM's methods in their plans with fortitude and firm adoption of the system to achieve their goals. Recent studies and researches have presented the positive impact of the TQM system integration with academic systems and being utilized in different instruments and hypothesis to legitimize employees, create a responsive environment, and highlight the needs of the students to bring out the excellence in them.

Outlining the definition of the TQM takes various metaphors; some can be as (James, 2002, p. 45; Zairi, 2002; De Miranda, 2003, p. 34):

a. A "total" approach:

The TQM is a long-term, significant, and all-encompassing method to management, including all members and activities into the quality improvement process, rather than being focused on limited features of the system. The total approach is about promoting a novel culture in the form of quality-based decision making (Berry, 1997, p. 58).

b. A customer-driven focus:

The TQM states a systematic management of relationships with stakeholders, which ensures sustainable progresses in quality performance (Murgatroyd and Morgan, 1993, p. 59).

c. Concept analysis:

The TQM is taken as a diversity of terms and concepts that are used to describe the quality management concept. The total management system can take any form of TQM, quality management system (QMS), systems management, and quality (Meyer, 1998). The perception of the TQM using this theme is more explained by Horwitz stating that the TQM can be sub-divided and described as follows (Horwitz, 1990, p. 56; Kachar, 1996, p. 2):

- Total quality management can be seen as a total process that recognizes quality as one form to the end product or service to the customer. This means that the system is unified as one unit and all should be contributing to the process leadership, operations, the classroom, the curriculum, etc. (Steyn, 1995, p. 16).
- Total quality management can be seen as a process that considers the quality as a global phenomenon for two reasons: The need to assure consistency and being the increasing need to compare outcomes on global base for quality competition purposes (Middlehurst and Gordon, 1995, p. 271).

Total quality management is a designed system that helps to capitalize the benefits out of the quality management and control systems; therefore, and mentioned above, the TQM is also concerned about evaluation processes. It includes forms of checks and inspections, such as examination systems. Inspectors who are involved might be from national or local community. Case like United Kingdom, where inspectors are selected depending on their promotion and ability to undertake a restrained observation of teaching so they can in turn decide which standards of teaching being met. They have privileges and authorities as well high trust of their experience of judging. This is a good practicing of a general coordination in academic standards. The general coordination system allows the general coordinator to request all teachers teaching a specific course to write their exam following general outlines and guidelines forwarded and agreed by the general course coordinator. Forming the exam in unified context and concept will give a fair chance to all relevant learners to have equal prospect of succeeding in this exam, and then the result is expected to reflect standards set by the course coordinator (from a national or local body). The same concept when we discuss the curriculum panel sets the examination and experts design marking schemes. Outcomes reflect quality achievements by learners. This scenario can be seen when we talk about higher education partnerships. When a well-established university decides to expand globally and open a number of the same educational system campuses in different countries, it needs to adopt this system of coordination to guarantee the quality of assessments in general and quality of and students' performance in specific.

We can conclude that the TQM is not a simple term as well; it is more taking a multi-focal model promoting approaches for managing purposes of overall the educational system. Academically, we can express by saying that the TQM is a management philosophy endorsing quality through systems and processes to meet and exceed the expectations of stakeholders (Rana, 2009).

4.3.2 Total Quality Management: Framework Fits Higher Education Institutions

To maximize the benefits from quality management and controlling systems, the philosophy of the TQM is used and deployed. The TQM concept promotes the educational system and enables it to handle all components, procedures, arrangements, and all who are influenced in any way with the quality of the education presented by this institutions. The TQM zooms particularly on total satisfaction for internal and external stakeholders by incorporating them into a management environment that is based on continuous enhancement cycles. The management environment framework manipulates the components of the higher education institution to work efficiently and effectively, on both sides, internal and external, as we can outline as follows:

1. Utilizing every staff member's ability and experience, by encouraging them to cooperate and exchange relative experiences and seek multi-functional teams more than singularity for promoting the improvement cycle and knowledge transfer.
2. Considering the system as "a total" means that TQM deals with the educational system and its processes as one unit seeking to enhance all parts and components to get the whole unit performing and producing as expected.
3. The TQM is more with flexibility to give more space for all parties to create and innovate for quality enhancing purposes.
4. The TQM is based on the understanding and the total adopting of quality culture, which raises the awareness, trains the parties, and integrates the entire workforce in order to achieve and attain the main objective, stakeholders' satisfaction.

The TQM can be considered as an accommodating tactic that helps to create a platform for a stable improvement process, focusing on superiority and excellence of QA and management systems. To be able to put this into practice, the TQM framework should be designed to suit the higher education system. There are five major steps to implement the TQM according to Napierala (2012), and can be discussed as follows:

1. Commitment and understanding from employee side: Employees are expected to be aware of the importance of such a mechanism and how the TQM mechanism is able to promote the quality concept inside the system.
2. Quality improvement culture: The culture of not accepting low level of performance or exceeding the expectation is wanted to empower the whole system to perform effectively.
3. Continuous improvement in process: The concept of quality at each level and fits for purposes keep the main contributors in active mode for a constant developing system.
4. Focus on customer requirements: Quality is fitness for purposes, and stakeholders' satisfaction and meeting their needs are the main objectives to take into consideration.

5. Effective control: Pretending to have quality is with no use, unless there is a well-formed QA structure to control this quality and ensure its sustainability. The TQM can be a great help in assuring the internal and external effectiveness.

Of course, if we are discussing the quality of education, there is an urgent need to discuss strategic planning, action processes, and assign experts as leaders to manage and control this quality. Leadership is based on commitment and the internal desire to identify mistakes and correct them. The whole process involves everybody in the educational system and empowers assigned experts and staff to make decisions to attain the goals and satisfy stakeholders' needs. Processes are expected to be continuously improved through the use of data assortment methods and extents. But yet, all what we have mentioned is not even close to be a suitable framework for a higher education system as long as we are not including processes to enhance the academic plans which are the backbone of any educational system and the plans where quality should be dominating as much as it is possible. Therefore, higher education in developed countries has formed a framework to inspire innovations in academic plans while maintaining and improving academic standards. Policy makers are playing crucial roles to experiment new academic QA strategies. Deming's concept of the TQM had outlined guidelines and principles for needed educational restructuring. These principles will be summarized next to draw a practice for supervision that goes beyond the control quality.

Principle 1: Synergistic Relationships

The total quality management is concerned on the system as one unit, where all are involved in the quality process. It encourages teamwork and collaboration seeking to utilize every single experience that can help to develop the system by developing each component of this system. Collaboration is a great chance to get more trained and a space for knowledge transfer in the purpose of achieving the system's objectives. For example, inside the classroom, teachers and students are working together as a team and expected to collaborate to enhance the teaching–learning process. The teacher practices quality of teaching and the students get a chance to learn and promote their abilities. In turn, learners by gaining the required knowledge are promoting the process of lifelong learning, which is an objective. The concept of synergy proposes that performance is enhanced by assembling the talent and knowledge of entities.

Principle 2: Continuous Improvement and Self-Evaluation

The total quality management system should be designed to focus on total dedication to continuous improvement cycles. To be able to encourage everyone's potential by dedicating ourselves to the persistent improvement of own skills, according to Deming, "no human being should ever evaluate another human being," and therefore, the total quality management underscores self-evaluation as an important step in the continuous improvement process.

Principle 3: A System of Ongoing Process

The total quality management is about identifying the total which means that the educational system is seen as one structure and the process within this structure is seen as an ongoing (unstoppable) process. If the theme of totality is framed, then all are involved and all are contributing and no one in particular is blamed when an unexpected failure occurs. Quality speaks to working on the system on a continuous basis. This means there will be a continuous evaluation and examination to identify and eliminate the imperfect processes that permit its contributors to fail.

Principle 4: Leadership

The total quality management's fourth principle applied to the educational systems is mainly about the theory of having a good quality practices at the top management level that should reflect the system's success. For examples, inside the classroom, teachers and educators are considered the top management level and they must innovate the context in which students can best achieve their potential through the continuous improvement that results from teachers and students cooperating together.

Many researches have tried to modulate more flexible framework and hypotheses of the TQM to be emerged in the higher education system. One of these studies entitled "Total Quality Management in Higher Education Industry in Malaysia" (Sabet et al., 2012) briefly reviewed how TQM can operate in the industrial field and clarified how the philosophy of the TQM may be translated into the educational field. The authors presented examples of how lecturers of universities implement the TQM procedures and concepts in their classrooms focusing on the link between these elements. The proposed model is concluded to present the TQM as Figure 4.3.

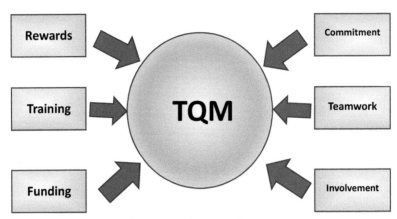

Figure 4.3   Model of total quality management.

"H1: Funding has a significant effect on TQM in higher education industry.
H2: Training has a significant effect
H3: Rewards and Recognition have a significant effect on TQM in higher education industry.
H4: Commitment has a significant effect on TQM in higher education industry.
H5: Teamwork has a significant effect on TQM in higher education industry.
H6: Involvement has a significant effect"
Reference to (Sabet et al., 2012).
Note: TQM is Total Quality Management

Executing the TQM system must be firm, secure, and visible based on values and strategies, not to be exclusively based on individual influence of the leader. Leaders and all staff must have a deep understanding and confidence toward the TQM in the educational system and they identify the benefits of the ongoing improvement. The process toward total quality in universities can be a slow process as adopting the change in the TQM requires time. Being optimistic, the change and the transformation can be attained with persistence, teamwork, and getting the right support of the top management level who is expected to have knowledge, belief, confidence, and skills towards the TQM.

4.3.3 Advantages and Disadvantages of Definition of Total Quality Management (TQM) Strategies

The effectiveness of the TQM depends on the knowledge, experience, and familiarity of assigned people. The TQM pre-requisite is to pay efforts to train and communicate with members about the importance of TQM concept for quality purposes and attaining objectives. It is truly essential adoption, for the endorsement of quality procedures and related measures to succeed. As we mentioned, knowledge transfer and experience exchange are important factors in the TQM as the more knowledge people acquire, the more they change their attitude and they become more effective. The adaption toward the TQM makes it possible, for example, to share and maintain university resources without conflicts that most of the time block the path of its success (Boothe, 1990). Gregory (1996) presented a good example of administered leadership succeeding in the academic institution's transformation to meet the higher education requirements and recommended four features of institutional leaderships: symbolic, political, managerial, and academic, which actually call to reconsider the TQM as a supervisory mechanism to overcome obstacles, promote strategies in the academic industry, and adapt to the standard of the education industry (Peat et al., 2005; Venkatraman, 2007).

The TQM principles and strategies have special themes that mean to help academic institutions at many clauses as:

1. The TQM has the theme of customer focus where the ideas of services to students are supported through staff training and development, which encouraged students' preference and independence.
2. The TQM has the theme of staff focus which appreciates the contribution of all members that create a platform of management structure with full of acceptance submission.
3. The TQM has the theme of service agreement positions and attempts to ensure excellence that meets the requirements of higher education.
4. The TQM has the theme of documentation as an important process as well as execution of systematic risk analysis and moderation along with the necessity to have emergency and back-up plans.
5. The TQM has the theme of refunding roles, purposes, and responsibilities in the system that involves all parties.
6. The TQM has the theme of planning for a comprehensive leadership and training for educators at all levels.
7. The TQM has the theme of creating staff development that helps to identify and recognize the attitudes and beliefs of school staff.
8. The TQM has the theme of enforcing the research and practice-based information to guide both policy and practice.

The TQM system can be viewed as a compact foundation of quality measures and it needs the ultimate commit from the the higher management to afford requirements to be properly initiated and implemented in the institutions quality policy. The TQM is one of the supreme effectual tools in the arena of teaching–learning process criterion. The mechanism works from scratch to involve each component of the system and get them incorporated and integrated to achieve the ultimate out of each. The TQM compromises the effectual development indorsing the quality of education institutions and affords the chance to students to be more proficient and competent on the one hand, and the academic institution as well performing with competitive features globally to face our profligate progressing world.

4.4 Accreditation

Accreditation is a type of a quality-oriented procedure performed under the supervision on the higher education system and executed by an external authorized body who has the full privileges to decide whether the academic institution (university) is meeting the requirements and attaining the standards of higher education or not. Accreditation is granted or denied based on this decision.

Accreditation can be simplified as a process of self-study taking place in parallel with other quality orientations such as external quality reviews to scrutinize higher education institutions such as universities and colleges for their commitment of QA and quality enhancement cycles. The external body is expected to recognize the level of quality inside the educational system against the standards required by the higher education. The process usually includes a self-evaluation, peer reviews, and site visits which optimistically end with being a recognized and accredited university. Accreditation is a bit complicated system that involves lots of procedures and evaluation systems. Audit is a related process that refers to reviewing an institution or a specific program in terms of internal components such as: curriculum, staff, and infrastructure. The inspection and the evaluation occur in relation to its own (the institution) mission and objectives, which leads to the institutional academic review which is actually a diagnostic self-assessment and evaluation system to examine teaching–learning process, academic standards, research, students' services, effectiveness of offered academic programs, as well as the quality levels of staff and faculty. Audit focuses on the accountability and reliability of the educational system managing processes to attain objectives and this why part of the audit process is at the institutional level, known as institutional review.

Institutional audit is a process based on facts, evidences of real practices of QA, enhancements, and control within the educational system itself giving space to identify quality gaps for enhancing purposes. The audit process should produce an audit report (known sometimes as an evaluation report) which is documenting for all quality assessments prepared by the peer review team who visited the university's campus (site) detailing measurements of institutional quality and academic quality standards with sufficient explanation of quality arrangements, impact, and effectiveness.

In Europe, the audit document is known as an evaluation report and sometimes known as an assessment report and most of the times it is prepared by the accreditation agency indicating the quality level and QA arrangements inside an educational system which actually reflects to the importance of the audit report that can affect the accreditation award or denial. Being aware of such importance, most of the universities have established an internal audit unit including expert staff and faculty members that can help the institution to exceed the expectations and meet the requirements of the accreditation agency. This internal unit has enough authority and open access to all local resources in order to promote the position of the institution toward accreditation.

In many countries, the term of accreditation is a result of a good quality practice in the higher education institution. And for an institution to be qualified as accredited from higher education systems, it is expected to go through the following common procedures:

1. A self-study and an accreditation proposal, prepared by the institution itself. The study is expected to outline quality practices inside the educational system along with the achievements that have been attained.
2. An external peer review that usually demonstrates the fact of quality practices and other important academic standards and indicators such as: curriculum, teaching and learning processes, research and university impact, and contribution to its community.
3. A site visit by an expert peer review team for site evaluations and inspections.
4. A site visit detailed report that focuses on quality standards, QA, and improvement cycles.
5. Accreditation decision based on previous steps.

At the end of the process, the university is awarded a specific status: Accredited, denied, or sometimes on hold status for missing documents or giving more time for enhancement cycles. This is called accreditation status which means that, if the university is accredited, then the institution has full privileges of this recognition nationally and globally. Accreditation and recognition add tremendous benefits in terms of institution prestige and reputation and raise the investment interest for fulfilling requirements of the higher education system.

The university has the right to have a copy of these reports and they are known as accreditation surveys for quality enhancement purposes in case of accreditation denial. The accreditation survey includes all evaluation documents and information provided by the institution itself and on-site observations by the peer reviewer team. The university is expected to utilize this feedback and other recommendations for the next cycle of quality enhancement process. In case of approval, the university needs to have proof of being accredited and this is known as portfolio for accreditation along with the formal accreditation licence. Portfolio for accreditation is a kind of accumulation of evidence, record of achievements of the educational system proficiency in managing and controlling high standards of quality.

Universities most of the times face problems in the field of accreditation by the higher education system; therefore, universities are expecting to have a simulation of what exactly the higher education request from their academic system. Such simulation and demonstration can compile key terms and definitions related to QA and accreditation such as benchmarking, recognition, external reviewers, site visits, and QA agencies as one incorporated system. The complication should play as a working tool and framework for universities and any higher education institution trying to understand QA concepts and how to develop it coming over obstacles and challenges preventing from getting accredited.

This section is dedicated to link QA to accreditation process and simplify the whole process with basic information needed for any higher education institution willing to be awarded with internally and externally accreditation. We want to emphasize that accreditation is a process performed by an independent body usually known as accreditation body running under the umbrella of the higher education. There are different types of accreditation bodies such as accreditation agencies, councils, or commissions and their main task is to evaluate the proposal submitted by the higher education institution for accreditation purposes. The accreditation process assesses the level of quality practices inside the educational system as one unit and/or assesses the academic quality standards of specific educational program against pre-determined standards (usually higher education standards or benchmarks).

Benchmark is a term to introduce here as it refers to a specific standard, or specific criterion the quality is measured against. The measurement involves evaluation of performance against outcomes of attained objectives. Accordingly, benchmarking is a process enabling means to describe the features and characteristics of academic institution's internal programs and academic standards in specific subject along with demonstration qualification standards and minimum expectations for the recognition process.

The process of accreditation as mentioned before involves three important steps: a self-evaluation proposed by the institution itself, a study visit done by external examiners, and the evaluation by the commission based on the evaluation report completed by the team of peer reviews resulting in a final decision. The mentioned accreditation main steps are actually done at two main levels:

1. Institutional level: The term being used is institutional accreditation. The process at this level refers to the accreditation of a whole entire institution system including all components: management levels, academic standards, academic programs, campus, infrastructure, methods of teaching–learning strategies without tracking, or implication to the level of quality performed within the educational system.
2. Program level: The term being used more often is the specialized accreditation as the process is concerned with individual components and mainly the academic units and programs. The evaluation is to examine the level of meeting and applying specific standards for curriculum and academic context.

Other countries such as the United States, they have a third level of accreditation called regional accreditation that granted to the higher education institution by a recognized accrediting association that conducts the accreditation process based on geographical distribution. The United States has around six regional accrediting commissions.

The accreditation process should be limited to an interval of time, known as accreditation duration as the validity of accrediting license is established by the accrediting body which holds the right to suspend or to renew the licence depending on the satisfactory resolving of identified problems.

Linkage and association between the QA systems and the accreditation process involve a re-defining of some QA terms to suit and integrate with the accreditation process. The QA terms are such as assessments, evaluation systems, quality control, outcome assessments, and quality assessments. Assessments are important systematic processes that help in augmenting and quantifying of relevant information required to examine the level of quality of instructional performance and effectiveness. They usually designed to measure quantities and qualities of educational activities, research outcomes, students' performance, and entire system's outcomes being able to attain the goals specified in the system's mission and objectives. The QA evaluation and enhancing systems within an educational institution are technically designed for evaluating the teaching–learning process and student's learning outcomes indicating to the teaching effectiveness and students' satisfaction level. Quality control and assessment are also related to assessment of individual qualifications concerned about faculty's qualification, competence spirit, innovation, and professionalism.

The accreditation process comes with a wide range of aims and objectives to raise the competence among academic institutions, and can be summarized as:

1. Promote a recognized measure of quality to attract students and investors, and of course to assure recognition of granted degrees internationally.
2. Secure the integrity of acceptable academic quality standards that form the reputation of higher education standards.
3. Endorsement of quality culture and practices as foundations for the educational system's success.
4. Stimulate the concept of academic performance in terms of teaching–learning process.
5. Empower the community with accredited graduates with high required skills into the marketplaces.
6. Protect the higher education sector from random "money seekers," when it comes to consider the academic institution just as an "investment" without ant commitment to the minimum requirements of quality.
7. Present reliable and accountable academic institutions which can run the education system properly.
8. Improve the quality enhancing cycles by stating specific standards to meet.
9. Recognize well-established universities' contributions and present them as a form of excellent and superiority.
10. Accreditation process is an important factor to achieve a sustainable higher education system.

The accreditation process is also subject to the change and massification of higher education as well as calls for flexibility. But, the accreditation concept will be always a crucial stage toward professionalism and excellence in higher education systems.

4.5 Globalization and Its Impact on Higher Education's Quality Assurance System

4.5.1 General Understanding

Globalization and internationalization are terms used interchangeably to form the higher education system being recognized globally. Knight and de Wit (1997) suggested that globalization "is the flow of technology, economy, knowledge, people, values, ideas... across the borders. Globalization affects each country in a different way due to a nation's individual history, traditions, culture, and priorities" (Knight and de Wit, 1997). This definition is the best practical and operational description when it comes to external growth and expansion of the higher education system. But the globalization comes with re-shaping pursuit that requires the higher education system to re-orient academic structures toward enlarging the scope of its provisions to be able to join the international competition and scope. The change and transform include in what way this higher education is aware to cope with global standards without impacting the internal QA systems.

Knowledge transfer and exchange are essential to promote a sustainable educational system in general, and development cycles in specific. There is a universal agreement that knowledge confrontation suspends the modernization as knowledge is the gear of growth and expansion progressions. On the other hand, affairs related to quality will be real concerns when it comes to educational systems that built on standards and principles to ensure missions and objectives of these academic systems.

This section will be keen to discuss and outline how the strategic plans within the system of higher education can be an important factor to attain the globalization with highly competitive attributes. Also the section will discuss the effect of the globalization on higher education quality systems and suggest plans how to overcome such obstacles taken into consideration that globalization is a demand now by all universities and this will be discussed in detail in Chapter 5.

4.5.2 Facts About Higher Education Globalization

The cope with globalization needs internal framework to broaden the higher education focus to the external norms instead of internal commitments and orientation. An effective mechanism of the QA system is to re-orient and re-assess the educational mechanisms and procedures to attain the satisfactory of international stakeholders demonstrating the practices of quality and generating the quality-oriented outcomes based on internationalization. How to start thinking about this is a real and existing challenge for the higher education system, while the excellence of the higher education system is based on globalization as main joint interconnecting many internal/external demands and action processes. Therefore, we would like to investigate the demands for deeper understanding of globalization trend in higher education, and this what we are outlining below:

1. Expansion of higher education is a fact which means that higher education system performs now quantitatively and qualitatively on the largest base it can be. Expansion includes the increasing number of academic institutions and the increasing number of students and their extensive expectations. The expansion is also caused by the diversity of offered academic programs, eagerness of stakeholders to invest in the higher education, inflammation of technology that has promoted the learning process to cross borders, and presenting miscellany of education delivery methods. All these demands represent in fact a burden on higher education to perform effectively and efficiently while being flexible and adoptable for change. All these factors caused the expansion and raised the needs for globalization.

2. E-learning revolution and the intentions to replace and limit the traditional learning process to reduce the cost and human and learning resources while widening the base of learners without limitations. Open learning and distance learning pedagogies offering diversity of programs, self-orientation, and learning necessitated going toward expansion and globalization for higher education systems.

3. Privatization of academic institutions that lead to the increasing number of higher education institutions to increase the resources retaining the accumulative number of students seeking university degrees and qualified education systems.

4. The concept of multination education which opened opportunities for higher education institutions partnerships with well-established universities in different countries usually, with excellent academic reputation, to offer mother's university programs locally.

5. Broadening and knowledge transfer caused the need for more related specialization to adopt this knowledge properly. The revolution in technology also emerged; new sciences were not taught before. The higher education system needs to cope with all these concerns guaranteeing

new opportunities to students and matching demands of marketplace and society.

6. Social responsibility assigned to each higher education to represent the internal nation and society in the best appropriate, attractive, and well-formed approaches reflecting the culture and demonstrating morals and ethics.

7. Academic collaboration and experience exchange programs are also steps to globalization and a step also in the enhancement and developing cycles.

What we have listed above was an attempt to understand the needs of going toward globalization and how it is important to attain for higher education. Benefits of globalization go beyond the external recognition, prestige, and academic reputation. Globalization is a tremendous scope of investing and financing prospective, as knowledge transformation is a real gain for both parties, the exporting and importing ones. The globalization makes it possible to have a global education network as we have mentioned in Chapter 1. Global knowledge network can reduce the cost by sharing the fundamentals and technology infra-structures as well. Globalization is a new trend as a problem solving tool by managing the experts to broaden their work and practice, benefiting others with experiments, and results can be adopted easily and properly to fit the internal society. It is time to mention the four keys of globalization mentioned by UNESCO regarding the higher education:

1. Permitting of knowledge transfer for globalization purposes.
2. Importance of availability, creativity, and innovation in terms of information and communication technologies (ICT).
3. Marketplace impact of the provision of higher education concerns.
4. Globalization needs for international framework for better practicing and attaining objectives.

4.5.3 Quality Assurance and Globalization

The main concern of coping with the trend of globalization is influence on the academic quality of the educational institutions taking into consideration that the definition of quality is a multi-dimensional concept that takes many forms to serve the objectives of the system it belongs to. Garvin (1987) proposed eight dimensions for quality as: performance, features, reliability, conformance, durability, serviceability, aesthetics, and perceived quality (Garvin, 1987). Patrick and Stanley (1998) defined QA as encompassing the entire educational system toward enhancing the teaching quality process in the higher education system as they specified that the indicators of quality are based on research and teaching.

We would like to enlighten two dimensions of terms "QA" here. First, the QA system is tangible which means that it comes with sufficient up-to-date equipment and facilities. It is expected to be ease of access and visually appealing environment supported with all infrastructure and support services. Second, the QA system is reliable which means that the system is trustworthy and being able to handle obstacles and attain goals. Actually, the two dimensions are perfect enough for globalization. Globalization is sort of a quality-oriented-based system supported with sufficient infrastructure along with features such as reliability and accountability. If the higher education is being able to coordinate the internal QA system and reflect it to cope the globalization needs with full commitments to academic standards in parallel with stakeholders' needs, then the higher education is global and deserves the prestige and reputation universally.

Global competitiveness may impact the quality orientation of a well-defined higher education in case of coping with the race without coordinating and mapping the quality to the new global framework.

Quality assurance agencies play crucial roles to edify the higher education system and get it customized for globalization. They effectively function, coordinate, and instrument for catalyzing the infusion of components that promote the system of the higher education to be truly international. The coordination is actually based on strategic planning theme in emerging the globalization thinking into mission and objectives of the higher education, and then consequently, all enhancing cycles of QA system will be re-oriented based on the new demand creating, innovating, and designing action processes that can come with the desired international fitting outcomes. Before pursuing with how to achieve the globalization, we need to address concerns and negative impact of globalization on the QA system.

Globalization has contributed to the rise; to be more accurate, it caused inflammation in life expenses not only in education, but also, in other sectors such as technology and health care. The knowledge transfer represents a real threaten if it causes a culture gap between education systems and society (roots). The main role of a higher education at the first to serve the environment it belongs to and attain the society's objectives of having graduates that can be leaders and make difference to their community. Therefore, imported knowledge should be customized to exert in the internal framework of quality in order to facilitate attaining the objectives. Globalization is expected to promote the quality frame of education as it requests effectiveness and efficiency in terms of teaching–learning progress, research, and outcomes. Globalization has also broadened the digital divide unfortunately, so students who are closer to the urban areas are more benefitting from higher educational internationalization policies and strategies. Globalization opened opportunities for global collaboration and encouraged higher education to seek international partners, providers, and stakeholders which on the one side have promoted the investment opportunities bring wide range of job vacancies and academic programs, but on the other hand, the higher education system will be getting weakened from inside structure depending on global partners and providers. Such huge investments and collaborations can be impacted with politics affairs, for example, which can impact the system negatively that can come to collapse cases.

The scenario we are trying to demonstrate here is related to the reliability and accountability, if the system loses the trust at one stage, which causes collapse for sure even in the long term. Similarly, globalization is still without a unified framework which makes it confusing and jeopardizing at the same moment.

4.5.4 How to Get to Globalization?

Higher education needs to re-orient the internal system to be able to achieve the globalization. Globalization requires the higher education to run a two-tier system, internally and externally as being able to integrate them fittingly. Globalization requires cycles of development progression that starts with the emerging of new higher education providers investing and funding the educational system for globalization race and competence. The providers can be local or foreign ones such as multi-national companies, who find interest in education sectors. New forms of education delivering are expected with globalization such as open and distance learning, increasing the mobility of students, educators, and programs and offering diversity of programs, certifications, and qualifications.

References and Suggested Literature

[1] Barrett, A. M., Chawla-Duggan, R., Lowe, J., Nikel, J., Ukpo, E., *The concept of quality in education: a review of the 'international' literature on the concept of quality in education*, EdQual Working Paper No. 3, UK, 2006.

[2] Schindler, L., Puls-Elvidge, S., Welzant, H., Crawford, L., Definitions of quality in higher education: A synthesis of the literature. *High. Learn. Res. Commun.* 5, 3–13, 2015.

[3] Dill, D. D., Quality assurance in higher education: Practices and issues, *The 3rd International Encyclopedia of Education*, eds B. McGaw, E. Baker, P. Peterson, Elsevier Publications, 2007.

[4] Soomro, T. R., Ahmad, R., *Quality in higher education: United Arab Emirates perspective, higher education studies*, Canadian Center of Science and Education, vol. 2, 2012.

[5] Salganik, L. H., and Provasnik, S. J., *The Challenge of Defining a Quality Universal Education Mapping a Common Core*, available at: https://www.amacad.org/pdfs/routledge20.pdf.

[6] Vlašić, S., Vale, S., Puhar, D. K., *Quality Management In Education*, available at: ftp://ftp.repec.org/opt/ReDIF/.../InterdisciplinaryManagementResearchV/IMR5a46.p.

[7] Batool, Z., Qureshi, R. H., *Quality Assurance Manual for Higher Education in Pakistan*. Higher Education Commission, Pakistan, 2007.

[8] *Total Quality Management in Higher Education: A Review*, available at: https://www.researchgate.net/publication/266394850_Total_Quality_Management_in_Higher_Education_A_Review [accessed July 31, 2017].

[9] Hayward, F. M., *Glossary: Quality Assurance and Accreditation, prepared by the Council for Higher Education Accreditation (CHEA) in February 2001,* available at: http://www.chea.org/international/inter_glossary01.html.

[10] Becket, N., Brookes, M., Quality management practice in higher education – What quality are we actually enhancing? *J. Hosp. Leis. Sport Tour. Educ.* 7, 40–54, 2008.

[11] Al-Azzah, F. M., Yahya, A. A., Quality procedures to review, mission, vision and objectives in higher educational institutions. *Eur. J. Sci. Res.* 45, 168–175, 2010.

[12] Mehrotra, D., *Applying Total Quality Management in Academics*, available at: https://www.isixsigma.com/methodology/total-quality-management-tqm/applying-total-quality-management-academics/

[13] Sabet, H. S., Saleki, Z. S., Roumi, B., Dezfoulian, A., A study on total quality management in higher education industry in Malaysia. *Int. J. Bus. Soc. Sci.* 3, 17, 2012.

[14] In'airat, M. H., Al-Kassem, A. H., Total quality management in higher education: A review. *Int. J. Hum. Res. Stud.* 4, 3, 2014.

[15] Hayward, F. H., "Quality assurance and accreditation of higher education in Africa," *Conference on Higher Education Reform in Francophone Africa: Understanding the Keys of Success*, 2006.

[16] Arab Network for Quality Assurance in Higher Education, *Survey of Quality Assurance and Accreditation in Higher Education in the Arab Region*, 2012.

[17] Erden, A., Understanding the quality management issues in education faculties: Experiences from faculty members. *The Anthropologist* 24, 1–15, 2016.

[18] Alalfy, H. R., Abo-Hegazy S. R. E, A suggested proposal to implementation quality management system ISO-9001 in Egyptian universities. *Am. J. Educ. Res.* 3, 483–489, 2015.

[19] Surbakti, F. P. S., Natalia, C., Inderawati, M. M. W., Sukwadi, R., Business process and quality management in higher education institutions, 2014, [Ronald et al., 1(3), 2014].

[20] Vlasceanu, L., Grünberg, L., and Pârlea, D., Quality assurance and accreditation: A glossary of basic terms and definitions, 2007.

[21] Amaral, A., Rosa, M., Recent trends in quality assurance. *Qual. High. Educ.* 16, 59–61, 2010.

[22] American Council on Education, *International Higher Education Partnerships: A Global Review of Standards and Practices.* Washington, DC, 2015.

[23] American Society for Quality. (n.d.). *Quality Glossary*, available at: http://asq.org/glossary/q.html

[24] Barker, K. C., *Canadian recommended e-learning guidelines. Vancouver, BC: FuturEd for Canadian Association for Community Education and Office of Learning Technologies, HRDC*, 2002.

[25] Biggs, J., The reflective institution: Assuring and enhancing the quality of teaching and learning. *High. Educ.* 41, 221–238, 2001.

[26] Bobby, C.L. (2014). The abcs of building quality cultures for education in a global world. *Paper Presented at the International Conference on Quality Assurance*, Bangkok, 2014.

[27] Borahan, N.G., Ziarati, R., Developing quality criteria for application in higher education sector in Turkey. *Total Qual. Manage.* 13, 913–926, 2002.

[28] Commonwealth of Learning (2009). *Quality Assurance Toolkit: Distance Higher Education Institutions and Programmes*, available at: http://www.col.org/

[29] Cullen, J., Joyce, J., Hassall, T., Broadbent, M., Quality in higher education: From monitoring to management. *Qual. High. Educ.* 11, 5–14, 2003.

[30] Danø, T., Bjørn, S., Still balancing improvement and accountability? Developments in external quality assurance in Nordic countries 1996–2006. *Qual. High. Educ.* 13, 81–93, 2007.

[31] Eagle, L., Brennan, R., Are students customers? TQM and marketing perspectives. *Qual. High. Educ.* 15, 44–60, 2007, *High. Learn. Res. Commun.* 5, 3, 2015.

[32] Ewell, P., Twenty years of quality assurance in higher education: What's happened and what's different? *Qual. High. Educ.* 16, 173–175, 2010.

[33] Harvey, L., *Analytic Quality Glossary*, available at: http://www.quality researchinternational.com/glossary/, 2014

[34] Harvey, L., Green, D., Defining quality. *Assess. Eval. High. Educ.* 18, 9–34, 1993.

[35] Harvey, L., Williams, J., Fifteen years of quality in higher education. *Qual. High. Educ.* 16, 3–36, 2010.

[36] Iacovidou, M., Gibbs, P., Zopiatis, A., An explanatory use of the stakeholder approach to defining and measuring quality: The case of a cypriot higher education institution. *Qual. High. Educ.* 15, 147–165, 2009.

[37] International Organization for Standardization (ISO). (n.d.). *Quality Management Principles.* available at: http://www.iso.org/

[38] Newton, J., A tale of two 'qualitys': Reflections on the quality revolution of higher education. *Qual. High. Educ.* 16, 51–53, 2010.

[39] Nicholson, K., *Quality Assurance in Higher Education: A Review of the Literature*, available at: http://cll.mcmaster.ca/, 2011.

[40] Amalia Venera Todorut a, The need of Total Quality Management in higher education, 2nd World Conference on Educational Technology Researches WCETR2012, Available online: https://ac.els-cdn.com/S1877042813012743/1-s2.0-S1877042813012743-main.pdf?_tid=e4a4a78f-87fd-4c06-ba28-14acc9bbbba5&acdnat=1525887622_ab9 6892e84fab9c3139cf2802dcd9e16

[41] Murad, A., and Rajesh, K., (2010). Implementation of total Quality Management in Higher Education, Asian Journal of Business Management, 2(1), 9–16.

5

Quality Assurance for Universities

5.1 University Quality System

5.1.1 Definitions

University quality system is described as the totality of characteristics and features of the university and its educational program claiming to have the required capability to satisfy stated standards by the higher education system. The quality system expression also implies to the attainment of skate holders' needs and producing outcomes exceeding the expectations in terms of critical issues such as students' services, students' academic performance, and prospective of their futures. According to an article of the World Declaration on Higher Education published by the United Nations, university quality is a multi-dimensional perception that embraces all its functions and activities mainly: teaching, academic programs, research and scholarships, staffing, students, buildings, faculties, equipment, services to the community, and the academic environment. University quality concept is also related to the processes of internal self-evaluation and external reviews for enhancing quality and endorsing quality management and control concepts.

Similarly, the expression "stated standards" is energetic in a variety of ways, ranging from being just basic definition of regulations to more comprehensive descriptions of good practice. And to be able to demonstrate these standards and create an evaluation system to help in estimating the university quality against these stated standards, we do introduce again the quality assurance system, but as a pursuit for universities.

Quality assurance is an ultimatum by all universities. Getting recognized and accredited by higher education is a challenge for most of the universities. And this challenge can be resolved by understanding the concept which is by itself deceive. Theoretically, it refers to an accepted level of quality of everything the university does, or will do, as per the local, regional, and international benchmarks and standards. This indeed covers all programs from the Foundation Program to all study programs at all levels in addition to the other operations that make the college or university keep going. And practically, quality assurance is a stimulating field. It is a science in addition to being a field work not based on ink and paper but based on what is actually being done and employed in the university. It necessitates highly qualified and authentic people with academic specialization in quality assurance as field involvement. It is very disappointing to witness that a university assigns the quality assurance's tasks and procedures to people who have no idea of how to plan for a university quality system. Quality absence inside the university leads for loosing accreditation opportunities and reputation later on.

The expression of "quality assurance" is a generic term in higher education which lends itself to many interpretations; it is not possible to use one definition to compromise all conditions. But what we want here is how to adapt the quality transformation in the whole educational system and outline a university fitting framework of quality as excellence practice at each level sustaining the quality orientation in higher education system. Bogue and Saunders (1992, p. 20) perceived quality assurance process in universities as primarily based on coordinating the mission and achieving the goal within a framework of publicly accepted responsibility and integrity. Such a definition makes firm assumptions for universities: it assumes (1) that the university should define the mission, (2) that the goals of the institution are explicit and achievable, and (3) that there are public and acknowledged standards which are promoted by the university. Harman and Meek (2000, p. vi) stated quality assurance for university as systematic management and assessment procedures adopted by a higher education institution (University) to perceive performance and to ensure achievement of quality outputs. Quality assurance aims to give stakeholders confidence about the management and control of quality and the outcomes achieved (Harman and Meek, 2000, p. vi).

University quality assurance should be present to support the basic role of quality "fitness for propose" and one of these purposes is attaining quality at each level and encouraging all to contribute which necessities the processes of quality assurance to be transparent. Transparency will ensure the quality of the academic standards and structural provision of courses and make it easier and more flexible for evaluation and review. The transparency term is expected in two steps: make university educational program transparent enough for all stakeholders and display the quality assurance processes with transparency for all. Students want to achieve the ultimate performance and get the recognition of their courses and programs which qualify and guarantee them to promote employment opportunities.

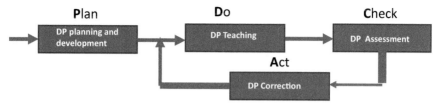

Figure 5.1 P–D–C–A Mechanism.

The framework of quality assurance affords two main methodologies of involving quality assurance within a university. The university can apply relevant processes to promote the quality assurance system for evaluation and quality indorsing. These methods can be summarized as:

1. Creating a fundamental quality platform reflecting basic values and principles of the university educational system focusing on the teaching–learning innovating process.
2. The extrinsic quality platform reflecting the capacities of the university to respond to the changing needs of the society with whom it cooperates. It concerns about the new demands directed toward the university educational program.

University quality assurance is impacted by the labor market. It exposes the role of academic research conducted for decades according to the do/plan/check/act (DPCA) cycle as shown in Figure 5.1. It shows the continuous improvement of quality at the classroom level not only at the institution level. One of the main purposes of a university quality assessment and assurance is meeting the stakeholders' satisfaction with system outcomes. Stakeholders can be students, graduates, and employers.

Quality of a DP is the level of acquiescence of the program's characteristics and outcomes with the requirements set for this DP (Gerasimov and Shaposhnikov, 2014). There are several methods to the assessment of quality in education described in the works of Damiechili et al. (2011), Lenn (1992), Haris (2013), and Eid (2014).

5.2 Quality Assurance General Framework for Universities

The university's common mission is to contribute effectively to the society through the pursuit of education, learning, and research at the highest international levels of excellence. This reflects one of its core values, which is to provide high-quality provision across all academic programs offered. The university's quality assurance procedures provide a framework within which its institutions can examine and enhance their teaching activities to ensure that they attain this aspiration of excellence. This framework is expected to present the quality culture as a set of shared, integrated patterns of quality usually known as principles to be endorsed within the system as well as presenting quality planning strategies referring to a set of actions applying the mechanism of the quality system taken into consideration the objectives of the university.

The university is accountable for the quality and standards of its provision and is required to participate in the activities of bodies such as the quality assurance agency and various professional bodies. The university's procedures for assuring quality in teaching, learning, and assessment are designed to reflect:

- The mission of the university
- The complexity, diversity, and federal structure of the university
- The university's nature as a community of scholars
- The conviction that academic staff and students are most effective in an environment that is supportive and participative rather than directive and managerial
- The collegiate nature of the university
- The delegated responsibility given to faculties and departments for their own quality assurance procedures
- A proportionate central approach to potential risks to learning and teaching provision and to assessment
- The general board's learning and teaching strategy.

5.2.1 Standards and Guidelines

There must be a set of standards and guidelines for the university internal and external quality assurance. They must be based on the quality assurance principles for the use of university education and the relevant quality assurance agencies covering key areas relating to quality and standards of the higher education system. The purpose of these standards and guidelines is to provide a source of assistance and guidance to both higher education institutions in developing their own quality assurance systems and agencies undertaking external quality assurance, as well as to contribute to a common frame of reference, which can be used by institutions and agencies alike to ensure that appropriate quality assurance mechanisms are in place and subject to independent reviews. Standards can be qualitative such as effectiveness, sustainability, or educational commitment to attain goals. Standards can also be quantitative for accreditation purposes. In general, university standards are the result of the benchmarking process and can be standards of excellence referring to highest level of quality attained. Standards of the university can be set and assessed in four categories:

1. The academic standards: They are related to the teaching–learning process. They are also known as educational standards and expected to be applied at all stages of the educational process promoting the relationship and integrating all these stage to work in harmony and more effectively utilizing learning resources, academic plans, and students' performance. These standards are included in the strategic planning and should be well stated and formed as they are more concerned about the process of learning in specific.

2. Standards of competence are related to the technical abilities of the institution and may involve students and educators. In most of the cases, these standards are used to manage and control the student's performance encouraging the self-centered learning and accumulating knowledge in parallel with the educational standards to attain at the end success of the teaching–learning process along with promoting private learning skills and own abilities. These standards are also a concern for staff and faculty as they usually seek the award and distinction.

3. Service standards: They are related to students' service afforded by the university. Educational and competence standards are with no use if they are not combined with leaning resources and facilities to support the teaching–learning process. University services and understanding of the importance of having a suitable standard of students and staff services will promote the quality in general as each component of the system will be enabled for creativity and innovation with full of awareness of university culture and requirements of quality assurance systems.

4. Institutional standards: They are related to general principles and procedures within the university. These standards are designed and planned to go along in harmony with the university's mission and objectives. The institutional standards are guards of achievements and specifications' required level of quality to the institution level to satisfy the stakeholders while still working in the frame of quality shaping the expected outcomes.

It will be very beneficial to mention here why these standards are important to be designed and planned during the strategic planning. The purposes of the standards and guidelines are:

- To improve the education available to students at the university.
- To assist universities in managing and enhancing their quality.
- To form a background for quality assurance agencies in their work.
- To make external quality assurance more transparent and simpler to understand for everybody involved.

A good quality assurance process in the university should be based on a number of fundamental principles which can be summarized as follows:

- The process should take into account the genuine interests of students, employees, employers, and the society more generally in good quality higher education.
- Universities are social educational institution and should be run as such. They are established to serve all the society not the self-centered interests.
- Universities as a higher education institution have the responsibility for the quality of their academic programs and how these programs are implemented.
- There need to be efficient and effective structures within the university which those academic programs can be provided and supported.
- Honesty in quality assurance processes is significant. Cheating and fake documents defeat the objectives of the process.
- Transparency is very important for any college or university to gain the confidence of the public. Hiding all the information and relevant data from the public in the name of "confidentiality" runs counter to the basic tenets of the QA process.
- Universities should build up and maintain a culture of quality among the management, students, teachers, and staff.
- Quality assurance processes should be developed through which higher education institutions can demonstrate their accountability.
- External quality assurance needs to be completely independent and objective and in line with international standards and guidelines.

Quality assurance agencies usually offer to help universities when they are at the start and provide them with all required information the university needs to launch a quality-oriented educational institution. They are supported to back-up universities not to play the role of criticizing only. Quality assurance agencies are really called to communicate with universities more frequent to have open discussion of the university's internal quality assurance system and quality standards. More issues can be discussed as well, such as effectiveness at the individual programs levels and how to submit careful to the quality transformation seeking to more truthful and reliable educational system. The student's performance is an important issue to discuss trying to improve quality cycles centered on promoting the teaching–learning process, satisfying marketplace with quality-oriented graduates and outcomes.

The uniqueness of offered programs grips the attention of students and manages to have them as direct customers in the market of educational services. And after completing their studies, students are distinctive at the labor market. The problem of teaching students demanded by the labor market should be solved in a systematic way through quality assurance of university education in competitive degree programs (DPs). A DP quality assurance should be based on the implementation of the PDCA (plan–do–check–act) cycle which provides a continuous improvement of the educational service offered by a university. Involving students in quality orientation serves the quality improvement cycle as well. Strategic planning should involve plans and action processes of using students' input and feedback concerning teaching–learning interactions. Students' and educator's feedback help the planning committee to understand and determine the compatibility between students' expectations and teaching quality and approaches, learning resources, and readiness of students for good learning practices. Usually, a student survey can help to complete this task, as students are asked to answer open-ended questionnaires about university services and academic policies. Such information helps a lot in the next quality enhancement cycles and gets it closer between efforts from the university side and students' expectations. It is a good practice of students' involvement in the quality assurance system.

5.2.2 Two Levels of Quality Assurance

1. Program Level

The program level quality assurance system is identified as essential and not to be replaced with institutional level quality assurance processes as standards vary among various academic programs even within the same institution. Thus, for the accuracy of information to standards, the specific program level quality assurance processes are given significant importance. Accreditation councils/professional bodies are responsible to assure that program level quality assurance processes are in place.

Study program and validation are the core modulator components of higher education including all activities such as: the design and the process of teaching, learning, and research. Study programs' features may vary from, such as, academic qualification, study mode, and field of knowledge specialization. Quality assurance system is achieved by a study program by taking into consideration the following: Quality-oriented curriculum design and managing, course syllabus detailing all related information, and specific knowledge field describing of the teaching and learning melodies. Which is not actually an easy procedure to designing quality-oriented activities that match the general purpose of the course and present the time-table distribution of these activities along with the expectation from such course and knowledge transfer. The quality assurance of a study program is expected to be supported with plans for validation which means that the program and its context will be assessed periodically against the requirements of higher education, students' satisfaction, and market-needs.

On the educator side, teachers' satisfaction is also a necessary need to achieve a stable quality assurance system and the most important obstacle that teachers are affected with is the workload. Workload is a quantitative measurement and indicator of learning activities needed to achieve learning outcomes as it may affect other assigned tasks such as researching, managing the assigned courses, and drifting the concentration to promote the quality of teaching because of the extra workload. Quality at the program level depends on the internal faculty cooperation and contribution. Faculty and staff need to communicate about educational policies and quality assuring consistency, especially with information afforded to students such as information about assessments, course syllabus, handbooks, regulations of teaching–learning process inside classrooms, sharing learning resources, and expectations from the program side.

More to add that faculty of each program should be able to confirm the following general procedures from their side meeting the requirement of the quality assurance system and strategic plans approved by the quality assurance institutional committee:

1. Induction: As each department and program management level should have induction and orientation to communicate the quality culture to the accepted students and recruited faculty. This helps to get them all involved and aware of program provision and expectation and hold their responsibilities.

2. Proper application of quality procedures in terms of course design, monitoring, and review. These procedures should be working in harmony with the general quality assurance system and strategic plans approved by the quality planning committee.
3. Running and managing the evaluation processes as excepted from the quality planning committee.
4. Including assessment procedures as an effective learning tool. Conducting assessments should be with full of security, consistency, and fairness. This also includes a transparent system of grading and feedback to promote students' learning.
5. Fair distribution of workload and link it with internal and external quality indicators.
6. Provide guidelines and afford open-channel discussion internally with students, educators, and staff.
7. Students' feedback mechanism should be working in a proper way as it is an important information gathering for next quality enhancement cycles. Feedback may also have categories of complains and appeals regarding the assessments' result, for example.
8. Quality at the program level is also a field of statistical information feeding the enhancement quality cycles with different kinds of data such as students' performance, educators' performance, and actually data on the effectiveness and impact of quality assurance plans and action procedures.

2. Institutional Level

The main quality assurance system at the institutional level is to assure quality on all stages focusing on quality in education, the student as the main stakeholder, and the expected outcomes. The institutional level quality assurance processes are required to develop an ultimate quality culture with the goal that quality is the central focus of the quality-orientation integral part of all academic practices. Institutional level quality assurance system defines quality as a prime responsibility of the entire university system by creating and enabling learning environment for the students which are actually the core of university's mission and objectives.

A well-functioning quality development system emphasizes the responsibility of each of institution's activities and involves all groups in the university community in the educational process. It is consequently a prerequisite and goal that all significant information about planned and implemented evaluation measures be made obtainable and communicated to the students and

Figure 5.2   Quality assurance developing system – University vision.

staff concerned. Quality assurance takes place in accordance with a plan that ensures continuity and an overview. Schematically, it can be described as a recurring process consisting of four phases. Expected learning outcomes must be communicated to each individual study program.

Quality at the institutional level is extended more than just running a quality assurance system. In addition of what was mentioned at the program level, institutional level of quality is still concerned with more details of quality assurance. Recruitment, selection, and admission to higher education policies should follow the quality assurance system and specified procedures. Learning and teaching processes will always be real concerns to the institution quality vision as being the main of educational system processes. Quality plans for enabling students' development and achievement, students' engagement, assessments, and the recognition of prior learning and knowledge transfer system are also concerns.

Institutional level is after a quality procedure to assure program monitoring, review, and managing higher education provision: setting quality-oriented academic standards, promoting the provision of learning opportunities, provision of information, and the enhancement of the quality of student's learning prospects. The quality at the institutional level should take the risk-based and evidence-based methods promoting the quality assurance system on greatest scrutiny and facts not only plans and passion. Action plans should address features of good practice and affirmation of quality commitment in order to achieve the internal and external recognition.

Quality assurance at the institutional and program levels is a prerequisite for accreditation. Higher education intuitions are constantly evolving and changing; accreditation is based on an evaluation done at a specific point in time, normally with reference to a specific area of the institutions, e.g., a course or facility. This normally leads to the awarding of certificate or recognition that the institution or part therefore meets certain standards. When accrediting, quality assurance should be the guarantee that the standard measured in the accreditation process can be upheld in the long term. Thus, accreditation cannot be said to be complete unless the three steps outlined in the quality assurance and accreditation policy are enacted and the process is seen as ongoing. The prominence of accreditation for students can be stated in three points:

1. Accreditation provides students with programs, which are clearly defined and appropriate. Accreditation provides added assurance that the program in which students are enrolled or are considering enrolling is capable of achieving what it sets out to do.
2. Accreditation facilitates the mobility of students because it provides the higher education institutions with independent approval of the various programs at other institutions where a student can come from. This can lead to the development of precognition of degrees.
3. Accreditation must facilitate the recognition of degrees in other countries and thus facilitates mobility of graduates.

University quality assurance system cannot be completely stated without introducing the concept of University Key Competence which is about "the ability to successfully meet complex demands in a particular context through the mobilization of psychosocial prerequisites (including both cognitive and non-cognitive aspects)" (Rychen and Salganik, 2003, p. 43). This definition incorporates three critical elements:

1. Integrating different experiences of individuals in different areas as it allows knowledge and experience transfer.
2. This definition of competence recognizes that a range of internal prerequisites combine to allow individuals to meet demands.
3. The role of context. Competencies are played out in the social and physical environment, and therefore their specifics, as well as the specifics of their internal components, are profoundly influenced by the individual's particular situation.

5.2.3 Evaluation and Assessment System for Quality Assurance

The evaluation system is meant to demonstrate the success of the quality assurance system being able to attain and achieve goals designed for. Evaluation and assessment system for universities is expected to have the three folds given below:

1. Internal quality assurance process that takes place within the academic program and meant to collect continued information in a systematic way about the quality being achieved inside the department/program. The self-assessment process produces reports and they are crucial to be described the corner stone of the whole university quality assurance system; therefore, the university pays efforts and attention in how to prepare these reports using the guidelines of the quality assurance agencies most of the times. The report should be transparent as it can be reflecting facts, weakness, and strengths to locate and identify quality concerns for enhancement purposes. It is a kind of comprehensive report about objectives' structure and content of the academic programs, learning, and teaching environments.

    The self-assessment process is an effective tool for quality academics assurance and provides feedback to administration to initiate action plans for improvement. Self-evaluation documents are usually used for accreditation purposes and being reviewed and validated by the peer review team

2. External quality assurance system that involves accreditation by respective accrediting bodies, peer-review. External quality assurance system needs certain standards and quality in all procedures to be followed in terms of international compatibility.

3. Meta quality assurance system by the government mainly through the higher education council for higher education institutions. This step is more to be described as an induction and preparation for institution for internal accreditation and international recognition as the role of the quality agencies and councils to lend a hand to help universities achieve the quality standards.

The universities are responsible for provision of quality education to the students through the self-assessing system of quality assurance and to work with a satisfactory system of external quality assurance. The independence of universities is predictable in terms of academics and governance; however, the autonomy is accountable to public, to the government, to present and prospective students, and to the society. The internal quality assurance and external quality assurance are strongly linked being complementary and integrated with each other. The internal quality assurance is essential for external quality assurance, while external quality assurance motivates internal quality assurance for future developments and improvements.

5.2.4 University Assessments

a. Assessment and accountability

Apparently, two trends have gained prominence throughout higher education: assessment and accountability. The first trend, "assessment for excellence," is an information feedback development to guide individual students, faculty members, programs, and universities in improving their effectiveness. The second trend, "assessment for accountability," is essentially a supervisory process, designed to assure institutional conformity to specified norms.

While the expressions "assessment" and "accountability" are often used interchangeably, they have imperative differences. Assessment is used when we assess our own performance, but when others measure our performance, it is accountability. Assessment is a set of initiatives we take to monitor the results of our actions and improve ourselves; accountability is a set of enterprises that others take to monitor the fallouts of our actions and to penalize or recompense us based on the outcomes.

"Assessment is not an end in itself but a vehicle for educational improvement" (AAHE, 1992). Assessment targets at the continuing improvement of student development, and is generally consistent with a "value-added" concept of education; note that the rationale for having better programs is to ensure better student outcomes. As shown in Figures 5.3 and 5.4, the collection of assessment information is only the first step in a four-part process. To be useful, it must be analyzed and reflected upon by appropriate decision makers, and then used to design and apply changes. In each iterative

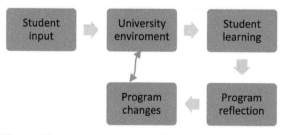

Figure 5.3   Assessment for excellence – feedback process.

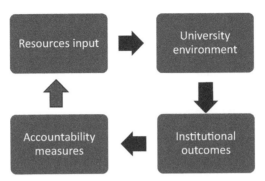

Figure 5.4   Assessment for accountability.

cycle, modified programs are then reassessed and readjusted, continually improving effectiveness.

Assessments are the most effective when it reflects an understanding of learning as multidimensional, integrated, and revealed in performance over time. Learning entails not only what students know but what they can do with what they know; it involves not only knowledge and abilities but values, attitudes, and habits of mind that affect both academic success and performance beyond the classroom. Assessment works best when the programs have clear, explicitly stated purposes.

Accountability measures are primarily concerned with resource allocation and financial efficiency. It is completely appropriate for students to evaluate critically what they get for their money from the education system.

Accountability aims at improving fiscal efficiency, but it is blind to issues of educational quality. Assessment aims at improving the quality of education, but it is necessarily constrained by budgets. Integration between assessments and accountability is the best solution to utilize both serving the teaching–learning progress. Accountability can be re-oriented and grounded in student learning to work with quality measurements in order to provide the most rational basis to measure university performance. The new aimed accountability measures should have two qualities: being unambiguous and they must be linked to indicators of quality. Such indicators, when combined with cost data, could also be used effectively as measures of change in institutional fiscal efficiency or overall performance.

b. Student learning outcome

It is the time to differentiate between "student outcomes" and "student learning outcomes." Student outcomes refer to aggregate statistics on groups of students and represent institutional outcomes, attempting to measure comparative institutional performance. They have generally been associated with accountability reporting. Unfortunately, student outcomes' statistics are often "output-only" measures (Astin, 1993). Student outcomes are computed without regard to incoming student differences and without regard to how different students experience the university environment. "Student learning outcomes," on the other hand, encompass a wide range of student attributes and abilities, both cognitive and affective, which are a measure of how their college experiences have supported their development as individuals.

Intellectual university outcomes embrace demonstrable acquisition of specific knowledge and skills, and affective outcomes are also of considerable interest. There are essentially three threads which must be interwoven into a program dedicated to the improvement of student learning: shifting curricular focus to student learning; developing faculty as effective teachers; and the integration of assessment into curriculum at several levels. To emphasize that students' learning outcomes are indictors for quality, we mention that "students and their learning should become the focus of everything we do starting from the instruction that we provide, to the intellectual climate that we create, to the policy decisions that we make" (Cross, 1998).

c. Quality assurance at all stages: The learning assessment cycle

Assessment can be expressed as the first phase in a continual learning cycle which includes measurement, feedback, reflection, and change. The purpose of assessment is not merely to gather information; the purpose of assessment is to foster improvement. Frequent assessment of students helps them to refine concepts and deepen their understanding; it also conveys high expectations, which further stimulate learning. "Students overwhelmingly reported that the single most important ingredient for making a course effective is getting rapid response" (Wiggins, 1997). Correspondingly, assessments of faculty teaching by students and faculty development consultants help teachers to improve their teaching and course organization. Program assessments keep departments and curriculum committee updated and informed of how well programs are meeting their objectives, and comprehensive university-level assessments provide feedback about how effectively university policies are contributing to the accomplishment of the university's mission and goals.

It is beneficial for universities if they start establishing some kind of "Faculty Development Center," which would provide confidential consultations, resource and technical support, and training to help faculty develop as teachers. University is expected to provide explicit support to improve both the quality of teaching and also the productivity of individual faculty, and provide incentives for teaching excellence.

5.2.5 Universities and the Fitting for Globalization

No doubts that most of universities as higher education institution seek the globalization, but selective universities are really committed to prepare the internal system to adopt the strategies of internationalization to get into the international competition and race, demonstrating that quality in outcomes is supposed to compete nationally and internationally. Again, our start is the strategic planning stage at the program level and institutional level. Strategic plans should include the globalization within the mission and main objectives of the educational system and raise the awareness of how internationalization is important to any education institution seeking a worldwide reputation and presence. Different universities may adopt different international strategies according to their environment, culture, and economic circumstances. Most of these strategies include that university is paying effort toward: research activities and publications provision, academic exchange programs and joint degrees, and re-orientation of the funding policies widening opportunities for international providers. The university should be able to create a coordination system, mapping the quality assurance system internally to reflect to the external quality system focusing on: supporting faculty, students' performance enhancing cycles, and offering more scholarships and international academic programs to attract international students and faculty as well.

To maximize the benefits from globalization and reducing the impact on quality, universities are encouraged to take the globalization as steps and stages ensuring that each re-orienting process is happening based on quality. It is very important for universities to take enough time to identify strengths and weakness of existing processes and measure the extend and the influence of globalization on these processes. Universities are also encouraged to discuss various mechanisms and coordinate with the quality assurance agencies that help and contribute with beneficial information and consulting issues to raise the chance for globalization objective. Globalization process and strategies should be a part of the quality assurance system cycles.

Quality assurance for universities comes with the theory part as we have explained in the previous section but being a working field, it should be supported with manual and specific steps and procedures about how to integrate and work within the system. Each university can design its own quality assurance system within the recommended framework by the higher education general system. In the next section, we give a general proposal that may fit for the university about how to launch the quality assurance system and start performing effectively.

5.3 Design a Manual for University Quality Assurance

5.3.1 Quality at All Stages

High quality in education is dependent upon a learning environment that stresses updated knowledge in all disciplines and the determination to achieve quality in teaching. Evaluation and the interpretation of evaluation measures are important preconditions for quality development. From the definition of good learning objectives via well-considered teaching and assessment methods through to achieved learning outcomes. A true commitment to student learning is a paradigm shift, but it does not have to happen all at once. The first recommendation is to define strategies for achieving such goals. Faculty within academic units must bear a particular responsibility for beginning a dialog about their own major programs, examining their willingness and ability to restructure the programs, courses, and assessment procedures to be consistent with improving learning outcomes. The university's quality assurance system should be established to maintain and further develop the quality of study programs in a manner that can be documented. Maintenance and further development of the quality in education can be achieved through a quality assurance system that:

- Systematizes knowledge of activities and improves the circumstances that permit students and employees better understand these activities and each other's views and perspectives.
- Encourages work on learning issues.
- Produces the information necessary to be able to propose and implement measures to improve the quality of education and study performance.
- Clarifies the responsibilities of students and staff to ensure that efforts to improve the quality of education succeed.
- Helps to ensure that sufficient resources are made available for study programs and support services.

Figure 5.5 shows a graphical scenario of entering students; the measurement of an array of student outcomes provides feedback about how well individual courses, programs, and the university as a whole are achieving their stated missions and goals.

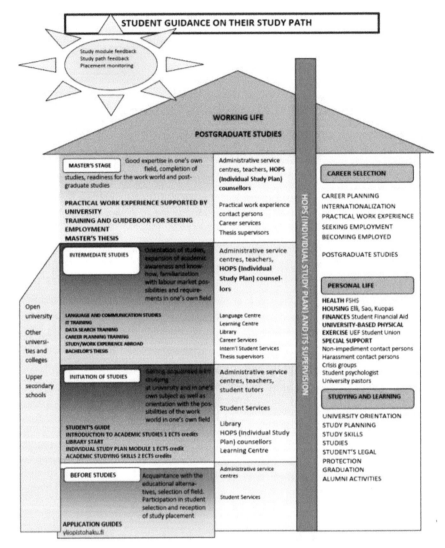

Figure 5.5   University of Eastern Finland Main Quality Manual.

Source: University of Eastern Finland Main Quality Manual
http://rritrends.res-agora.eu/uploads/22/UEF_plk_in_English_2.pdf

Strategic planning stage is actually the main key to launch the quality assurance system for universities. Including the main concept of evaluating and enhancing are also crucial keys in quality assurance systems (plan–do–check–act). The institutional effectiveness manual of quality assurance systems start by identifying vision, mission, and main goals that can be a part of the strategic planning phase. We present in the following practical step of how to start a quality assurance system:

1. *Vision*

The university vision can be ambition to be among the leading learning centers' region, by achieving international quality standards in teaching, research, and community engagement.

2. *Mission*

The mission of the universities has multi-stages and phases. Please note that all the following phases have a solid relationship with quality concept.

Phase 1: The university always strives to be a learning center of excellence that responds to market needs and prepares graduates who possess the scientific and technological competencies.

Phase 2: The university plays an active role in the creation of knowledge through quality teaching and research.

Phase 3: It values community engagement and always contributes to the environment it belongs to.

Phase 4: Accreditation and recognition.

Phase 5: Globalization.

3. *Strategic plan* *(objectives): in the following examples of these goals.*

Goal 1: Commitment of providing a quality-oriented education system with features of equality, consistency, transparency, and a high level of direct interaction between faculty and students. The education system is expected to promote curricular reforms and innovations in all areas of the academic program.

Goal 2: Contribute to society and the local community in particular, by addressing their educational, cultural, social, and economic interests. Contributing to society requires the university to start the support from inside by enhancing the professional stature of the faculty body and improve teaching and learning at all levels. Promote internal and external activities that cultivate positive perceptions of university and reflect this to the society. The university's collaboration with other academic institutions in the areas of teaching, research, consultancy, and training programs will contribute to the society and open more opportunities for students.

5.3.2 Main Keys in University Quality Assurance

a. Key concept and processes

1. Assigning the planning committee to experts and well-formed knowledge of business lunching and strategic planning and quality assurance systems to cope with higher education standards.
2. Recognition of stakeholders and interested part needs.
3. Recognition of environment and marketplace expectation.
4. Funding plans.
5. Mission, vision, and objectives.
6. Creating a quality-oriented university system including the structure, educational system, and administration.
7. Quality and education policy specifying the quality objectives.
8. Recruiting experts and professional faculty in the same domain of the university serving the university objectives and being aware of how to achieve them.
9. Admission standards.
10. Internal quality assurance system.
11. Creating a quality enhancement unit.
12. University, implementing and documentation of quality management:

    (a) Action processes
    (b) Documentations.

13. Assuring quality at the institutional level.
14. Assuring quality at the program level.
15. Graduates.
16. Promoting internal and external quality assurance system.
17. Accreditation.
18. Self-documentation and contribution to the society.

b. Core processes

1. Educational system processes:

    (a) Assuring quality at the program level.
    (b) Applying academic standards.
    (c) Teaching–learning processes.
    (d) The academic structure for each program.

(e) Knowledge transfer system.
(f) Academic cooperation and collaboration.
(g) Assessments and developing cycles.

    i. Course continuous assignments and assessments and students' achievements' system such as examination and projects.
    ii. Assessments' quality assurance and management system ensuring equality and fair opportunities.
    iii. Assessment of education policy, instructing, and learning procedures.
    iv. Audits experience: Internally and externally.
    v. Assessment evaluation process in order to measure the attained knowledge and feedback.

(h) Students' performance and expectations.
(i) Academic outcomes.
(j) Students' feedback and evaluation.
(k) Educator's feedback and evaluation.
(l) Quality culture awareness and involvements.
(m) Students' induction and quality orientation.

2. Research activities and requirements.
3. Social interaction and educational objectives and reflection.

c. Support processes

1. Resources:

(a) Human resources policy and strategies.
(b) Recruiting.
(c) Health safety.
(d) Induction and orientation.
(e) Working hours and load.
(f) Developing units.

2. Learning recourses:

(a) Suitability of learning process.
(b) Availability.
(c) Accountability and reliability of learning resources to help the learning progress.
(d) Training how to employ them.
(e) Creativity and innovation space for educators and students.

3. Equipment and research needs.
4. Information infrastructure and safety.
5. Laboratory quality systems.
6. Strategic management and performance management.
7. Administrative support.
8. Management reviews and developing procedures.
9. Management of complaints and appeals.
10. Management system of promotions and appraisal system.

d. Accreditation process
1. Self-evaluation documents.
2. Enhancing cycles for accreditation.
3. Committee for accreditation applying.

e. Creating a system for external quality and globalization process
1. Coordination with internal and external quality assurance bodies.
2. Globalization strategies.

References and Suggested Literature

[1] Bogue, G. E., Saunders, R. L., *The Evidence for Quality: Strengthening the Test of Academic and Administrative Effectiveness*, San Francisco: Jossey-Bass, 1992.

[2] *European Association for Quality Assurance in Higher Education, 2009, Helsinki, 3rd edition*, available at: http://www.enqa.eu/

[3] Eurydice, *Key Data on Education in Europe 2012. Education, Audiovisual and Culture Executive Agency,* 2012.

[4] Frazer, M., "Quality Assurance in Higher Education," in *Quality Assurance in Higher Education: Proceedings of an International Conference Hong Kong, 1991, Alma Craft (ed.),* Hong Kong Council for Academic Accreditation, Taylor and Francis e-Library, 2005.

[5] Harman, G., Meek, V. L., *Repositioning Quality Assurance and Accreditation in Australian Higher Education.* Centre for Higher Education Management and Policy University of New England, 2000.

[6] Harvey, L., A history and critique of quality evaluation in the UK. *Qual. Assur, Educ,* 13, 263–276, 2005.

[7] Lindop, N., *Academic Validation in Public Sector Higher Education.* London: HMSO, 1985.

[8] Mizikaci, F., A system approach to program evaluation model for quality in higher education. *Qual. Assur. Educ.* 14, 37–53, 2006.

[9] *QAA-Strategy-2014–17 UK*, [available online]. Retrieved 25 November 2016.

[10] Reisberg, L., *Quality Assurance in Higher Education: Defining, Measuring, Improving It. Boston: Boston College*, 2010.

[11] Roberts, V., Global Trends in Tertiary Education Quality Assurance: Implications for the Anglophone Caribbean. *Educ. Manage. Admin.* 29, 425–440, 2001.

[12] Stanley, E., Patrick, W., "Quality Assurance in American and British Higher Education: A Comparison," in *Quality Assurance in Higher Education: An International Perspective*, ed. G. Gaither, *New Directions for International Research*, 25, 39–56, 1998.

[13] Stimac, H., and Katic, S., (n.d.). "*Quality Assurance in Higher Education*," [available online]. Retrieved 25 November 2016.

[14] UNESCO, *Guidelines for Quality Provision in Cross-border Higher Education*, Paris, 2005.

[15] Woodhouse, D., Quality assurance: International trends, preoccupations and features. *Assess. Eval. High. Educ.*, 21, 347–356, 1996.

[16] Salganik, L. H., Provasnik, S. J., *The Challenge of Defining a Quality Universal Education Mapping a Common Core*, available at: https://www.amacad.org/pdfs/routledge20.pdf

[17] Vlasceanu, L., Grünberg, L., Pârlea, D., *Quality Assurance and Accreditation: A Glossary of Basic Terms and Definitions*, Bucharest, 2007.

[18] Popescua, F., "South African globalization strategies and higher education," *International Conference on Education, Reflection, Development, ERD 2015*, Cluj-Napoca, Romania, 2015.

[19] Bakhtiari, S., Shajar, H., *Globalization and Education: Challenges and Opportunities*, available at: file:///C:/Users/sharif/Downloads/3461-Article%20Text-13841-1-10-20110217.pdf.

[20] Singh, M. K., *Challenges of Globalization on Indian Higher Education, Sr. Research Fellow, Apeejay-Stya Education Research Foundation, New Delhi*, available at: http://www.aserf.org.in/presentations/globalization. pdf.

[21] Adapa, P. K., *Strategies and Factors Effecting Internationalization of University Research and Education*, available at: http://homepage.usask. ca/~pka525/Documents/Internationalization.pdf

[22] Edrak, B., Nor, Z. M., Maamon, N. M., *Internationalization of Higher Education: Key Factors Attracting International Students to Study in*

Private Higher Education Institution in Malaysia, available at: http://jespnet.com/journals/Vol_2_No_4_October_2015/19.pdf, *2015*.

[23] University of Oxford, *Education Committee Quality Assurance Handbook*, available at: http://www.admin.ox.ac.uk/media/global/wwwadminoxacuk/localsites/educationcommittee/documents/QA_Handbook.pdf, *2008*.

[24] Shahidi, N., Seyedi, S. M., *The Impact of Globalization in Higher Education on the Universities' Educational Quality: A Regional Project on Shiraz Universities*, available at: https://pdfs.semanticscholar.org/eb10/959e7fb432d5149c41eb0227111a39716271.pdf, *2012*.

6

Quality Assurance Piloting
the E-Learning Sector

6.1 Introduction

In Chapter 1, we introduced the e-learning concept and here we would like to emphasize that e-learning has become widely spread in education systems regardless of being traditional or non-traditional as we explained that even the traditional systems have adapted to e-learning and depend on it to build their e-learning centers. E-learning has become a part of the continuing education, the adult education, and the corporate training, simply because it is flexible, rich, allows for resource-sharing, and less cost-effectiveness. The quality of e-learning education reflects the relationship and interaction between learning seen as a result, a process, or as an education system adopting e-learning and the demands of technology transfer, goals, standard regulations, and requirements set in e-learning framework decided by individuals, businesses, and stakeholders in general.

According to Jung and Latchem (2007), most institutions realize quality culture in their e-learning systems as for the other modes of delivery. Endean et al. (2010) pointed that e-learning by its design presents an appreciated level of quality assurance by internal designed and developed quality procedures. It is by nature in electronic tools. The implementation of a set of quality standards that convey some wider recognition discourses the need for internal progressions.

This chapter also considers the role of quality assurance in e-learning, reflecting on the conditions necessary for successful e-learning in higher education. It assesses the international work on quality assurance in this area and drives to deliberate the conducts in which the quality of a process or activity can be assessed focusing on the use of benchmarking and specification of standards.

There are several phases when we develop academic quality culture in e-learning sector. According to Ehlers et al. (2004), models for quality approaches are classified by nature to determine the situation they run. Classifications are most related to the way that they were designed and the goals they should meet. The focus, the method, and the approach are examples of these types and goals where the focus is for quality inheres in the process, and the method is about evaluating or managing the approach, while a quality model or instrument is the approach itself (Ehler et al., 2004).

When we aim toward quality, we should think seriously about "evaluation." Evaluation plays a great role in e-learning application as much as the used methodology is stable, powerful, and well integrated into the e-learning system. Evaluation should cover criteria as course content, student's performance, and social interventions such as strategic plans and program designs. Six areas had been identified to evaluate an e-learning usage in academic systems, according to United Kingdom's quality assurance agency guidelines for distance learning, which included e-learning evaluation subject (QAA, 1999). These areas are as follows:

1. Academic standards and how these standards can be formed with technology.
2. Program delivery management and control by admins, tutors and students via e-learning platforms
3. System design and representations e-learning is expected to be system built to overcome obstacles faced by traditional methodologies
4. Interaction and communication with students via this e-learning system which is a cycles of teaching, instructing, assessing and evaluating.
5. Student's evaluations and assessment which are the crucial for any learning process.

Evaluation of e-learning system that guarantees quality assurance is expanded beyond the above areas. Dedicated budget that can be expanded while working with different technologies, admission standards and contnous cycles of strategic planning to present the excat teaching-learning intended quality that involves up to some level an interaction between educators and learners. The way of delivery and communication do not affect only teachers and learners, it also covers and affects the spectrum of institutional processes to support learners outside the frame of e-learning. Stake and Schwandt (2006) presented quality in e-learning evaluation that distinguishes between "quality-as-measured" and "quality-as-experienced": "Evaluation studies are fundamentally a search for claim about quality"

(Stake and Schwandt, 2006). The comparison and balance were deployed to: using quantity techniques and experiencing as using perceptual and qualitative techniques. This is a framework that deploys harnessing small details together through participative inquiry, interview, and quantity measurements such as surveys and statistical reports. Quality presented by the delivery and design is more effective when it matches the understanding of users' preferences. In the same frame, an investigation was done by Benson (2003) who presented the different quality-related perceptions that stakeholders communicate and discuss on, when you have e-learning activities. She came up with the conclusion that quality in e-learning is no longer related to certification, but it is related to effectiveness and cycles of development processes that end with quality of the pedagogy. Frudenberg in (2002) as mentioned in Ossiannilsson's (2010) study had a review of quality models in e-learning as chunked quality elements constructed and integrated to work in the following nine dimensions: *"the administrative commitment, the technological infrastructure, the student services, the educational planning and development, the educators and educational services, the financial prosperity, the distribution of programs, the legal and regulatory requirements, and the program evaluation"* (p. 65). A recent chapter with the title "Frameworks for Assessing and Evaluating e-Learning Courses and Programs," by Martin and Kumar (2017), presented an overview of different frameworks, benchmarks, and instruments that can be used to investigate and evaluate "quality" and "effectiveness" in e-learning products for academic purposes. Elements and presented multiple frameworks are integrated together and identify seven quality indicators: "institutional support, technology infrastructure, course design, learner and instructor support, learning effectiveness, faculty and student satisfaction, and course assessment and evaluation in the "What you need to know" section (Martin and Kumar, 2017).

We end this section; we need to emphasize that quality assurance in e-learning services is a must if this e-learning is a platform for academic purposes. The quality in e-learning is not limited to the course/program content; it should be expanded to cover criteria in the whole electronic system as the content of the website should be taken in parallel and it will also influence users' perception and satisfaction. A study with the title "E-learning service quality" investigated concerns arise on how the factors of e-learning service quality affect the quality of e-learning systems in education and extended to conclude that the satisfaction of pillars is influencing on e-learning system quality (Rahman and Hamid, 2017).

E-learning has lots of trends being added to the e-learning services and the way of implementations. The next section will go through clouding in e-learning and how the quality assurance is the main factor in implementing this trend. Later on, the book will discuss the artificial intelligence (AI) embedding into the e-learning system and as well how to integrate these trends along with the "quality" concept.

6.2 E-Learning on the Cloud

E-learning is the theme providing learning by electronic mediums – usually the Internet – and requires virtual environments for better integration and integration with users' preferences. The e-learning systems come with lots of characteristics and attitudes with diverse functionality such as e-mail, web pages, forums, and learning platforms in order to support the academic platform and the procedure of teaching–learning. The cloud computing is a trend in computer science created for many reasons but having an atmosphere of natural podium makes it a perfect combination to provide provision to e-learning systems. The characteristics added by clouds are efficiently used for academic institutions to support the learning process and afford an atmosphere of a natural podium to provide provision to e-learning systems and also for facilitate maintaining and manipulating big data for data-mining practices and analysis for attaining more knowledge can be used and utilized for further enhancement procedures. Clouding has more benefits in energetic adaptation by providing a scalable system for changing necessities along time. Cloud computing technologies empower organizations lacking technical expertise to get support to their own infrastructure and to get access to computing on demand which reduces installation costs in labs. Cloud computing makes it possible for almost anyone to install tools that can scale on demand to serve as many users as desired. On the side of service providers, clouding affords greatly simplified software installation and maintenance and centralized control over versioning consumers can access the service anytime and anywhere. Clouding allows data sharing and affords perfect environment for collaboration more easily, and keeps customers' data stored safely in the infrastructure. On the customer side and relevant user, clouding as a structure is invisible, hiding all irrelative details and costly requirements. It is more toward getting a service and advantage of applications, regardless of the used technology to support them. For many organizations, cloud computing evolution introduced a cost-effective solution to the problem of how to provide services, data storage, and computing power without investing a huge amount of money in physical machines that need to be maintained and upgraded on a regular basis.

With huge benefits and as expected, higher education institutions were the first to adopt the technology of clouding and taking the advantage of existing applications to enable their users to perform tasks comfortably with the least cost. For hosting a service on the cloud, it usually requires site licensing, installation, and maintenance of individual software packages.

Recent features in computing, multimedia, and communication technology have added great characteristics and gave a chance to build a self-growing unit sharing virtual environment for teaching and learning. This syndicates a wide range of technologies and tools for education with access content across the Internet independently without reference to the underlying hosting infrastructure.

6.2.1 Cloud Computing

The cloud computing technology has been inspired by the network diagram in Internet clouds. The technology is based on establishing huge data-centers and using of existence ones to provide services and a unique access point for all requests coming from all clients over the world. Cloud computing is constructed with three layers:

1. Infrastructure as a service (IaaS): Information technology infrastructure can be offered as a service without physical requirements.
2. Platform as a service (PaaS): Virtual platform to enable clients to develop and use applications.
3. Software as a service (SaaS): Clients access these layers depending on the requirements and services they choose. It allows accessing the application on demand.

In order to enhance the educational system, many universities have adopted the cloud computing technology in their e-learning systems. This integration has been realized as an efficient way of promoting the leaning and increasing the number of students, services, and education content as the cloud service affords more resources and tools to enhance the learning process. The educational process itself is based on different philosophies and teaching methods, which makes it important to identify which stages are the crucial for developing learners' skills. The educational philosophy introduced education as "a synthesis of realist and idealist world views, with a primary focus on performance" (Horvath and Bruner, 1960). The call for adopting

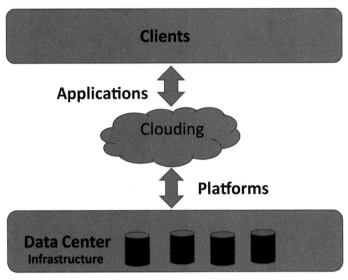

Figure 6.1   Cloud computing and big data.

cloud computing in universities as the basic for, is shown in Figure 6.1 which helps to create a knowledge-based system associated with a large volume of available historical data and expertise to support decision makers in achieving their goals and enhancing the learning process (Khedr and Idrees, 2015). Recently, cloud computing has contributed to technology by accelerating the innovation for the computer industry in general and for e-learning in specific that we can say cloud computing brought a new educational model, where the frame of services provided is widening such as resources that can be shared and utilized for the teaching–learning enhancement cycle. The new model is widely accepted and used today due to its key advantages:

- The cost is low or even free in some cases such as hardware upgrades.
- For some applications (like spreadsheets), it can be used even in the offline mode to give more space and Internet-free work time. Clients are then able to synchronize offline work to update data with one on the cloud.
- The client can reach the same work result with minimum software requirements.
- Devices with minimal hardware requirements could be successfully used as cloud clients.
- Internet connection is the only requirement to work on the cloud.

- The cost of licensing different software packages includes upgrading when new patches are released.
- No data lost if the user has a crash problem because everything is stored into the cloud.

The cloud computing model helped a lot the educational institutions who have financial problems by providing resources that can be either externally owned being public cloud such as provided by Google and Amazon or internally owned which is a private cloud. Public clouds offer access to external users who are typically billed on a pay-as you-use basis. The private cloud is built for the access within the enterprise where the users can utilize the facility without any charge.

The methods of meeting challenges such as user interface, task distribution, and coordination are explained and evaluated in Praveena and Betsy (2009) who have described the application of clouding in universities. Delic and Riley (2010) assessed cloud-based enterprise knowledge management and how cloud computing would turn into a more global and efficient infrastructure. They have discussed the architecture as well as applications. Figure 6.2 identifies differences between traditional computing and cloud computing.

Integration between cloud computing and e-learning sector has expanded to continuous education, company trainings, academic courses, and any other academic activities that more into reducing cost and get unlimited services. The cloud-based e-learning is still a system that needs a quality-oriented

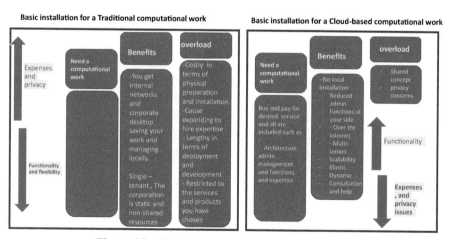

Figure 6.2 Cloud computing compared to traditional.

procedure to ensure quality and customer satisfaction. There are at least two main entities involved in a cloud-based e-learning system: the students and the educators. Each has own set of activities and cloud-based e-learning is expected to successfully fulfill these requirements:

1. The students: Common activities such as taking online courses, taking exams contributed with feedback, and submitting homework and projects.
2. The educators: Common activities deal with content management, preparing tests, being able to assess tests, design, and grade homework, and designing and managing projects taken by students. Educators would be able to provide feedback and communicate with students via this e-learning system.

The main cloud computing disadvantage is completely being dependent on the Internet connection speed that may affect the overall performance, especially when we indicating to e-learning services. Other disadvantages can be on a long-term basis that the data center subscription fee may be more expensive than buying the hardware and the service quality is crucial for enhancing purposes. Clients should be able to have backups locally for data security.

6.2.2 Virtual Learning and Cloud Computing

Virtual learning environments (VLEs) are electronic platforms that can be used to provide and track e-learning courses and enhance a face-to-face teaching–learning process with online tools and components. Primarily, they automate the teaching–learning process with full automated administration practices and allow archiving and recording learner activities for quality purposes and usage of enhancing cycles. VLEs have evolved quite differently for formal education and corporate training to meet different educational needs. VLEs are the dominant learning environments in higher education institutions. Also known as learning management systems (LMSs) and course management systems (CMSs), their main function is to simplify course management aimed for numerous learners. The content within VLE is developed by teachers, who are mainly experts of a special domain. VLEs provide an easy-to-use system for flexibly delivering learning materials, activities, and support to students across an institution. For administrator, a VLE as simple as shown in Figure 6.3, provides a set of tools which allow course content and students to be managed efficiently and provide a single point of integration

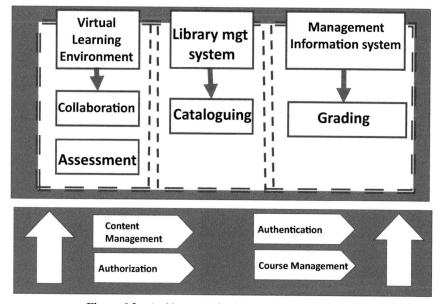

Figure 6.3 Architecture of a simplified learning system.

Reference: Masud2012AnES, An E-learning System Architecture based on Cloud Computing, M. Masud and Xiaodi Huang A World, 2012.

with student record systems. For the tutor, a simple set of integrated tools allows the creation of learning content without specialist computer skills, while class administration tools facilitate communication between the tutor and individual learners. One major drawback of existing VLEs is that it is content-centric. Many instructors simply move all their teaching materials to the system. The materials are presented uniformly to all learners regardless of their background, learning styles, and preferences.

A study in 2012 entitled "An E-learning System Architecture based on Cloud Computing," by Masud and Huang (2012) introduced the characteristics of the current e-learning and described the architecture of cloud computing platform by combining the features of e-learning. The authors tried to build an e-learning cloud and make an active research of the following aspects: architecture, construction method, and external interface with the model. The proposed e-learning cloud architecture investigated the architecture, the method of constructing, and the external interface. The model according to Masud and Huang (2012) can be divided into the following layers: "*Infrastructure layer as a dynamic and scalable physical host pool, software resource layer that offers a unified interface for e-learning developers, resource management layer that achieves loose coupling of software and hardware resources, service layer, containing three levels of services (software as a service, platform as a service and infrastructure as a service), and application layer that provides with content production, content delivery, virtual laboratory, collaborative learning, assessment, and management features"* (Masud and Huang, 2012). We will discuss each separately:

A. Infrastructure layer

The layer is composed of information infrastructure and teaching resources. Information infrastructure provides all details in terms of Internet/Intranet, system software, information management system, and some common software and hardware. While teaching resources, the phase is built by accumulating the traditional teaching model and then distributing in different domains. This layer is located in the lowest level of cloud service middleware; the basic computing power like physical memory, CPU, and memory is provided by the layer. The use of virtualization technology, physical server, and storage and network forms virtualization group for being called by the upper software platform. The physical host pool is dynamic and scalable; a new physical host can be added in order to enhance the physical computing power for cloud middleware services. Figures 6.4 and 6.5 depict this in a clearer view.

B. Software resource layer: The layer is built up by the operating system and the middleware. The layer presents a variety of software resources being integrated to provide a unified interface for software developers to develop applications based on software resources and embed them in the cloud, making them available for cloud computing users.

C. Resource management layer is a very important layer being a key to achieve loose coupling of software resources and hardware resources. The aimed integration of virtualization and cloud computing allows for scheduling strategy and distribution of on-demand free software over various hardware resources.

D. Service layer has three levels of services: SaaS (software as a service), Paas (platform as a service), and IaaS (infrastructure as a service). In SaaS, the service is provided to customers and users use software via the Internet by simply paying a monthly fee.

E. Application layer: This layer is to integrate the teaching resources into the cloud computing model including interactive courses and sharing the teaching

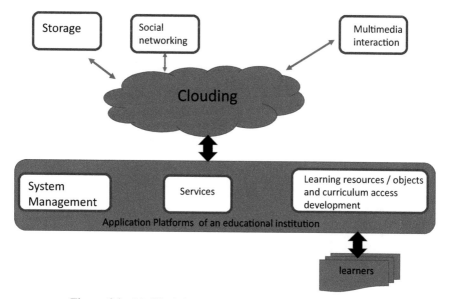

Figure 6.4   Modified cloud based e-learning system architecture.

http://clipartstation.com/laptop-clipart-black-and-white-1/

resources taken full advantage of the underlying information resources after finishing made, and the course content as well as the progress may at any time adjust according to the feedback, and can be more effectiveness than traditional teaching. The layer allows for sharing of teaching resources that include teaching material resources, teaching information resources (such as digital libraries and information centers), as well as the full sharing of human resources. This layer mainly consists of content production, educational objectives, content delivery technology, assessment, and management component.

The above study presented a cloud model which is a successful attempt to identify an architecture which will be suing cloud computing within higher education. And in the following, we discuss the expected benefits from the architecture according to Masud and Huang (2012):

1. Powerful computing and storage capacity: The cloud based e-learning architecture traces data in a large number of distributed computers to afford huge data storage space and enable students to practice computing on cloud-based compilers.

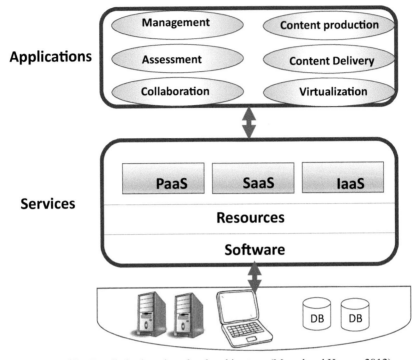

Figure 6.5   Detailed e-learning cloud architecture (Masud and Huang, 2012).

Reference: Masud2012AnES, An E-learning System Architecture based on Cloud Computing, M. Masud and Xiaodi Huang A World, 2012.

2. High availability: The integration of mass storage and high-performance computing power provides a higher quality of service as being able to automatically detect the node failure and exclude it, without affecting the normal operation of the system.

3. High security: Through the model, data are storied intensively using one or more data center with automatic reliable real-time monitoring to guarantee the users' data security.

4. Virtualization: It is an important characteristic that each application deployment environment and physical platform is not related as it is managed through a virtualization platform.

5. The architecture provides easy access to costly software running on high-performance processors to rural students at institutions which lack considerable facilities. Figure 6.6 presents the Connectivity scenario of the institutions in cloud environment as (Masud and Huang, 2012) proposed.

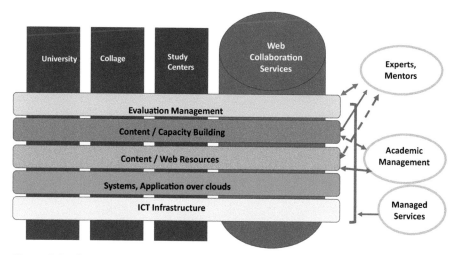

Figure 6.6 Connectivity scenario of the institutions in the proposed architecture (Masud and Huang, 2012).

Reference: Masud2012AnES, An E-learning System Architecture based on Cloud Computing, M. Masud and Xiaodi Huang A World, 2012.

6. In the classic e-learning system, teachers assign teaching tasks, conduct regular lectures, or train students' skills. The students attend the online autonomous learning act and cooperative learning sessions, or accomplish teachers' assignments. But in the proposed architecture, teachers also answer students' questions and offer essential teaching to major and difficult points. In addition, teachers can also use multimedia to enhance the teaching content. Students work out their own learning plans, determining learning methods autonomously. They conduct online autonomous learning when they study each unit, finish its test via the Internet, and do some statistics to the test results. Teachers also encourage students to cooperate with each other to finish simple learning tasks or complex group-based projects. Through cooperative learning, students cannot only acquire knowledge, their team spirit and coordination will also be fostered, skills in dealing with people will be improved, and abilities to express themselves will be enhanced. Thus, the learning and teaching will be more interactive which the demand of the age is. The interactive mode of the proposed architecture is furnished in Figure 6.7.

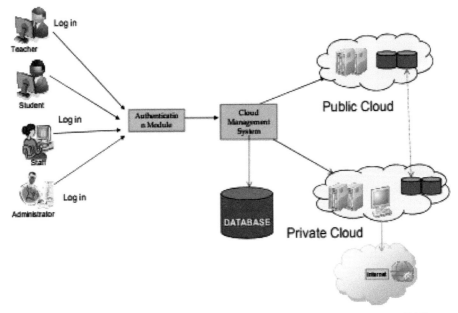

Figure 6.7   Interactive mode of the proposed architecture (Masud and Huang, 2012).

Reference: Masud2012AnES, An E-learning System Architecture based on Cloud Computing,
M. Masud and Xiaodi Huang A World, 2012.

At the end of this section, we would like to emphasize that the development of e-learning solutions cannot ignore the cloud computing trends. But it cannot also ignore trying to build a quality-oriented cloud-based e-learning system. The trends using cloud services have affected the ICT sectors and changed it a lot to have more economic nature that in turn led to many socio-cultural transformation impacting user's behavior asking and expecting more from the service provider (Upadhyaya and Ahuja, 2017). Unfortunately, there are no standards to define the quality of services provided by providers and being able to evaluate the service whether to meet the user's references or not. Upadhyaya and Ahuja (2017) raised the concern for developing a metric model for enhancing quality of services in cloud computing for e-learning sector, especially when it is applied to higher education applications. Especially that clouding is based on the idea of sharing resources and achieving coherence. According to Maguad (2007), the future success of online learning services will increasingly depend on the proper identification of the issues and variables related to quality of service. There are many benefits from using the cloud computing for e-learning systems.

6.3 The Future of Artificial Intelligence in e-Learning Systems

The artificial intelligence, which is also known as machine intelligence (MI), is apparently intelligent behavior by machines, rather than the *natural intelligence (NI)* of humans and other animals. In computer science, AI research is defined as the study of "intelligent agents": any device that perceives its environment and takes actions that maximize its chance of success at some goal. Typically, the term "artificial intelligence" is used when want to describe a machine that mimics "cognitive" functions that humans associate with other human minds, such as "learning" and "problem solving"(Russell and Norvig, 2009, p. 2).

The scope of the AI is not limited to any specific category of applications simply because our life is now considered a machine (technology)-based as machines become increasingly capable and most of tasks now require "intelligence." There has been a phenomenon known as the AI effect, leading to the quip "AI is whatever has not been done yet." For instance, optical character recognition is frequently excluded from "AI," and has become a routine technology. Capabilities generally classified as AI as of 2017 include successfully understanding human speech, competing at a high level in strategic game systems (such as chess), autonomous cars, intelligent routing in content delivery networks, military simulations, and interpreting complex data. The scope of AI is open for designing more intelligent software that can analyze its environment and make intelligent choices for online learning. But what exactly could be the future of AI in e-learning?

Artificial intelligence-based research engines made the accessing to all the information on the Internet and big data analytics faster and more efficiently but being a more complex process. Educators may find themselves in the position of simply feeding results into databases and developing theories and algorithms for AI to validate or dismiss. When it comes to e-learning, the best instructor is to have an introduction to AI to help promoting the teaching–learning process. The benefit of AI comes from its ability to evaluate learning with a dynamic strategy that is supported with features of e-learning tools.

While the current evaluation of educational techniques indicates that face-to-face instruction from a human tutor, if available, leads to a better understanding than self-learning via online lessons. But it is not the case everywhere and every time. Some students need to go online for accumulating knowledge as they cannot afford being tutored in classrooms because of their work times and limited resources to afford traditional tutoring. AI in e-learning can be a problem-solving agent as the future of AI lies in its potential for making the most of all the elements that make e-learning so promising. AI, previously, was more suited to logic such as mathematics which focused on antiseptic problem solving and dedicated to find ways to adapt any kind of instruction to relate them with relationships, abstract concepts, and real-world use.

One of the greatest benefits of e-Learning is that it allows students for self-learning and self-development at their own pace and explores new materials turned up by simple searches. Designing an AI tutor to be present to support the e-learning system means that students are left free to start e-learning by their own by exploring topics in depth and test their knowledge in complex scenarios rather than straightforward answers as YES/NO. Some detractors may maintain that a computer is less relevant, but a computer's ability to reproduce human images and voices is child's play even now. AI instructors that are more dedicated, more knowledgeable, and less error-prone than human instructors are coming. And given that personality is not a prerequisite for classroom instructors, AI programs could be more relatable, after all.

A. Artificial Intelligence in Education

Artificial intelligence plays crucial roles to automate educational activities, like grading, which can be a very consuming task for large student enrollment courses. Educational software can be adjusted according to students' needs and preferences. AI units can contribute a lot through the application of greater levels of individualized learning and adaptive learning programs, such as games and software. These systems cope with the needs of the students, put greater emphasis on certain topics, repeat things that students have not understood, and generally promote self-learning and encourage work at own pace. These AI units can identify the weakness of courses and what to develop. They can also play the role of being a teacher assistant to help and support teachers. Teachers most of the times are being overloaded and may not always be aware of gaps in their lectures and educational materials which impact the teaching–learning process; here, AI will support and offer a way to solve that problem. Coursera, a massive open online course provider, is a noteworthy example and is already applying this side of practice. In Coursera intelligent system, the system can alert the educator when a large number of students submit the wrong answer to a homework assignment, and gives future students a customized message that offers hints to the correct answer. This is a good practice of acting as a teacher assistant who fills in the gaps in explanation that can occur in courses and helps to ensure that all students are building the same conceptual knowledge. Rather than waiting to hear back from the professor, students get immediate feedback that helps them to understand a concept and remember how to do it correctly the next time around.

Students could get additional support from AI tutors. Some tutoring programs based on AI already exist and can help students through basic mathematics, writing, and other subjects. These programs can teach students fundamentals, but so far they are not ideal for helping students learn high-order thinking and creativity. AI-driven programs can give students and educators helpful feedback: AI can not only help teachers and students to craft courses that are customized to their needs but also provide feedback to both about the success of the course as a whole. These kinds of AI systems allow students to get the support they need and for professors to find areas where they can improve instruction for students who may struggle with the subject matter.

It is altering how we find and interact with information: We rarely even notice the AI systems that affect the information we see and find on a daily basis. Google adapts results to users based on location, Amazon makes recommendations based on previous purchases, Siri adapts to your needs and commands, and nearly all web ads are geared toward your interests and shopping preferences. These kinds of intelligent systems play a big role in how we interact with information in our personal and professional lives, and could just change how we find and use information in schools and academia as well. It could change the role of teachers: There will always be a role for teachers in education, but what that role is and what it entails may change due to the new technology in the form of intelligent computing systems. AI systems could be programmed to provide expertise, serving as a place for students to ask questions and find information or could even potentially take the place of teachers for very basic course materials. In most cases, however, AI will shift the role of the teacher to that of the facilitator. Teachers will supplement AI lessons, assist students who are struggling, and provide human interaction and hands-on experiences for students. In many ways, technology is already driving some of these changes in the classroom, especially in schools that are online or embrace the flipped classroom model.

Artificial intelligence can make trial-and-error learning less intimidating. Trial and error is a critical part of learning, but for many students, the idea of failing, or even not knowing the answer, is paralyzing. Some simply do not like being put on the spot in front of their peers or authority figures like a teacher. An intelligent computer system, designed to help students to learn, is a much less daunting way to deal with trial and error. AI could offer students a way to experiment and learn in a relatively judgment-free environment, especially when AI tutors can offer solutions for improvement. In fact, AI is the perfect format for supporting this kind of learning, as AI systems themselves often learn by a trial-and-error method.

Data powered by AI can change how schools find, teach, and support students: Smart data gathering, powered by intelligent computer systems, is already making changes to how colleges interact with prospective and current students. From recruiting to helping students choose the best courses, intelligent computer systems are helping to make every part of the college experience more closely tailored to student needs and goals. Data mining systems are already playing an integral role in today's higher education landscape, but AI could further alter higher education. Initiatives are already underway at some schools to offer students AI-guided training that can ease the transition between college and high school. Who knows but that the college selection process may end up a lot like Amazon or Netflix, with a system that recommends the best schools and programs for student interests? AI may change where students learn, who teaches them, and how they acquire basic skills: While major changes may still be a few decades in the future, the reality is that AI has the potential to radically change just about everything we take for granted about education. Using AI systems, software, and support, students can learn from anywhere in the world at any time, and with these kinds of programs taking the place of certain types of classroom instruction, AI may just replace teachers in some instances (for better or worse). Educational programs powered by AI are already helping students to learn basic skills, but as these programs grow and as developers learn more, they will likely offer students a much wider range of services.

More to discuss here is about the ability of AI units to analyze data in terms of software. Imagine what is involved in making chess programs that can defeat a grandmaster, and you see what AI is capable of making data-driven analysis and decisions faster than a human can. AI is already making a big impact on medicine and transportation, and is about to play a major role in education. Also AI plays well in teaching software sectors which are now with AI are more adaptive which means that learning as well is adaptive. Adaptive learning has always been a part of the online experience in evaluating via pre-programmed tests. Adaptive learning is more about programs that branch out into different subroutines based on user responses. Real AI software will do much more than that. AI emphasizes areas that need improvement: Some more recent AI teaching software is able to identify areas where students are deficient and focus on that content. Advanced versions can generate new problems from source materials. These online systems actually generate better materials and more comprehensive testing than typical classroom curriculum. Flexibility and adaptability are actually with the quality-oriented concept as the cycle of developing should include procedures to assure the quality of this software.

B. Artificial intelligence and e-LEARNING: An Intelligent Tutoring System Integrated with Learning Management Systems
1. Definition

As one result of integrating e-learning with AI, researcher presented the intelligent tutoring system (ITS) which is a computer-based system designed to educate students and tutor them by providing immediate and customized instructions or feedbacks to learners without the intervention from a human tutor. ITSs have the common goal of enabling learning in a meaningful and effective manner by using a variety of computing methodologies. An ITS typically aims to replicate the demonstrated benefits of personalized tutoring, in contexts where students would otherwise have access to one-to-many instruction from a single teacher, or no teacher at all. ITSs are often designed with the goal of providing access to high-quality education to each and every student.

There is a close relationship between intelligent tutoring, cognitive learning theories, and design, and there is a continuing research to improve the effectiveness of ITSs. There are many specimens of ITSs being used in educational systems and in which they have demonstrated their proficiencies and competences.

2. Structure

Intelligent tutoring systems are built using four basic components based on a general compromise among researchers (Nwana, 1990; Freedman, 2000; Nkambou et al., 2010).

i. The domain model

It is the intellectual model or expert knowledge model that is built on a theory of learning; more specifically, this model "contains the concepts, rules, and problem-solving strategies of the domain to be learned. It can fulfill several roles: as a source of expert knowledge, a standard for evaluating the student's performance or for detecting errors, etc." (Nkambou et al., 2010, p. 4).

Another approach for developing domain models is based on Stellan Ohlsson's theory of learning from performance errors, known as constraint-based modeling (CBM). In this case, the domain model is presented as a set of constraints on correct solutions.

ii. The student model

It is an overlay on the domain model and considered as the core component of an ITS paying special attention to student's cognitive and affective states and their evolution as the learning process advances.

iii. The tutor model

It accepts information from the domain and student models and makes choices about tutoring strategies and actions. At any point in the problem-solving process, the learner may request guidance on what to do next, relative to their current location in the model.

iv. The user interface

Component "integrates three types of information that are needed in carrying out a dialog: knowledge about patterns of interpretation (to understand a speaker) and action (to generate utterances) within dialogs; domain knowledge needed for communicating content; and knowledge needed for communicating intent" (Padayachee, 2002, p. 3).

3. Eight principles of intelligent tutoring system design and development

Anderson et al. (1987) outlined eight principles for intelligent tutor design and Corbett et al. (1997) later elaborated on those principles highlighting an all-embracing principle which they believed governed intelligent tutor design; they referred to this principle as:

Principle: An intelligent tutor system should enable the student to work to the successful conclusion of problem solving.

- Represent student competence as a production set.
- Communicate the goal structure underlying the problem solving.
- Provide instructions in the problem-solving context.
- Promote an abstract understanding of the problem-solving knowledge.
- Minimize working memory load.
- Provide immediate feedback on errors.
- Adjust the grain size of instructions with learning.
- Facilitate successive approximations to the target skill.

4. Design and development methods
i. The first stage
The goal is to specify learning goals and to outline a general plan for the curriculum; it is imperative not to computerize traditional concepts but develop a new curriculum structure by defining the task in general and understanding learners' possible behaviors dealing with the task and to a lesser degree the tutor's behavior. In doing so, three critical scopes need to be dealt with:

1. The probability a student is able to solve problems;
2. The time it takes to reach this performance level;
3. The probability the student will actively use this knowledge in the future.

ii. The second stage
Cognitive task analysis is a detailed approach to expert systems programming with the goal of developing a valid computational model of the required problem-solving knowledge. The key methods for evolving a domain model include:

1. Interviewing domain experts;
2. Conducting "think aloud" protocol studies with domain experts;
3. Conducting "think aloud" studies with novices;
4. Observation of teaching and learning behavior.

iii. The third stage
Initial tutor implementation involves setting up a problem-solving environment to enable and support an authentic learning process. This stage is followed by a series of evaluation activities as the final stage which is again similar to any software development project. The fourth stage, evaluation, includes

1. Pilot studies to confirm basic usability and educational impact;
2. Formative evaluations of the system under development;
3. Parametric studies that examine the effectiveness of system features;
4. Summative evaluations of the final tutor's effect: learning rate and asymptotic achievement levels.

5. Applications
In recent years, ITS has begun to move away from the search-based to include a range of practical applications. ITSs have expanded across many critical and complex cognitive domains, and the results have been far reaching. ITS systems have cemented a place within formal education and these systems have

found homes in the sphere of corporate training and organizational learning. ITS offers learners several affordances such as individualized learning, just in time feedback, and flexibility in time and space.

There are now many applications found in education and in organizations. ITSs can be found in online environments or in a traditional classroom computer lab, as well as in universities. There are a number of programs that target mathematics but applications can be found in health sciences, language acquisition, and other areas of formalized learning. Reports of improvement in student comprehension, engagement, attitude, motivation, and academic results have all contributed to the ongoing interest in the investment in and research of these systems.

6. Modern intelligent tutoring systems

In the late 20th century, Intelligent Tutoring Tools (ITTs) were developed by the Byzantium project, which involved six universities. The ITTs were general-purpose tutoring system builders and many institutions had positive feedback while using them (Kinshuk, 1996). This builder, ITTs, would produce an Intelligent Tutoring Applet (ITA) for different subject areas. Modern-day ITSs typically try to imitate the role of a teacher or a teaching assistant, and increasingly automate pedagogical functions such as problem generation, problem selection, and feedback generation. However, given a current move toward blended learning models, topical work on ITSs has been initiated concentrating on customizing these systems to perform more. Also, recent work has employed ethnographic and design research methods to examine the ways that ITSs are actually used by students and teachers across a range of contexts, often revealing unanticipated needs that they meet, fail to meet, or in some cases, even create.

Different teachers shaped the ITAs and built up a large inventory of knowledge that was accessible by others through the Internet. Once an ITS was created, teachers could copy it and modify it for future use. This system was efficient and flexible. However, Kinshuk and Patel (1997) believed that the ITS was not designed from an educational point of view and was not developed based on the actual needs of students and teachers.

C. The Future of Artificial Intelligence in E-learning

Virtual learning environments are online platforms designed mainly to deliver courses and training and manage information online. AI played a crucial role in promoting these LMSs to be more substantial redesigns (Ford, 2008) including the addition of the ability to simulate emotions and participate in complex social interactions (Payr, 2005). Emerging of AI in education and business has released a version of a scientific community that looks to industry as animated interfaces and guidelines which in turn improve learner's engagement with marketplace. This emerging has a profitable design consideration and ability of presenting that lasts for many years (ACM, 2004).

Artificial intelligence integration with the idea of building semantic web will for sure promote the learning as well from the stage of directed learning to the space of thinking of what learner is learning as Boyle (1998) stated that educators and parents must never forget that in most cases, it is more important to teach students how to think and not *what* they should be thinking about (Boyle, 1998). Adaptive, AI-based structures with agents can be teaching, tutoring, and interaction supported with animated user interfaces that afford an extraordinary opportunity to improve the knowledge experience by engraving it to the learner, and the perfection in student motivation, ownership, and rendezvous warrants further exploration.

With the vast wealth of information available publicly on the web, what may be required is the addition of a learning layer which integrates all of these data (Andersen, 2011) and could evolve organically using the same model of public contribution which has made Wikipedia the largest encyclopedia in the world. This idea can be further developed by integrating social networking into the system, so that a student consumes only educational content which has been marked as useful by people or organizations that they trust.

By this point, you have a globally accessible source of multimedia learning content filtered by sources you know to be trustworthy, which Anderson (2011) says can facilitate the below paradigm change in the learning process:

"The truly fascinating shift is that you wouldn't necessarily start by consuming the media that goes with the questions. Instead, you would simply start answering the questions in your bank. As you encounter learning questions that you can't answer, you could dive into the content at those points in time—this is the exact point between boredom (with things you already know) and frustration (with things you don't know), the point to engage in learning." (p. 16)

Researchers of e-learning and LMSs have been innovating and suggesting new features along utilizing the characteristics of artificial against to make the online learning more enjoyable and more interesting. One of these notable suggestions to improve student motivation is to add a game layer and make learning the goal of the "game" (Biswas et al., 2005; Chin et al., 2010; Priebatsch, 2010). It is a fact that at least one US Corporation is already trying to add a game layer to every action people engage in as a commercial enterprise (SCVNGR, 2011). On the other hand, other researchers were not that impressed with the While this excellent way to improve engagement, as some studies have shown that not all educational games do really encourage the constructive reasoning required to learn as the expectation must be (Conati, 2002).

We have recently noticed that the educational game concept has been invaded not only the educational platforms but also the social networks, famously the Facebook. This invasion comes with advantages and also impacted the education process negatively. Education game-based process has more than the purpose of "fun to learn" concept being used in higher education, it is more adopted to accumulate a quality-based knowledge that can sustain and support the graduate in the marketplace, while gaming concept lacks standards and most likely limited to serve with a high level of quality. For the transform in learning process, the scientific community needs to agree on a standard set of ontologies, or the framework required to create these standards. Ontologies, not unlike concept maps, provide the conceptual structure for describing and linking knowledge for a given field. Ontologies can be used to outline a learner's development and choose learning tracks for a student using AI agents (Deliyska and Rozeva, 2009). The diagram (Figure 6.8) shows a potential architecture for such a system (Chen, 2009).

Such proposals and systems are open for further improvements that could be to record and process information on the students' memory capacity and incentive (Jeremić et al., 2009) and attempt to adapt to their unique learning styles and abilities (Villaverde et al., 2006).

Avatars or what we call them animated computer characters, have been emerged into the teaching–learning process and actually have added a remarkable value to the education system, especially with the concept of encouraging the self-learning, student-centered learning process. Avatars have made it further supporting the education system as experience of "valuecosm," which is a system that permits educational digital avatars to make automated decisions on our behalf, based pre-trained data concept inside the AI units to match with learner's values, thoughts, and considerations. This would allow learners to delegate non-important decision making functions to their avatar and afford them more free time to pay attention to the important work of learning and meta-cognition (Andersen, 2011; Chin et al., 2010). Finally, the movement toward the "semantic web" may mean that intelligent systems become more capable of understanding and processing data

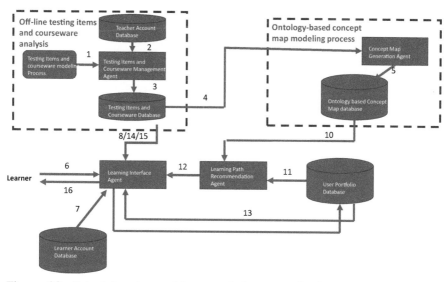

Figure 6.8 Potential system architecture of the personalized e-learning system with ontology-based concept maps.

independently. This would move AI and educational research into previously unimaginable areas (Welham, 2008).

D. A Learning Management System or an Intelligent Tutoring System?

To choose between LMSs or ITSs, we need to high point the metamorphoses, benefits, and opportunities of each.

1. Comparing the systems

Learning management systems and ITSs are e-technologies that came with AI base to promote the learning process. They both store and manipulate electronic content and convey an interactive experience to the learner. The advantage of performing in the digital world means that they are capable of presenting any number of content interactions, from basic gap fill exercises to sophisticated simulations, evaluating user input and offering assessment and grading based on their performance.

The key difference between the LMSs and the ITSs is that the latter provides direct customized feedback to learners based on their input into the system. It employs AI agents meaning that it can employ a variety of AI techniques to understand, inform, and direct the user after completing exercises.

It seeks to replicate the role of a tutor, effectively guiding and coaching the student through the content. So, the ITS plays the role of a private tutor who responses to each action in order to assure the learning accumulation per step, not at the end of the processes of learning or assessment as the LMS activities work. The structure of the ITS is a bit complex that than the LMS as it is built compromise missing the four modules: the student interface provides the interface with which the student interacts with the system, the domain model containing a description of the knowledge, or behaviors that represent expertise in the subject-matter domain the ITS is teaching. While the student model defines the knowledge and behavior of a sample student of the subject, the tutor module is designed to provide feedback to students. The fact that the ITS is based on AI units means that the system has already a knowledge-base and has lots of data-training procedures to build the required experience to make decisions such providing certain feedback that should work exactly as a human tutor in such situations would work, suggest and do.

The ITSs sound superlative and ideal for educational systems, or cases where the system lacks expert teachers to teach a specific course; an ITS can act as this teacher supported with a well-formed and designed pursuit of the required knowledge and way of interacting. Apparently, this means that ITSs are intensive in terms of the expert knowledge and resources required to build it. This increases if the subject is qualitative rather than quantitative, with qualitative knowledge containing a greater degree of implied knowledge. ITS timeframe of design, train, and put into function is longer than the LMS's as it needs to cover all aspects of content analysis, student, and tutor roles, but yet if successfully designed, it can offer a rich student experience as well as being an accurate barometer of subject knowledge.

Alternatively, the LMS performs with less complexity as it does not act like private tutoring systems. It can use individual exercises to offer paths through the learning experience. Content can be easily amended in a variety of different formats. Logical rules do not have to be associated with content and users are not going through a complete AI learning experience. The LMSs can provide correct answers to learners when each exercise has been completed, but cannot offer intelligent feedback like the ITSs can. In case of feedback, LMS uses customs of feedback or requires the cooperation from a human tutor to interact with the students and guide them. An ITS could offer greater learning value where the requirement is to learn faster within a shorter space of time, with direct customized feedback offering the learning provider a tangible way of delivering their learning. The outcomes of ITSs are more accurate and ITSs come with expectations more than the LMSs.

The comparison can continue for longer times as both have lots of advantages and features and they are ideal for different cases of learning. But to maximize the benefits of both systems, an integration in between is expected to be feasible, though this would depend on a coherent and focused e-learning vision within an organization. The integration included re-orienting and re-designing of some functions such as student communication and assessment could be managed within the LMS, while maintaining the ITS could deliver the learning experience, for instance.

Comprehensive technical analysis would be a prerequisite to providing the opportunity of integration, and at the same time, strategic benefits would have to be recognized prior to this. Effective semantic analysis of the subject scope would be critical to be able to fully exploit the capabilities of the AI system. With the LMSs, it is simpler as just adding course digital content without the need of intensive experience to do so while adding content to the ITSs, administrators would have to consider context in addition to student and tutor outcomes. Migrating content between ITSs, or between ITSs and LMSs, would be riskier, with capabilities required for any logical rules and associated sequencing relevant to ITS content.

Both types of systems are blended learning solutions, not seeking to replace the teacher but to aid the teacher both in and out of the classroom, and function as an aid in the delivery of the curriculum. ITSs would have an obvious advantage in student testing, as direct customized feedback which has more value student assessment. The integration in between as described for is the ideal learning environment for the teaching–learning process involving the idea of AI. When faced with the choice, business considerations would have to be supreme. Using an ITS would require considerably more resources and investment, depending on the level of customization. Recent researches have contributed with a new suggestion to integrate both or at least to embed the AI unit into LMS as agents for specific tasks. Studies' abstracts can be as follows:

1. Study entitled with: "An Intelligent Tutoring Systems Integrated with Learning Management Systems," for Cecilia E. Giuffra P., Ricardo Azambuja Silveira, and Marina Keiko Nakayama, *Part of the Communications in Computer and Information Science* book series (CCIS, vol. 365), 2013 has contributed with the following abstract:

"Learning Management Systems (LMSs): LMSs are used in distance learning and classroom teaching as teachers and students support tools in the teaching–learning process. Teachers can provide material, do activities, and

create assessments for students. Nevertheless, this procedure is done in the same way for all the students, regardless of their performance and behavior differences. The work proposes an ITS integrated with LMS to provide adaptability to it, using Moodle as a case study, taking into account student performance on tasks and activities proposed by the teacher and student access on resources."

2. Study entitled with: "Enhancing Performance of Learning Management Systems (LMSs) through Intelligent Agents," for Maryam Ghanilou, Rozita Jamili Oskouei, *MAGNT Research Report (ISSN. 1444–8939), Vol. 2 (3). PP:192–198* has contributed with the following abstract:

"Intelligent agent is an entity that has sensors for perceiving its environment and acting through effectors. Each intelligent agent has some internal characteristics such as autonomy, reactivity, goal-based, and learning/reasoning and several external characteristics such as communication, cooperation, and mobility capabilities. E-learning is using electronic communication techniques and the Internet to deliver learning materials for learners in various distance locations. It is a very important way for providing facilities for disabled people or anyone who has problem for learning through traditional learning models. LMSs are proposed to create virtual classrooms. One LMS allows learners to connect and use these systems from anywhere and anytime based on their desires. Some of the important facilities that are providing LMSs are including: delivering learning materials to learners and tracking the learners' learning process, making quizzes and exam administration, supporting learners' collaboration, managing learners' profile, scheduling events, providing facilities for registration and prerequisite and screening and cancellation notification, etc. Since several intelligent agents are proposed for increasing the performance of LMSs. The paper presents a comprehensive study about various intelligent agents that are proposed for enhancing the learning capabilities of learners in LMS environments and their functionalities."

E. Artificial Intelligence-led Quality Assurance for E-learning Concept

Currently, learning new subject and making training classes for employees of educational environments is one essential task. However, all managers or employers are intended to make training facilities for all their workers with the best quality and less money charging. E-learning can be a help to these employers or managers to deliver the desired information and knowledge to their employees with low cost and high quality. LMSs provide facilities for the quality assurance purpose. Further, it can help the managers for tracking the employees' performance and attendance in these virtual classes. Intelligent agents (IA) are providing facilities for adding personalized learning contents and delivering these contents for learners through the virtual LMSs. Since personalization is one of the main challenges nowadays, several applications of intelligent agents specially related to supporting personalization in e-learning and LMS are under research and discussion. The approach on AI/machine learning (ML)-based quality assurance is design-based complying with specific steps:

1. Explore and research: Create smart assess using data repositories including defects and logs that can be used for analytical and predicate purposes.
2. Learn: Identify relation between test assets such as defects and software requirement document for developing insights.
3. Sense: Predict the occurrence of an incident, impact, and do analytics and insights.
4. Respond cycle: Respond to an incident, input, and result for continuous learning.
5. The knowledge base stores and builds patterns of learning which in turn helps in learning and making decisions and responding to defects and sudden actions, which actually formulates a quality-oriented AI unit in e-learning.

The e-learning system built using this idea should train the system on predicting the next and building a self-learning confident to solve problems and comes with a new solution in times of unavailability of scientific methods. The AI unit has strategic decisions in terms of allocating resources, for example, and aborting missions in risky situations. Quality assurance procedures are life cycles of continuous assessing, predicating, and coming with solutions to produce the quality we seek in education which matches with internal and external requirements. Exactly with similar scenario as we have described above that fits perfectly with artificial intelligent units for piloting e-learning and blended learning concepts, which in turn comes with results for higher education institutions who have already adopted e-learning and blended learning. Figure 6.9 presents the basic of a quality-oriented artificial intelligent unit can be deployed for quality assurance purposes.

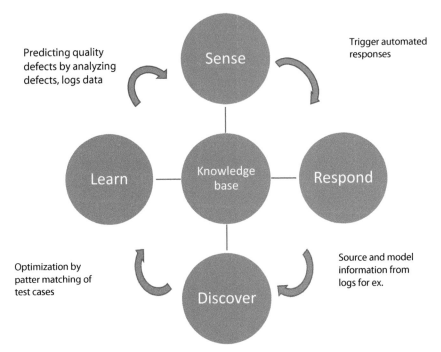

Figure 6.9   Quality-oriented artificial intelligence unit.

References and Suggested Literature

[1] Anderson, T., and Dron, J., Three generations of distance education pedagogy. *Int. Rev. Res. Open Distr. Learn.* 12(3), 80–97, 2011.

[2] Bloxham, K. T., *Using Formative Student Feedback: A Continuous Quality Improvement Approach for Online Course Development.* All Graduate Theses and Dissertations, 801, 2010.

[3] Bonk, C. J., and Graham, C. R., *The Handbook of Blended Learning: Global Perspectives, Local Designs.* Hoboken, NJ: John Wiley and Sons, 2012.

[4] Charmaz, K., *Constructing Grounded Theory: A Practical Guide through Qualitative Analysis.* London: SAGE Publications Ltd., 2006.

[5] Contreras, S., Torres, J., Palominos, P., and Lippi, L., "Quality Management Systems in Educational Contexts: A Literature Review." In *ICERI2015 Proceedings*, Sevilla, 1790–1796, 2015.

[6] Corbin, J. M., and Strauss, A., Grounded theory research: Procedures, canons, and evaluative criteria. *Qual. Soc.* 13(1), 3–21, 1990.

[7] Decreet van 08/05/2009. (n.d.). Retrieved December 17, 2014, available at: http://www.ond.vlaanderen.be/edulex/database/document/document. asp?docid=14129

[8] Decreet van 15/06/2007. (n.d.). Retrieved February 9, 2015, available at: http://www.ond.vlaanderen.be/edulex/database/document/document. asp?docid=13914

[9] Deepwell, F., Embedding Quality in e-learning Implementation through Evaluation. *Educ. Technol. Soc.* 10(2), 34–43, 2007.

[10] Deming, W. E., Elementary Principles of the Statistical Control of Quality. *JUSE,* 1950.

[11] Dumont, B., and Sangra, A., "Organisational and cultural similarities and differences in implementing quality in e-learning in Europe's higher education." In *Handbook on Quality and Standardisation in ELearning* (Berlin: Springer), 331–346, 2006.

[12] Dzakiria, H., Illuminating the Importance of Learning Interaction to Open Distance Learning (ODL) Success: A Qualitative Perspectives of Adult Learners in Perlis, Malaysia. *Eur. J. Open, Dist. E-Learn.* 2012.

[13] Ehlers, U.-D., Quality in e-learning from a learner's perspective. *Eur. J. Dist. Open Learn.* 2004.

[14] Ehlers, U.-D., Quality Literacy-Competencies for Quality Development in Education and e-Learning. *Educ. Technol. Soc.* 10(2), 96–108, 2007.

[15] Ehlers, U. D., Understanding quality culture. *Qual. Assur. Educ.* 17(4), 343–363, 2009a.

[16] Ehlers, U. D., Web 2.0 – e-learning 2.0 – quality 2.0? Quality for new learning cultures. *Qual. Assur. Educ.* 17(3), 296–314, 2009b.

[17] Ehlers, U.-D., and Pawlowski, J. M., "Quality in European e-learning: An introduction." In *Handbook on Quality and Standardisation in E-Learning* (Berlin: Springer), 1–13, 2006.

[18] Frydenberg, J., Quality Standards in eLearning: A matrix of analysis. *Int. Rev. Res. Open Distr. Learn.* 3(2), 2002.

[19] Garrison, D. R., and Kanuka, H., Blended learning: Uncovering its transformative potential in higher education. *Inter. High. Educ.* 7(2), 95–105, 2004.

[20] Graham, C. R., "Blended learning systems: Definition, current trends, and future directions." In C. J. Bonk and C. R. Graham (eds), *Handbook of Blended Learning: Global Perspectives, Local Designs* (San Francisco, CA: Pfeiffer Publishing), 3–21, 2005.

[21] Graham, C. R., and Robison, R., "Towards a Conceptual Frame-work for Learning in Blended Environments." In A. Picciano and

C. Dziuban (eds), *Blended Learning: Research Perspectives* (United States of America: the Sloan Consortium), 83–111, 2007.

[22] Graham, C. R., Woodfield, W., and Harrison, J. B., A framework for institutional adoption and implementation of blended learning in higher education. *Inter. High. Educ.* 18, 4–14, 2013.

[23] Grifoll, J., Huertas, E., Prades, A., Rodriguez, S., Rubin, Y., Mulder, F., European Association for Quality Assurance in Higher Education (ENQA), *Quality Assurance of E-learning*. ENQA Workshop Report 14. ENQA (European Association for Quality Assurance in Higher Education), 2010.

[24] Hansson, H., *E-learning quality. Aspects and criteria for evaluation of e-learning in higher education*, 2008. Available at: http://www.diva-portal.org/smash/record.jsf?pid=diva2:283764

[25] *TOJET: The Turkish Online Journal of Educational Technology*, 2017, 16(3), The Turkish Online Journal of Educational Technology, 177.

[26] Harroff, P. A., *Dimensions of Quality for Web-based Adult Education*. Unpublished Doctoral Dissertation, University of Georgia, Georgia, 2002. Available at: https://getd.libs.uga.edu/pdfs/harroff_pamela_a_200 208_edd.pdf

[27] Inglis, A., Quality Improvement, Quality Assurance, and Benchmarking: Comparing two frameworks for managing quality processes in open and distance learning. *Int. Rev. Res. Open Dist. Learn.* 6(1), 2005.

[28] Inglis, A., Approaches to the validation of quality frameworks for e-learning. *Qual. Assur. Educ.* 16, 347–362, 2008.

[29] Jara, M., and Mellar, H., Factors affecting quality enhancement procedures for e-learning courses. *Qual. Assur. Educ.* 17(3), 220–232, 2009.

[30] Jung, I., The Dimensions of E-Learning Quality: From the Learner's Perspective. *Educ. Technol. Res. Dev.* 59(4), 445–464, 2011.

[31] Kear, K., Rosewell, J., Williams, K., Ossiannilsson, E., Covadonga, R., Paniagua, Á. S.-E., Mellar, H., *Quality Assessment for E-learning: a Benchmarking Approach*, 3rd edn, Maastricht, The Netherlands: European Association of Distance Teaching Universities (EADTU), 2016. Available at: http://excellencelabel.eadtu.eu/images/ Excellence_manual_2016_third_edition.pdf

[32] Korres, M. P., Karalis, T., Leftheriotou, P., and Barriocanal, E. G., "Integrating adults' characteristics and the requirements for their effective learning in an e-Learning environment." In M. D. Lytras, P. Ordonez, de Pablos, E. Damiani, D. Avison, A. Naeve, and D. G. Horner

(eds), *Best Practices for the Knowledge Society. Knowledge, Learning, Development and Technology for All* (Berlin: Springer), 570–584, 2009.

[33] MacDonald, C. J., and Thompson, T. L., Structure, Content, Delivery, Service, and Outcomes: Quality eLearning in higher education. *Int. Rev. Res. Open Dist. Learn.* 6(2), 2005.

[34] McLoughlin, C., and Lee, M. J., The Three P's of Pedagogy for the Networked Society: Personalization, Participation, and Productivity. *Int. J. Teach. Learn. High. Educ.* 20(1), 10–27, 2008.

[35] Moskal, P., Dziuban, C., and Hartman, J., Blended learning: A dangerous idea? *Inter. High. Educ.* 18, 15–23, 2013.

[36] Ossiannilsson, E., and Landgren, L., Quality in e-learning – a conceptual framework based on experiences from three international benchmarking projects. *J. Comp. Assist. Learn.* 28(1), 42–51, 2012.

[37] Ossiannilsson, E., Williams, K., Camilleri, A., and Brown, M., *Quality models in online and open education around the globe. State of the art and recommendations.* Oslo: International Council for Open and Distance Education – ICDE, 2015.

[38] Scheerens, J., School Effectiveness Research and the Development. *Int. J. Res. Policy Pract.* 1, 61–80, 1990.

[39] Scheerens, J., School Effectiveness Research and the Development of process indicators of school function. *Int. J. Res. Policy Pract.* 1(1), 61–80, 2006.

[40] Shea, P., "Towards a conceptual framework for learning in blended environments." In A. G. Picciano and C. Dzuiban (Eds.), *Blended Learning: Research Perspectives* (Needham, MA: The Sloan Consortium), 2007.

[41] Srikanthan, G., and Dalrymple, J. F., Developing a Holistic Model for Quality in Higher Education. *Qual. High. Educ.* 8(3), 215–224, 2002.

[42] Stodel, E. J., Thompson, T. L., and MacDonald, C. J., Learners' Perspectives on what is Missing from Online Learning: Interpretations through the Community of Inquiry Framework. *Int. Rev. Res. Open Distr. Learn.* 7(3), 2006.

[43] Ubachs, G., Brown, T., Williams, K., Kess, P., Belt, P., van Hezewijk, R., Riegler, K., *Quality Assessment for E-learning: a Benchmarking Approach,* 1st ed. European Association of Distance Teaching Universities (EADTU), 2007.

[44] Volungeviciene, A., Tereseviciene, M., and Tait, A. W., Framework of quality assurance of TEL integration into an educational organization. *Int. Rev. Res. Open Distr. Learn.* 15(6), 2014.

[45] Williams, K., Kear, K., and Rosewell, J., *Quality Assessment for E-learning: a Benchmarking Approach, 2nd edn.* Heerlen: European Association of Distance Teaching Universities (EADTU), 2012.

[46] Zhang, W., and Cheng, Y. L., Quality assurance in e-learning: PDPP evaluation model and its application. *Int. Rev. Res. Open Distr. Learn.* 13(3), 66–82, 2012.

[47] Available at: https://www.infosys.com/IT-services/validation-solutions/service-offerings/Documents/machine-learning-qa.pdf

[48] Available at: https://www.learningsolutionsmag.com/articles/45/a-learning-management-system–or-an-intelligent-tutoring-system

[49] Al-Zoube, M., E-Learning on the Cloud. *Int. Arab J. Inform. Technol.* 1(2), 2009.

[50] Pocatilu, P., Alecu, F., and Vetrici, M., Using cloud computing for E-learning systems. In *Proceedings of the 8th WSEAS international conference on Data networks, communications, computers* (pp. 54–59), *World Scientific and Engineering Academy and Society (WSEAS)*, 2009.

[51] Masud, M. A. H., and Huang, X., *An E-learning System Architecture based on Cloud Computing*, World Academy of Science, Engineering and Technology, 6(2), 2012 International Science Index, Information and Communication Engineering, 6(2), 2012.

[52] Babateen, H. M., The role of virtual laboratories in science education. In *5th International Conference on Distance Learning and Education IPCSIT*, IACSIT Press, Singapore, 12, 100–104, 2011.

[53] O'Connell, R., *Artificial Intelligence in e-Learning Systems*, 2011, available at: https://richardoconnell.wordpress.com/2011/01/27/artificial-intelligence-in-e-learning-systems/

[54] *Web 3.0: Technology and the Valuecosm*

[55] ACM, Pedagogical Software Agents. *Communications of the ACM*, 47(4), 47, 2004.

[56] Andersen, M. H., The World Is My School: Welcome to the Era of Personalized Learning. *Futurist* 45(1), 12–17, 2011.

[57] Bargh, J. A., and Schul, Y., On the cognitive benefits of teaching. *J. Educ. Psychol.* 72(5), 593–604, 1980.

[58] Biswas, G., Leelawong, K., Schwartz, D., and Vye, N., Learning by teaching: A new agent paradigm for educational software. *Appl. Artifi. Intell.* 19(3–4), 363–392, 2005.

[59] Bodenheimer, B., Williams, B., Kramer, M. R., Viswanath, K., Balachandran, R., Belynne, K., and Biswas, G., Construction and

Evaluation of Animated Teachable Agents. *J. Educ. Technol. Soc.* 12(3), 191–205, 2009.

[60] Boyle, M., Has Minsky anything to say for education? *J. Comput. Assis. Learn.* 14(4), 260–267, 1998.

[61] Cassell, J., Towards a model of technology and literacy development: Story listening systems. *J. Appl. Dev. Psychol.* 25(1), 75–105, 2004.

[62] Chen, C., Ontology-based concept map for planning a personalised learning path. *Br. J. Educ. Technol.* 40(6), 1028–1058, 2009.

[63] Chin, D., Dohmen, I., Cheng, B., Oppezzo, M., Chase, C., and Schwartz, D., Preparing students for future learning with Teachable Agents. *Educ. Technol. Res. Develop.* 58(6), 649–669, 2010.

[64] Conati, C., Probabilistic assessment of user's emotions in educational games. *Appl. Artif. Intell.* 16(7–8), 555–575, 2002.

[65] Deliyska, B., and Rozeva, A., Multidimensional Learner Model In Intelligent Learning System. *AIP Conference Proceedings*, 1184(1), 301–308, 2009.

[66] Eppler, M., and Mengis, J., The Concept of Information Overload: A Review of Literature from Organization Science, Accounting, Marketing, MIS, and Related Disciplines. *Information Society.* 20(5), 325–344, 2004.

[67] Forbes-Riley, K., Litman, D. J., Silliman, S., and Tetreault, J. R. (2006). "Comparing Synthesized versus Pre-Recorded Tutor Speech in an Intelligent Tutoring Spoken Dialogue System." In *FLAIRS Conference*, 509–514, 2004.

[68] Ford, L., A new intelligent tutoring system. *Br. J. Educ. Technol.* 39(2), 311–318, 2008.

[69] Gratch, J., Émile: Marshalling Passions in Training and Education. In *Proceedings of the fourth international conference on Autonomous agents, AGENTS'00* (New York, NY: ACM), 325–332, 2000.

[70] Jeremić, Z., Jovanović, J., and Gašević, D., Evaluating an Intelligent Tutoring System for Design Patterns: the DEPTHS Experience. *J. Educ. Technol. Soc.* 12(2), 111–130, 2009.

[71] Kinchin, I. M., and Hay, D. B., How a qualitative approach to concept map analysis can be used to aid learning by illustrating. *Educ. Res.* 42(1), 43, 2000.

[72] Leelawong, K., and Biswas, G., Designing learning by teaching agents: The Betty's Brain system. *Int. J. Artif. Intell. Educ.* 18(3), 181–208, 2008.

[73] Payr, S., Not Quite an Editorial: Educational Agents and e-learning. *Appl. Artif. Intell.* 19(3/4), 199–213, 2005.

[74] Pek, P., and Poh, K., Making Decisions in an Intelligent Tutoring System. *Int. J. Inform. Technol. Decis. Making* 4(2), 207–233, 2005.

[75] Peso, J. D., and de Arriaga, F., Intelligent E-Learning Systems: Automatic Construction of Ontologies. *AIP Conference Proceedings,* 1007(1), 211–222, 2008.

[76] Priebatsch, S., *Seth Priebatsch: The game layer on top of the world |Video on TED.com*, 2010. Available at: http://www.ted.com/talks/seth_priebatsch_the_game_layer_on_top_of_the_world.html

[77] Rishi, O. P., and Govil, R., DCBITS: Distributed Case Base Intelligent Tutoring System. *AIP Conference Proceedings,* 1007(1), 162–176, 2008.

[78] SCVNGR., *SCVNGR*, 2011. Available at: http://scvngr.com/

[79] Shaw, K., The application of artificial intelligence principles to teaching and training. *Br. J. Educ. Technol.* 39(2), 319–323, 2008.

[80] Sundberg, P., *Learning Theories Concept Map,* 2011. Available at: http://www.ed.uiuc.edu/courses/edpsy317/sp03/learning-maps/sundberg-learning-theories.gif

[81] Van den Brande, L., *Flexible and Distance Learning.* New York, NY: John Wiley and Sons, Inc., 1993.

[82] VanLehn, K., The behavior of tutoring systems. *Int. J.* Artif. Intell. *Educ.* 16(3), 227–265, 2006.

[83] Villaverde, J. E., Godoy, D., and Amandi, A., Learning styles' recognition in e-learning environments with feed-forward neural networks. *J. Comp. Assis. Learn.* 22(3), 197–206, 2006.

[84] Weibelzahl, S., and Weber, G., "Evaluating the inference mechanism of adaptive learning systems." In *Proceedings of the 9th international conference on User modeling, UM'03* (Berlin: Springer-Verlag), 154–162, 2003. Available at: http://portal.acm.org/citation.cfm?id=1759957.1759984

[85] Welham, D., AI in training (1980–2000): Foundation for the future or misplaced optimism? *Br. J. Educ. Technol.* 39(2), 287–296, 2008.

[86] Available at: https://en.wikipedia.org/wiki/Intelligent_tutoring_system

[87] Available at: http://meta-guide.com/robopsychology/artificial-intelligence-in-learning-management-systems-lms

[88] Available at: https://www.learningsolutionsmag.com/articles/45/a-learning-management-system–or-an-intelligent-tutoring-system

[89] Available at: https://createlms.com/knowledge-base/blogs/role-artificial-intelligence-ai-learning-management-systems

[90] Available at: https://www.researchgate.net/publication/318900824_Valid ation_of_a_Conceptual_Quality_Framework_for_Online_and_Blended_ Learning_with_Success_Factors_and_Indicators_in_Adult_Education_a_ qualitative_study

[91] Jung, Insung; Latchem, Colin., Assuring Quality in Asian Open and Distance Learning, Open Learning, v22 n3 p235–250 Nov 2007. Open and distance learning (ODL) is enjoying phenomenal growth in Asian higher ..., https://eric.ed.gov/?id=EJ776781

7

Teaching–Learning Process and Quality Assurance in Higher Education

7.1 Overview

7.1.1 Introduction

The higher education institutions (HEIs) as universities have addressed quality assurance for their teaching provision in a variety of methods, and the acknowledgement given to the high standards of teaching has been evidence of their educational systems' effectiveness. Various factors to be taken into these provisions, such as the augmented size of the university, the greater diversity of courses and the variety of teaching and assessment methods, technology interaction means most of institutions have adopted to such as e-learning platforms and the need to interrelate and follow learning process step by step with learners to avoid system failure, all together, along with the growth of these educational institutions requiring to meet standards of quality have added burdens to the teachers and educators who are the frontline of the educational systems. This urged all institutions and education initiates to call for more systematic and coordinated approaches to quality assurance in general and quality of the teaching–learning process in specific.

Higher education institutions who have being performing in the academic field for time with notable high levels of quality of teaching and learning are still having difficulties to demonstrate this level of quality to outside world. These well-established higher education institutions have been forming emergency strategic planning from time to time to articulate famed academic concerns and keep the enhancement cycles to be in line with new transforming and changing plans, launching new academic programs, adopting to globalizations. Emergency calls can be also when the strategic planning underperforming are not well-formed for specific students and so, then that they lack systematic mechanisms for quality control from a while to while. What we want to convey here is that meeting the standards of

191

higher education can't be once for academic life. Huge responsibilities are attached to these weel-established institutions to keep top position and top ranked they have achieved. Emergency calls we have mentioned just above should always concerns as well about the quality of teaching–learning as the quality teaching–learning in higher education matters for student learning outcomes such as graduates and meeting the stakeholder's expectations. So, the quality education is not a random process, it requires well-educated and trained teachers and is expected to give further consideration to the quality of education process as a unit and quality of teaching in specific to ensure that the system is effective on both sides' qualitative and quantitative standards.

This chapter presents quality of teaching as it has never been thought of before and discusses some recent academic research to highlight the importance of quality and also to suggest some promising solutions to overcome obstacles that have been encountered by teachers and educators.

7.2 Quality of Teaching–Learning Process

7.2.1 Quality of Teaching–Learning Process

One definition of teaching is as simple as: teaching is basically a cultural activity that has started since earliest days of humanity, and mainly to transfer the gained knowledge to someone else in order to get him/her knowledgeable. But now it is a bit complicated, not only teaching but also we investigate the definition of teaching quality which is the use of pedagogical performances to harvest the best learning outcomes for learners (students). It involves several scopes, starting from the effective design of curriculum and course content, a diversity of learning frameworks including, for example, steered independent learning, project-based learning, combined learning, investigation, and research methodologies. Teaching is no long a single term, it is usually associated with learning indicating that no use of teaching if we could not reach the quality level of this teaching and present the best learning experience. The teaching–learning process comes with important features such as: asking questions, inquiring more information, using feedback, and effective assessment of learning outcomes. It also involves well-adapted learning environments and student support services. But in terms of teacher quality, definition is still no simple task, though, because the criteria for doing so vary from person to person, from one community to another, and from one era to the next.

There is much debate within the higher education community on how teaching or teaching effectiveness may be defined (Braskamp and Ory, 1994). For instance, Centra (1993) defined effective teaching as "that which produces beneficial and purposeful student learning through the use of appropriate procedures" (p. 42) and Braskamp and Ory (1994, p. 40) included both teaching and learning in their definition, defining effective teaching as the "creation of situations in which appropriate learning occurs; shaping those situations is what successful teachers have learned to do effectively." HEIs' experiences to foster quality assurance and apply it at the level of teaching, proves that quality is not a simple term or a concept applied to one level. One important levels of quality, is "quality of teaching" which as well, is not a simple or one-step procedure. On the contrary, it is iterative multi-levels of endeavor which embed the quality of teaching as a core of the inter-dependent level. We emphasize on this inter-level to be with no/limited factors, impacts, and influences which can distract the teaching–learning processes. We are no calling for a stand-alone teaching–learning process but we are trying to build this architecture of learning with a quality-oriented concept and base. This quality-oriented base of teaching is after the support from all parties who are included and contributed to this learning process.

Starting from students (learners) who can on spot get constructive feedback to help enhancing the learning process, teachers who are educators and in charge of designing the concept of learning with different tools and experiences. The institution has a basic role to assure the quality of teaching concept which should give the right space, healthy environments, and right/requested tools and teaching equipment (e-learning and blended learning tools and concepts) to inspire this architecture to rise up and survive as long as the teaching process is going on. To give formal positions for this concept and let it comes to true, quality of teaching should take place at four inter-dependent levels:

1. At the institution level: The institution's strategic plans and policy design should support to sustain the internal quality assurance systems. Institution plans should include endorsing quality culture and orientation.
2. At the program level: Comprising actions to measure and enhance the design, content, and delivery of the programs within a department.
3. Individual level: Including initiatives that help teachers achieve their mission, encouraging them to innovate and to support improvements to the teaching–learning process.

4. Quality assurance's agencies' level: These agencies have introduced teaching as a real unit that should be taken into consideration when it comes to evaluate the educational process.

Teachers and learners should be included as main parties in the simplest procedure of these agencies, that their voices should be heard from broad agents rather than the institutional body. For instance, agencies' site-visits to any university should include meetings with teachers and students. These meetings are meant to not only hear their voices or complains but further to exchange educational experiences with external examiners and give a space and time for exchanging experiences, training how to achieve quality of teaching, as well as getting the training they seek for to be up to the level, and also, to give a fair chance to participate in the quality assurance for external standards.

Supporting quality teaching at the program level is a key to ensure improvement in quality teaching at the individual level and across the institution. Support for quality teaching can be demonstrated over many eclectic assortments of activities that are likely to improve the quality of the teaching process, of the program content, as well as the learning conditions of students.

Essentially, visiting the history back, we find that a number of factors have brought quality teaching to the forefront of higher education policies. Practically, every education system has experienced significant growth of student numbers in contemporary decades and the student profile has become more diverse. At the same time, higher education faces greater challenges from students, parents, and employers to account for their performance and demonstrate their teaching quality. Talking more practically and supported with latest statistics and feedbacks, we can summarize the reason why to follow the trend of "quality" at the teaching level:

1. Institutions should foster quality teaching and consider it an essential to sustain the quality within the entire system taken into consideration that the quality of teacher is related to the most important process in the educational system, teaching–learning process.

2. To respond to the growing request for evocative and relevant teaching skills and abilities. Students as well as employers want to ensure that their education will lead to rewarding employment and will prepare them with needed skills to evolve professionally over a lifetime.

3. Institutions would like to have the chance and demonstrate that they are reliable providers of good quality higher education, while operating in a complex setting, with multiple stakeholders, each with their own

expectations (ministries, funding agencies, local authorities, employers, etc.).

4. Achieve the stability of teaching performance that can last along with research performance.
5. Plan for mechanisms to effectually contend for students against the increase of higher tuition subscriptions and greater student mobility.
6. To escalate the efficiency of the teaching–learning process to attain the expected outcomes.

All what we have mentioned above can be categorized as challenges at institutional level, but add to this burden, higher education challenges in terms of internationalization and globalization that ask for well-skilled outcomes of this higher education that can compete the workplace and external education systems, rapid change of technology that affects even the satisfaction of all parties in this learning–teaching units, and engagement of graduates in the community which reflect the right choice for higher education to approve programs that have been offered by HEIs. So, teaching quality is not limited by local issues, even if they look like, nevertheless, quality of teaching throughout the world is also influenced by contextual shifts within the higher education environment, not the local higher education policies but also the external fleet of higher education and quality requirements.

It is important to distinguish teacher quality from teaching quality. If universities are not well organized and supportive, it is possible that even good teachers will not be successful (Raudenbush et al., 1992). The quality of teaching depends on many factors, including the level of instructional resources available, faculty qualifications, continuing professional development, and support from administrators and parents (Johnson, 1990). To learn how institution's level might promote teacher quality, by adopting a hybrid forms including ingenuities such as:

1. The institution involves plans to promote teaching and create centers for teaching and learning development, including activities and iterative cycles of training utilizing old and provisions of experiences.
2. The university system should be able to permit faculty to have a private channel to speak aloud inside the institutions and permits for professional development activities. Sabbatical years of research give a space to this faculty to seek more experience that they could not find in their environment without getting worried about losing their credentials and positions.

3. The system has to have funds for teaching and innovation as well as research fostering and funding teaching excellence awards.
4. The university system should be able to afford healthy teaching conscription criteria and give more opportunities to pioneer pedagogy.
5. The university system should appreciate any student achievement. Actually, in teaching–learning linkage, teachers should always get credentials for this achievement because they are the first line in teaching progress. The fostering can be both for the teacher as a supervisor and student which can be counseling, career advice, or mentoring.
6. Teacher contribution and right to know about the students' evaluation if they are not directly related to their teaching. This kind of knowledge builds more teaching confidence of being aware of thoughts and expectations of students.
7. Affording multi-ways of communication between teachers and institutions that involve self-evaluation of experimentations, peer-reviewing, and benchmarking of practices.
8. Teachers being more engaged with community service. This point may also include a development-based program for self-developing.
9. Competence-based quality of teaching assessments and evaluations. Competence is a nature, and should be used wisely and in the positive direction.

7.2.2 Quality Assurance System Is a Personal Touch When it Comes to Teaching–Learning Process

Defining teacher quality is fundamental to understand the quality role and being able to promote it. Quality of teaching examines teacher quality which may include the knowledge, skills, abilities, and dispositions of teachers. These features are used to discuss current conceptions of teacher quality and to develop these standards for advanced certification, describing what accomplished. Standards can be a very wide aspect, but yet represent contemporary views of teacher quality and are relied on, in part, what teachers need to know and do to promote student learning.

The quality assurance system should be able to monitor the course of teaching and learning processes and should be able to evaluate on periodical base being extremely important, not only to regulate the teaching and learning process, following the quality assurance orientations at a national and international level, but also to reflect and share teaching practices that enhance the whole academic experience, for students, teachers, and researchers' perspectives. This monitoring should be able to touch the base to categorize problems and good practice situations identified by the students' survey and reports feeding back about the teaching process. Teachers are encouraged to be part of such practices. The quality assurance systems are to help teachers to personify virtue and practice these criteria for quality of teaching purposes with high moral characters. Reliably, teachers are often expected to play good role models for students and to represent the highest standards of social propriety. This view of teacher quality was especially widespread in the early 1900s, when teachers were often positioned on pedestals, to speak, as were ministers. When a teacher entered a room, people stopped talking and became self-conscious and embarrassed.

Let us dig more about quality and re-ask the question, why did quality teaching become an issue of importance? What we are trying to do here is to fetch more reasons and consequences of quality in education in general and in teaching in specific. Expanding our awareness toward the need behind quality paves the road to our goal toward the excellence of the higher education systems. We can add here that nowadays the number of students has increased and so we have more pupils to be taught than ever before. The challenge of number is along with the recent trends in higher education that have increased toward the attention given to the quality of the teaching offered to the students. Let us try to go through changes and raise the need and try to conclude the best definition again:

- First, the advent of mass higher education that produced a shift in the conception of the role of universities. In fact, according to Coaldrake and Stedman (1999), teaching was the major function of universities. But the export of the German model of research and teaching to the United Kingdom and the United States led research to become the first and during the 20th century, whereas teaching was often perceived as a second-class activity. Academic institutions could not hold more because of the expansion of the higher education sector, the importance of teaching is now being re-examined and re-assessed.
- Second, the funding structure of many universities is changing according to priorities and increased the focus on the quality of teaching as the higher education itself is being seen as an investment that should contribute to national success in the long term.
- Third, quality assurance culture and structure in higher education have also become a focus of attention for private universities (Jones, 2003). Students might now be described as "clients" of HEIs (Telford and

Masson, 2005). Which give them the right to be actually concerned about the quality of the lectures they pay for?

- Fourth, the learner is changeable and their expectations are growing up, and teaching methods are too to cope with the new requirements.
- Fifth, globalization has influenced the higher education system and universities are more to get the best students on a national scale and internationally. So, teaching faculty are also involved in this globalization as many professors are now teaching international students, and consequently must promote their quality of teaching.
- Sixth, teaching methods have also developed. Educators who wish to integrate their teaching with e-learning and blended learning need to become familiar with new pedagogical methods.
- Seventh, quality may be seen as "fitness for purpose," the purpose being that of the institution is always attached and related to the teaching–learning quality.

Accordingly, quality of teaching and teacher quality, and also diversity of arguments have raised more concerns such as what constitutes a "good" teaching and how a "quality culture" in higher education supports quality of teaching. The definition of quality teaching is contingent on the meaning one preferences to give to the concept of quality. As Biggs (2001) points out, "quality" can alternatively outline an outcome or a process, and therefore it is not surprising that the phrase "quality teaching" has been given several definitions. In this chapter, we try to build a strong base of quality of teaching to pave the way toward the excellence of higher education. Then it will be more logically if we associate the "excellence" performance to the components of this HEI and reflect this on the procedures and processes that have been adopted to. Let us start by saying that quality is the "excellence" which is the traditional conception of quality and it is the leading and dominating one in many of old elite HEIs.

Skelton (2005) stated that because there is no one single understanding of what constitutes "teaching excellence," there are various types of quality cultures. Skelton presented four meta understandings of teaching excellence in higher education: traditional, performative, psychologized, or critical. An institution might want to change its quality culture, as "in any given culture, understandings of teaching excellence may change over time" (Skelton, 2005). But before trying to foster change, an HEI should previously consider what it currently regards as teaching excellence and review how the institution works (Skelton, 2005). Indeed, it is easy to understand that the quality of an educational system cannot exceed the quality of its teachers. Webbstock

(1999) underlines that good teaching is a type of teaching that correlates with the educational institution's mission statement. Taylor (2003) lists 13 abilities needed for quality teaching and learning:

1. Engagement locally and globally,
2. Engagement with peers and colleagues,
3. Equity and pathways,
4. Leadership,
5. Engagement with learners,
6. Entrepreneurship,
7. Designing for learning,
8. Teaching for learning,
9. Assessing for learning,
10. Evaluation of teaching and learning,
11. Reflective practice and professional development,
12. Personal management,
13. Management of teaching and learning.

7.2.3 Quality Assurance Leads to Quality of Teaching–Learning Process

As globalization continues, the national and international competition for the best students is likely to increase among HEIs, so reinforcing pressure for quality teaching and quality assurance. There are more and more students who study at various universities, benefitting from opportunities like international scholarships. These students are likely to compare the quality of the teaching received at these different institutions. Therefore, it is important that the university education, and the approach of learning, will requisite to formulate students for entry to such an atmosphere and equip them with fitting skills, knowledge, values, and attributes to thrive in it.

There is a strong drive to shape and create knowledge together with an understanding of working life and reformulate the notion of knowledge in education situations. If we can release the imagination a little bit and reformulate the teaching–learning process instead of being a student-centered process, it can be formulated to be an iterative loop of the knowledge-based system that accumulates the knowledge for the profit of both teachers and learners. The smart system needs procedures to follow and methods to control. Figure 7.1 demonstrates a new thinking of knowledge-based system combining teachers and students in a centered unit to build the educational system.

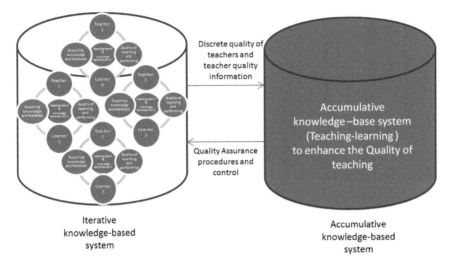

Figure 7.1   Quality assurance system to build and control the quality of teaching–learning and teacher quality.

Figure 7.1 actually is still missing factors that directly affect this accumulative enhancing unit. The teaching–learning process is not that easy or can be reduced in this simplicity. Lots of out factors and duties can affect and negatively impact the relationship of not cautiously dealt with, if we look closer to the factors which can be actually classified: relative and irrelative to this kind of relation between teachers and learners. Figure 7.2 gives a close-up view.

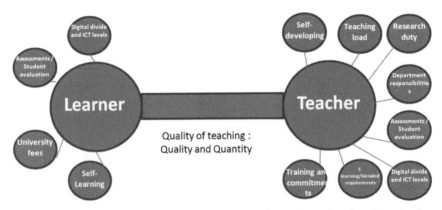

Figure 7.2   Quality of teaching–learning measures quality and quantity of teaching that have been transferred to the learner and vice versa in terms of feedback and satisfaction.

We can understand that the relationship in between is a channel that cannot take/handle lots of arguments. Quality and quantity of information that have been proved/transferred and exist can be badly impacted if this channel is overloaded. This can be on a teacher–learner direct interaction and the impact would of course be accumulative once the problem is not solved at this level. Quality of teaching is not a step from the teacher side. It is also a task that the learner should contribute to help his/her self at the first and then be a part and more engaged in the teaching–learning processes. Diversity of learners in the classroom raises issues of teaching and learning and actually was discussed in many previous books and went through different analysis and was of processing.

A book with a title "Teaching for quality at university: what the students does" was written by McGraw-Hill Education (United Kingdom) which is available on Amazon.com presented a very well example about engaging students' learning activities appropriately minimizes differences of ethnicity between students as far as learning itself is concerned. Let us look at two students attending a lecture. Susan is academically committed, she is bright, interested in her studies, and wants to do very well. She has clear learns; she goes about it in an "academic 'way. She comes to the lecture with sound, relevant background knowledge, possibly some questions she wants answering. In the lecture, she finds an answer to a preformed question: it forms the keystone for a particular arch of knowledge she is constructing. Or it may not be the answer she is looking for and she speculates, wondering why it is not. In either event, she reflects on the personal significant of what she is learning. Students like Susan virtually teach themselves and they do not need much help from teachers. Now, Robert is at university not out of a driving curiosity about a particular subject, or a burning ambition to excel in a particular profession, but to obtain a qualification for a decent job.

Students like Robert are in higher proportions in today's classes. They need help if they want to reach acceptable levels of achievement. To say that Robert is "unmotivated" maybe true, but it is unhelpful. All it means is that he is not responding to the methods that work for Susan, the likes of whom were sufficiently visible in most classes in the good old days to satisfy us that our teaching did work.

That was an excellent example of student's engagement to evaluate the quality of teaching as I suppose that the feedback from Susan to the teaching–learning process will be for sure different from the feedback that can be received from Robert. Figure 7.3 is a chart of academic performance of Susan and Robert that measures their engagements levels in the teaching-learning

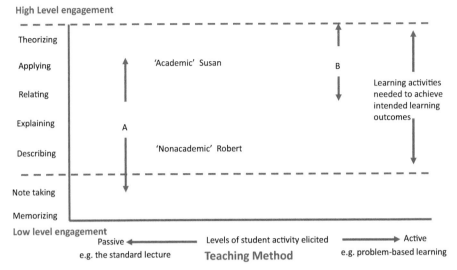

Figure 7.3   Students' orientation, teaching method, and level of engagement [McGraw-Hill Education (United Kingdom), p. 6].

Reference: "Teaching for quality at university: what the students does" [McGraw-Hill Education (United Kingdom), p. 6].

process according to McGraw-Hill Education (United Kingdom). We are not pushing students to give positive feedback, but we would expect that feedback from Susan will be based on logical and scientific methods and acquiring information from her side will be more beneficial for the teaching–learning process. Such a student will add lot to the quality-teaching evaluation process. So, this brings us to the next question "are we doing teaching/learning right?" and how we can enhance the process if we are just concentrating on the teacher quality only. The academic institution is also asked to take a serious step here and contribute to spread the awareness of quality and make it as a culture, student can live, have experience and then encouraged to contribute. Students can take quality-oriented concept classes besides lectures. Students along with teachers may notice that definitions of quality in higher education as a process, an outcome, or a property are not necessarily in conflict, but can potentially be used by HEIs as complementary.

Just like "quality culture," "quality culture" is also a fashionable word these days. But Harvey and Stensaker (2007) pinpoints that if we always assign a taken-for-granted meaning to it, we are not helping those who want to enhance their work and consequently not contributing to the essential processes of teaching and learning. Culture is one of the two or three most complicated words in the English language" (Williams, 1983). The conclusion comes to all after that quality is a never-ending process of reduction of defects. Hau (1996) argues that quality in higher education, and quality teaching in particular, comes from a never-ending procedure of reduction and removal of defects. Argyris and Schön (1974) determine that quality enhancement in HEIs should be a nested cycle of quality enhancement.

Integrating all definition of quality were mentioned above, along enhancement process with points that were presented by the European Universities Association (EUA) "quality culture" project (2002–2006), we conclude that every quality culture was based on two distinct elements: a set of shared values, beliefs, expectations and commitment toward quality and second element is a structural element with well-defined processes that enhance the quality. Quality culture must be deployed to enhance the quality assurance at the teaching–learning level to reach to what we are looking forward to achieve and must not be considered as a concept capable of answering challenges, but as a notion that helps to recognize challenges.

7.2.4 Responsibilities for Quality Assurance in Teaching and Learning

The integration between quality assurance and quality of teaching has resulted in generating guidelines and outlining frameworks for institutions to set up reliable quality assurance systems. Those guidelines do not address the quality of teaching as such, since they all rely on the principle that quality lies first within the institutions and that there is no one-size-fits-all way of achieving quality. However, they all advise institutions to take the attention of quality teaching by ensuring that quality of teaching and research are possible by the quality of faculty and the quality of their working conditions. They also advise that other relevant instruments be taken into account by all institutions and providers to support good working conditions and terms of service, shared governance, and academic freedom (UNESCO Recommendation concerning the Status of Higher Education Teaching Personnel).

a. National regulatory framework

Agreements between higher education and quality assurance agencies (internally and externally) may differ from a country to another but here we can call for and highlight major arrangements to ensure the continuing of the high quality of programs and academic standards in terms of teaching–learning process. Agreements assure evaluating systems designed for teaching and learning processes and outcomes, including related supporting services, for example, monitoring staff and faculty's performances in terms of teaching–learning processes by ensuring a systematic monitoring and reviewing system of the academic management of offered courses. Support individual staff development, performance improvement, and opportunities for innovation in teaching as well as recognizing and rewarding teaching excellence. Quality of teaching also needs an infrastructure to support a high-quality learning environment and services to enrich and support the teaching–learning process.

One of the action processes in terms of teaching–learning quality that universities are recommended to formulate an academic committee to be responsible for the supervision and development of all academic activities of the university, including the maintenance of high standards in teaching and research, as well as for communication with the academic community through the faculties and departments. The academic committee achieves its work through a series of committees and meetings such as:

- Academic Programs' Committee (APC): It develops academic policies specially the teaching and learning practices.
- Selection Procedures' Committee (SPC): It develops policies relating to the admission of students into university courses and the principles on which they are based.
- Teaching and Learning Development Committee (TLDC): It develops policies to encourage excellence and innovation in order to enhance the quality of teaching and learning.

Also the university is also encouraged to establish a teaching and learning quality assurance committee (TLQAC) with particular responsibility for quality assurance. It plays a key role in evaluating teaching and learning in the university by:

- Providing advices to the university community on quality assurance policy and processes for teaching and learning processes.
- Having particular responsibilities in relation to quality assurance of award courses' assessment and examination policies.
- Developing appropriate qualitative and quantitative measures of performance of teaching and learning,
- Evaluating quality in teaching and learning processes.

Academic units with heads can contribute to the quality assurance systems as well. Academic units are executive agents and have responsibilities for teaching and assessment offered through their unit. These responsibilities include quality assurance of all subjects, maintaining documentation relating to subjects, monitoring staff performance, appraisal of teaching staff, and providing opportunities for individual staff development.

Quality of teaching and teacher quality acknowledge the diversity of the student population in a way not previously done, as it asks for a level of instruction that is more intellectually difficult and meaningful than has the traditional cases. What we have emphasized above is less concerned with teachers' character traits and it is more concerned with teachers' ability to teach students and get them involved in meaningful activities.

To end this section properly, we need to emphasize that quality of teaching by nature is a quality assurance at infrastructure level of the teaching–learning process in HEIs. To be able to foster this concept and build blocks of this quality-oriented bricks that shape the quality assurance system, we need to do lots of iterative loops that keep the quality alive from inside the educational system. The report "Fostering Quality Teaching in Higher Education: Policies and Practices" which is an IMHE (Institutional Management in Higher Education) Guide for HEIs by authors, Fabrice Hénard and Deborah Roseveare (2012), has stated key elements to consider in fostering quality teaching as follows:

- The ultimate goal of quality teaching policies is to improve the quality of the learning experiences.
- Teaching and learning are inherently entangled and this necessitates a holistic approach to any development initiative.
- Sustained quality teaching policies require long-term plans and calls for institutional commitment.
- Quality teaching policies should be designed consistently at institutional, program, and individual levels.
- Establishment of horizontal linkages and creating synergies is an effective way of supporting the development of quality teaching.
- The temporal measurement in quality teaching means that what can be done at a certain point of time cannot be repeated.
- The institution is primarily responsible for the quality of its teaching and should set the standards within the system.
- Internal quality teaching should express the institutional quality culture and reflect to be a global quality approach.
- Quality teaching happens first in the classroom and should pay all efforts toward enhancing this process at the classroom first.

7.3 Quality of Teaching–Learning Effectiveness and Challenges

7.3.1 Quality of Teaching–Learning as Indicator

Performance indicators help and assess the progress within an educational institution, and they are four groups of performance indicators: input, process, output, and outcome indicators (Borden and Bottrill, 1994; Cave et al., 1991; Chalmers, 2008; Richardson, 1994). The input indicator imitates the resources involved in processes within the institution. These resources can be human, physical, or/and financial (Chalmers, 2008). Imitation means how far the system is the best selecting the input to its processes. Process indicators are more keen about the performance and delivery of the educational system sensing the effectiveness in each activity and procedure and such indicators are the best to measure the performance as they evaluate the real actions and activities on the spot with highest transparency reflecting the quality of teaching and learning, provide an understanding about an institution's quality practices, and detect errors where the enhancement quality procedures can be applied. This makes them the most practical and appropriate indicators. Output indicators are to indicate the quality at the final or as a "product" and "These… can be immediate measurable results or/and direct consequences of activities implemented to produce such results" (Bruke, 1998). Outcome indicators indicate the quality of educational program through its outcome as the graduate status and their employment opportunities. They assess the educational system to illustrate how close the results are to what is expected. They are actually meaningful measurements and can be employed to enhance and improve the entire system being aware of all challenges these indicators have referred to. Universities need such indicators to survive by evaluating the performance from different corners. Other reason of using such indicators can be as (Chalmers, 2008; Kember, 1997; Rowe, 2004) stated:

- To monitor their own performance;
- To enable assessment and evaluation of operations;
- To provide information and reports for external quality assurance audits and accreditation;
- To ensure ongoing enhancement of the institution (Chalmers, 2008; Kember, 1997; Rowe, 2004).

Input and output indicators are responsible for the quantitative measurement of a desired result or system transform. When performance is more related to teaching–learning processes, we need to specify more accurate and closer indicators testing and showing the quality of teaching indicators. Qualitative indicators can provide deeper interpretation and understanding.

The quality of teaching by nature measures "effective design of curriculum and course content, a variety of learning. It also involves well-adapted learning environments and student support services" (p. 192). Therefore, quality of teaching can be used as an indicator in order to enhance the quality and/or make it more effective. Quality of teaching indicator can assess to measure the quality. Figure 7.4 shows the relation between indicators and teaching–learning quality.

Figure 7.4 indicates that the indicators must go through iteration loops, where they are assessed for validity and practicality, using a set of criteria that can be agreed by the quality assurance committee. Indicators are effectively selected to gather the most relevant information. According to Chalmers (2008), indicators should be:

- Specific enough to classify what they are measuring.
- Measurable, which means being sensitive to what is measured and verifiable.
- Attainable to gather clear and valid information.
- Relevant to the intended outcome or output.
- Trackable, to follow information back to the source.

Quality indictors may be used to assess an institution to recognize performance of the quality teaching, teaching enhancement, and a teaching culture within the educational system. Quality of teaching indicator reflects if the

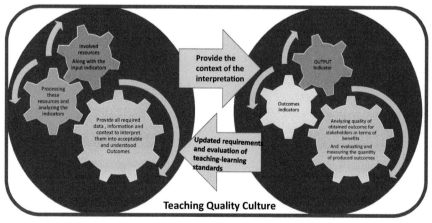

Figure 7.4 Relationship between indicators and teaching–learning quality.

teaching quality is a priority or not. Quality teaching may be viewed at three inter-dependent levels: the institution level, the program level, and the individual level (see Appendix 3; Chalmers, 2008).

1. Institutional level indicators: The indicator reflects the university supporting the quality of teaching by many phases such as: including the quality of teaching within the mission statement; the existence of a teaching and learning center; quality of learning resources; and technology-based teaching environments.

2. Program level indicators: The indicators reflect the quality of teaching within the department. They may include: promoting a balance between the evaluation of teaching and learning and research performance; communication with faculty, encouraging the faculty to be updated with pedagogical teaching and learning best practices; and fostering research and innovation. And the most important is the balance of the workload.

3. Individual level indicators: The indicators highlight initiatives and programs meant to help faculty to achieve their institutional mission, encourage them to be involved in the quality enhancement cycles, and promote their teaching level by using different teaching methods.

7.3.2 Quality Teaching and Research

Teaching and research are vital to the delivery of higher education and stability of an educational system. The affiliation between teaching and research is fundamental in defining the distinctive nature of the university as an institution" (Taylor, 2007). It is also possible that teaching promotes abilities of teaching faculty to be better researchers. Quality of teaching initiatives has impact on teaching and on research as teachers have become more aware of the objective pursued by teaching outside their own knowledge area; they recognize their role as individuals and as contributors to a collective mission.

In particular, quality teaching initiatives boost information technology in pedagogy enhancement and scrutinizing student–teacher interfaces. In institutions that are autonomous in program design, quality teaching initiatives support teachers and leaders to get to the aims and content of the program. Instruments and policies that foster quality teaching are expected to be expedient to research activities. On the research side, research feeds the theoretical background of quality teaching and teachers. The researches for example promote thinking of some cases such as why some teachers are hesitant to get involved in initiatives on quality teaching because of perceived weakness or the absence of sustaining theory. The challenge here is about commitment to quality of teaching and research. Stability and working in parallel give research and quality of teaching enough renovation will provide institutions with an opportunity to test the conduit between research and teaching and enjoy the outcome of such collaboration.

Unfortunately, research is still a challenge for faculty, and therefore, institutions are responsible to act and promote quality of teaching and research activities by, for example, giving the staff chances to contribute and explore every opportunity to develop the quality of teaching, and present their research experiences, to be a part of (re-)accreditation procedures, to be in the committee of institutional audits, promotions, appraisals, and so on. Institutions along with higher education have the ability to get their workforce and other institutions engaged in national, regional, and international networks to share best research practices not only in quality of teaching but also in diversity of eras get the staff interested and more willing to give and contribute. Higher education can also arrange for national or regional events (e.g., conferences) giving exposure to institutional achievements on quality of teaching and other research aspects. HEIs are also keen to assemble for a research propensity through the promotion of research-teaching linkages, such as: demonstration of how research supports and inspires teaching and vice versa, promotion of students' research skills, and the best is to develop a research-based teacher education. In teacher education programs, research-based approaches have lately received increasing interests both among researchers of teacher education and in communal dialog. This is based on the notion that the knowledge base of the study program is dynamic and that student teachers are active processors of knowledge (Zeichner and Conklin, 2005).

7.4 How Quality Assurance Should Communicate about Quality Teaching?

Launch quality teaching enhancing and culture spread of teaching–learning process should be and expected to be under direct supervision and control of quality assurance agencies the role of the department is always central in enabling these enterprises to be successful. Departments may also adopt to implement quality enhancement processes on a top–bottom basis. This may be stress-free to do at the department level as the department's closeness to the teachers is greater than that of the central administration of the university.

Indeed, academics recognize first with their discipline and then with their department (Hannan and Silver, 2000). However, in order for these top–bottom teaching enhancement processes to be well accepted by the professors, internal communication between the various levels of the association must be of great quality. This can be achieved by quality management, principles of Van der Wiele (1995), using the instruments of the total quality management framework:

1. Transparency of the organization.
2. Employees' involvement in the decision-taking process.
3. Quality policy deployment: The quality policy must be integrated in the mission and business strategy.
4. Communication about the importance of quality assurance.

Teaching faculty must work in collaboration with the university's system to attain a quality culture and enhance the quality of teaching. Gibbs (1995) and Patrick and Lines (2004) claim that the quality system must be integrated in the university's core goal and mission: "For the quality system to be effective, it must be acknowledged by the assortment of constituencies within the institution while at the same time framing the strategic direction and nature of change for the whole university" (Patrick and Lines, 2004).

Education development units in particular have a part to play as to enhance the quality of the teaching in HEIs. Education development units can profit the university by inducting initiatives and innovations in teaching. They may get involved in comprehensive and systematic implementation of teaching and learning initiatives. Education entities should create and facilitate communities of learning, and support bottom-up engagement and management on educational initiatives (Chalmers and O'Brien, 2004).

Chalmers (2007) asserts that an indicator of great quality reflects to several principles including validity, reliability, bearing to mission and policy, potential for disaggregation, timeliness, coherence transversely diverse sources, simplicity, and transparency with respect to known limitations, accessibility, and affordability.

Evaluation of teachers' portfolio is also a way to communicate about teaching quality as a possible method to assess teaching quality and identify best practices to increase the awareness of being able to evaluate the vital issue here, quality teaching. The teacher's portfolio evaluation is a valuable technique because it is based on multiple sources of evidence and multiple levels of scrutiny (Webbstock, 1999). However, as it was noticed that it is problematic to reach agreement on which substances should be included in the portfolio and on how much each of these items should be given a score. The question remains whether quantitative weighs should be attributed to each item of the portfolio to increase the transparency of the process or whether this would transform the portfolio evaluation process into a mechanical task.

Quality assurance agencies can spread the quality culture and communicate about quality control and managing by involving faculty and teachers in the quality assurance agencies processes. And actually this is what is ongoing, as no one better than teachers and professors to evaluate and contribute not only internally but also externally. External examiners are equipped with knowledge and practical methods to help quality assurance agencies to practice their tasks and mainly to be able with high level of communication and smartness to grip the attention of learning systems to work harder to achieve the quality-oriented design that supports the educational system. External examiners are a teaching faculty in most times and being called by the agencies to serve in the quality assurance management and control theme. Quality assurance site-visits should include faculty for discussion in terms of exchanging experiences and expanding quality culture.

Quality assurance agencies can also involve the faculty to develop a report template that institution would use for internal quality processes and recommend suite to help institution choose practices to enhance their teaching culture and quality of teaching. This kind of contribution moves the quality culture beyond the framework of regular quality assurance agencies. It helps to develop the internal awareness of quality-oriented environments within the education systems rather than just focusing on how to assess the quality process. Training faculty, adopting successful quality culture from abroad, and trying to deploy the faculty's experience in quality of teaching are the necessary tasks by both quality assurance agencies and institutions, under the umbrella of higher education. This kind of support for quality of teaching usually generates awareness of the responsibility of teachers in the learning process and justifies the institutional need for helping them to fulfill their mission.

References and Suggested Literature

[1] Guest, R., Duhs, A., Quality Assurance and the Quality of University Teaching. *Austr. J. Educ.* 47, 40–57, 2003.

[2] Defining Teacher Quality. National Research Council, *Testing Teacher Candidates: The Role of Licensure Tests in Improving Teacher Quality*. Washington, DC: The National Academies Press, 2001.

[3] Anderson, G., Assuring Quality/Resisting Quality Assurance: Academics' responses to 'quality' in some Australian universities. *Qual. High. Educ.* 12, 161–173, 2006.

[4] OECD, *Learning our lesson: Review of Quality Teaching in Higher Education, IMHE*, available at: http://www.oecd.org/edu/imhe/quality teaching

[5] Hénard, F., Roseveare, D., Fostering Quality Teaching in Higher Education: Policies and Practices, *An IMHE Guide for Higher Education Institutions*, 2012.

[6] *Responsibilities for Quality Assurance in Teaching and Learning*, available at: http://about.unimelb.edu.au/__data/assets/pdf_file/0004/861223/responsibilities.pdf

[7] Available at: https://evolllution.com/opinions/teaching-assessment-quality-assurance-higher-education/

[8] Henard, F., Leprince-Ringuet, S., *The Path to Quality Teaching in Higher Education*, Paris: OCED Publication, 2008.

[9] Hare, D., Heap, J. L., Effective teacher recruitment and retention strategies in the midwest: Who is making use of Them? *ERIC* 95, 2001, available at: https://eric.ed.gov/?id=ED477648.

[10] García-Gallego, A., Georgantzís, N., Martín-Montaner, J., Pérez-Amaral, T., (How) Do research and administrative duties affect university professors' teaching? *Appl. Econ.* 47, 4868–4883, 2015.

[11] Available at: https://chss.gmu.edu/faculty/teaching-load-standards-tt-faculty

[12] Kustra, E., Meadows, K. N., Dawson, D., Hondzel, C. D., Goff, L., Wolf, P., et al., Teaching culture indicators: Enhancing quality teaching, *Centre for Teaching and Learning Reports*, 2014.

[13] Cishe, E. N., Fostering quality teaching and learning in higher education through academic staff development: Challenges for a multi-Campus university. *Medit. J. Soc. Sci.* 5, 272–277, 2014.

[14] Henard, F., Cros, M., Thanh, J. T., *Institutional Policies, a Key to Foster Quality Teaching*, available at: http://learningavenue.fr/wp-content/uploads/2015/11/Leadership-in-Quality-Teaching-Learning.pdf

[15] Roger, E., (Ed.). Quality assurance for university teaching. *ERIC* 1993, https://eric.ed.gov/?id=ED415735

[16] Hoecht, A., *Quality Assurance in UK Higher Education: Issues of Trust*, Control, 4 Professional Autonomy and Accountability. *High. Educ.* 51, 541–563, 2006.

[17] Felder, R. M., Brent, R., How to Improve Teaching Quality. *Qual. Manage. J. 6*, 9–21, 1999.

[18] Angelo, T. A., Cross, K. P., *Classroom assessment techniques: A handbook for college teachers*, 2d ed. San Francisco: Jossey-Bass Publishers, 1993.

[19] Beaver, W., Is TQM appropriate for the classroom? *Coll. Teach.* 42, 111–114, 1994.

[20] Bellamy, L., Evans, D., Linder, D., McNeill, B., Raupp, G., Active learning, team and quality management principles in the engineering classroom. In *Proceedings of the 1994 Annual Meeting of the American Society for Engineering Education* (Washington, DC: ASEE), 1994.

[21] Bloom, B. S., *Taxonomy of educational objectives. Chapter 1. Cognitive domain*. New York: Longman, 1984.

[22] Bonwell, C. C., Eison, J. A., *Active learning: Creating excitement in the classroom*. ASHE-ERIC Higher Education Report No. 1 (Washington, DC: George Washington University), 1991.

[23] Boyer, E. L. *Scholarship reconsidered: Priorities of the professoriate*. Princeton, NJ: Carnegie Foundation for the Advancement of Teaching, 1990.

[24] Brent, R., and Felder, R. M., Writing Assignments – Pathways to Connections, Clarity, Creativity. *Coll. Teach.* 40, 43–47, 1992.

[25] Burke, K., *The mindful school: How to assess thoughtful outcomes*. Palatine, IL: IRI/Skylight Publishing, 1993.

[26] Campbell, W. E., and Smith, K. A., (eds). *New paradigms for college teaching*. Edina, MN: Interaction Book Company, 1997.

[27] Deming, W. E., *The new economics*, 2nd edn. Cambridge, MA: MIT Center for Advanced Engineering Studies, 1994, Cited in Latzko, 1997.

[28] Ewell, P. T., National trends in assessing student learning. *J. Engr. Educ.* 87, 107–113, 1998.

[29] Felder, R. M., Any questions? *Chem. Engr. Educ.* 28, 174–175, 1994a.

[30] Felder, R. M., The myth of the superhuman professor. *J. Engr. Educ.* 82, 105–110, 1994b.

[31] Felder, R. M., A longitudinal study of engineering student performance and retention. IV. Instructional methods and student responses to them. *J. Engr. Educ.* 84, 361–367, 1995.

[32] Felder, R. M., Brent, R., *Cooperative learning in technical courses: Procedures, pitfalls, and payoffs*. ERIC Document Reproduction Service, ED 377038, 1994.

[33] Felder, R. M., Brent, R., Navigating the bumpy road to student–centered instruction. *Coll. Teach.* 44, 43–47, 1996.

[34] Felder, R. M., Brent, R., Speaking objectively. *Chem. Engr. Educ.* 31, 178–179, 1997.

[35] Felder, R. M., Felder, G. N., Dietz, E. J., A longitudinal study of engineering student performance and retention. V. Comparisons with traditionally-taught students. *J. Engr. Educ.* 87, 469–480, 1998.

[36] Glassick, C. E., Huber, M. T., Maeroff, G. I., *Scholarship assessed: Evaluation of the professoriate*. San Francisco: Jossey-Bass, 1997.

[37] Grandzol, J. R., Gershon, M., Which TQM practices really matter: An empirical investigation. *Qual. Manage. J.* 97, 43–59, 1997.

[38] Gronlund, N. E. *How to write and use instructional objectives,* 4th edn, New York: Macmillan, 1991.

[39] Jensen, P. A., Robinson, J. K., Deming's quality principles applied to a large lecture course. *J. Engr. Educ.* 84, 45–50, 1995.

[40] Johnson, D. W., Johnson, R. T., Smith, K. A. *Active learning: Cooperation in the college classroom*, 2nd edn. Edina, MN: Interaction Press, 1998.

[41] Latzko, W. J., Modeling the method: The Deming classroom. *Qual. Manage. J.* 5, 46–55, 1997.

[42] McKeachie, W., *Teaching tips*, 10th edn. Boston: Houghton Mifflin, 1999.

[43] Meyers, C., Jones, T. B., *Promoting active learning*. San Francisco: Jossey-Bass, 1993.

[44] Millis, B. J., Cottell, Jr., P. G., *Cooperative learning for higher engineering faculty*. Phoenix: Oryx Press, 1998.

[45] NISE (National Institute for Science Education), *Collaborative learning: Small group learning page,* 1997, available at: *http://www.wcer.wisc.edu/nise/cl1/*

[46] Panitz, B., The student portfolio: A powerful assessment tool. *ASEE Prism* 5, 24–29, 1996.

[47] Rogers, G. M., Sando, J. K., *Stepping ahead: An assessment plan development guide.* Terre Haute, IN: Rose-Hulman Institute of Technology, 1996.

[48] Rogers, G. M., Williams, J., Building a better portfolio. *ASEE Prism* 8, 30–32, 1999.

[49] Shelnutt, J. W., Buch, K., Using total quality principles for strategic planning and curriculum revision. *J. Engr. Educ.* 85, 201–207, 1996.

[50] Shuman, L. J., Atman, C. J., Wolfe, H., Applying TQM in the IE classroom: The switch to active learning. In *Proceedings of the 1996 Annual Meeting of the American Society for Engineering Education* (Washington, DC: ASEE), 1996.

[51] Stedinger, J. R., Lessons from using TQM in the classroom. *J. Engr. Educ.* 85, 151–156, 1996.

[52] Summers, D. C. S., TQM Education: Parallels between industry and education. In *Proceedings of the 1995 Annual Meeting of the American Society for Engineering Education* (Washington, DC: ASEE), 1995.

[53] Braskamp, Larry A., Ory, John, C., Assessing Faculty Work: Enhancing Individual and Institutional Performance. Jossey-Bass Higher and Adult Education Series, 1994, https://eric.ed.gov/?id=ED368305

8

Assessments in a Quality Assurance Suit

8.1 Introduction

Ramifications of the higher education shift from elite to mass are count-less. Higher education entry and participation in this shift had formulated a growing heterogeneity of students in terms of their socio-economic con-textual, academic aptitude and attentiveness, career predications, incentive, and rendezvous. This broadening reflects the increasing social demand for education and the subsequent greater participation (OECD, 2008). Students' demands are also fluctuating in terms of learning they get and assessments they attend. Learners increasingly look for courses that enable them to update their knowledge throughout their working lives. In addition, as learners seek to acquire particular knowledge or skills to satisfy labor market needs, more and more prefer to pick and choose courses from the most suitable providers, rather than studying a traditional clearly defined program at one institution (OECD, 2008). This kind of change still needs to work inside the framework of quality-oriented higher education strategic plans that plan for an effective and high level of teaching–learning process on both parties, the teacher (the educator) and the student (the learner). These strategic plans will be formulating an atmosphere in which educators are expected to be part of continuous review/evaluating cycles in terms of their competition, contri-bution, and performance and enthusiastically backing peer-observation. The importance of executing the strategic plan is to recognize the teacher's skills at earliest stages and help to outline a formative practice and demonstrate an educational community that goes along with learning and practice, also has a chance for sharing learning objectives among educators, and some of formative feedbacks that have been proved prodigiously affirmative in terms of their prospective improvements in teaching–learning process.

Assessments can be defined simply as the instance of making a judgment about a specific performance in a specific field. Assessments are often used interchangeably with examinations, but not limited to tests and examination practices. They are more likely to focus on the discrete learner. We are more promoting assessment with different approaches to be a quality-oriented risk management policy for learning process. Assessments are techniques which are being altered to fit-for-purposes of education process and they play a crucial role in the teaching–learning process. Assessments' formulation depends on the amount assessed and what to expect. In education process, assessments are vital for authenticating the teaching–learning process and for enhancing cycles. The assessment process includes a variety of methods that educators use to evaluate and document the academic readiness, skill acquisition, educational requirements, and students' needs to reach the required academic readiness to learn and accumulate the required knowledge.

The review of educational assessments usually should present a wide variety of forms of assessment in the education sector and summarize purposes why education needs such a concept. Educational progress cannot achieve results without including assessments that take the attention and consideration of everyone who is involved in this progress and interested in the outcomes, starting from students and parents, teachers, business and industry, government, and society. Without exceptions, each party would like to be involved and aware of the teaching–learning progress, and assessments help a lot on this side, as they are methods, well-designed actions, and techniques to monitor and evaluate learning in order to improve it (Brady and Kennedy, 2009). Assessments are also a way to open discussion as each party has own view on the purpose of any evaluation process and afford a chance to clear ambiguity and answer questions related to the educational process as well as fulfilling different stakeholders' needs.

Eisner (2001) has established a distinct ground of assessments and their purposes, as he believes that there are five major purposes of assessments. Many other researches have been modifying assessments' purposes to better fit the new educational systems. In the following, we will try to lay out some of the most important purposes of assessment in any educational system:

1. Assessments per program are expected to be designed to determine whether the desired outcome of this program is achieved or not as the ultimate goal of any educational progress is to "educate" and accumulate knowledge. Assessments are also expected to facilitate learning and increase awareness not to complicate the educational progress challenging students' and teachers' abilities and skills. The best assessment is the one designed to help the learning process, not for accumulating data

about student's or educator's performance. This kind of assessment has a "formative" purpose: it helps to shape what lies ahead rather than merely to instrument and record past achievements.

2. Assessments are required over the learning process to indicate how effective teaching–learning process and technique have been. This review of assessments is a teacher-oriented purpose where evaluation is keener to monitor teacher's performance and ability to educate. The assessment goes beyond evaluation as it holds strategic plans to decide whether the performance is satisfactory or needs modification to better suit students' needs. And this is actually a third purpose of any assessment as it will be explained in the next point.

3. Assessments are meant and designed to provide feedback to all parties at different levels. Students get feedback on their performance and recommendations how to improve, as well as their parents want to know how students are performing and desire to be involved to lend further assistance to students in the right time. Teachers are expected to harvest benefits out of these feedbacks for self-development purposes.

4. Assessments and evaluations have purposes to direct students along a certain pathway to attain the required knowledge and get the right experience preparing them for space market after graduation. In general, business and industry are more interested in the amount of contribution and creative mentality the graduator affords, "They are concerned with knowledge and skills that can be applied immediately to specific work requirements" (Brady and Kennedy, 2009, p. 5). Assessments and evaluations are expected to be designed to identify strength and weakness, a learner may have in certain skills, as well as providing opportunities to gain practice and overcome obstacles. Also our chapter attempts to identify two sets of guidelines, for policymaking and practice.

5. Assessments have general purposes that include parties such as higher education itself and government. Assessments should follow strategies and play a crucial role in terms of their consistency with higher education policies that present that they have been tamed by quality assurance procedures and processes. Assessments with quality-oriented strategic policies should give clear information to the educational community and the public about the views on assessment which are to be promoted or prioritized in an education system, in order to avoid misperception and overlapping of approaches among experts in the teaching–learning process. And to reach this far, assessments are customized in a way to assure quality in each step in the educational system which aimed and

goaled to assess actually the teaching–learning process by evaluating and putting efforts to understand the outcomes.

6. We think that we have a sixth purpose here. Any educational systems as a well-defined and recognized institution internally and externally should extend further bonds with students by asking them to contribute with suggestions and recommendations. Some of these trials come in an assessment framework, surveys, or questionnaires to motivate students to be creative and motivated to contribute, which in turn become indicators or key factors that come along with input for designing next teaching–learning step procedures. This should revive the teaching–learning process and assure to find-and-fix approach and act dynamically to avoid any failure.

Despite the fact that assessments are not the only way to know that students have accumulated knowledge or have learnt, investigating the benefits of assessing a learning process drives facts about whether quality assurance really happens at the assessment level of the learning process or not. In traditional classes, assessment and instruction are highly integrated parts of the classroom experience; students are examined only on the materials directly covered in a class while in current classrooms such as the ones in e-learning and blended learning, assessments can take different concepts and by deploying online tools and assessments can cover more requirements than the traditional classes do as these concepts of learning come with different quality orientations and requirements of knowledge to cope with. The above purpose points and discussion actually encourage and propel to re-define the educational assessment to be based more on quality orientation and more linked to higher education standards. Therefore, we can say that assessments are integral parts of instruction for most important interpretation processes to understand the prime mission of the teaching–learning process and determine whether or not the goals of education are being met as educators discover the extent to which students have learned whatever it is they are required to know to meet their learning purposes. Assessments are designed based on methodical processes of authenticating and using pragmatic data on the knowledge, skill, and attitudes to augment programs and recover student learning for quality assurance purposes and enhancements. And as they play vital roles of evaluating learning process, they should take the form of being continuous to give time and more opportunities for enhancing and following up. Educators who evaluate in this way are anxious not just to confirm and verify what their students have learnt, but also to help their students and themselves apprehend what the next steps in learning should be and how they might be undertaken.

Assessment data can be attained from directly analyzing student work to assess the achievement of learning outcomes and they take more than one method such as formative/summative assessments, research-based assessments, self-assessments, a simple form of a questioning method, requiring a feedback, or any way put the learning process under conversation and evaluating in any setting is an assessment.

Before moving to the next section and get in details with assessment types and design, it is fair to mention that depending on assessments is a bit a hectic as we have approved that not only assessments are being used to look for quality in the teaching–learning process. To be realistic, we would like to mention that during the teaching–learning process, it is a bit difficult to measure the quality of student's learning and accumulating knowledge at the level of teaching itself. Teaching and educating happen at the instructing level, which is the actual time of attaining knowledge and providing/affording and giving information. Teaching–learning process is equipped with different delivery teaching tools and supported with materials using all learning resources available, but at the end of each teaching session, yet instructors have no idea if their students are actually learning, regardless of how good their lectures or other instructional materials might be, until they test students' understanding, mastery, or knowledge. Teaching–learning interaction and a live discussion can help to avoid misunderstanding at an information level, but when we want truly to measure the concept learning and knowledge estimation, assessments help a lot. Regardless of the process of assessment like objective exams, written assignments, projects, or, for example, just demonstrations, if assessments are well designed and firmly regulated to reflect what a student has learned, it is the assessments that constitute evidence of student learning.

When it comes to educators and teachers, assessments' results might disturb their decisions about grades specially when educators are not satisfied about students' achievements, assessments design and in some cases, about the curriculum as well. This and as a sequence raises questions about teaching–learning acquiring skills and methodologies and, in some cases, they do affect funding has been expected from interested parties who feel not that contented with the assessments' process for instance. Such scenarios would stimulate educators to raise questions and urge to be honest to answer in order to indicate and identify weaknesses. Questions can take different forms trying to touch the base: Questions about the teaching performance itself and whether they can deliver the right knowledge or not. Questions related to students and their abilities to attain the right and expected knowledge or not and, of course, more hard questions about whether there are new teaching technologies educators must learn and use or what they are doing and the way they are teaching should be more beneficial. Today's students want to know not only the basic reading and arithmetic skills but also abilities that will tolerate them to face a world that is incessantly changing. They must be able to contemplate censoriously, to analyze, and to make inferences.

Changes in the skills' base and knowledge our students need necessitate new learning goals; these new learning goals change the relationship between assessment and instruction which necessitates in turn new teaching–learning methodologies. Teachers need to take an active role in making decisions about the firmness of assessments and the content that is being considered.

"*Assessment should be deliberately designed to improve and educate student performance, not merely to audit as most school tests currently do.*" Grant Wiggins, Ed.D., President and Director of Programs, Relearning by Design, Ewing, NJ, United States

References to educating and teaching studies and experiences, have concluded that assessment works best when they are designed with principles, it does the following:

1. Assessments are more beneficial when they are on-going process and take place continuously to keep students occupied with the learning process in an effective way that is ready to be tested and evaluated regularly.

2. Assessments by design are on-going process that provide mores more than one opportunitiy for learners to promote their understanding of assessments and how they are expected to perform meeting the requirements to attain the knowledge persisting that purpose of assessment is to facilitate the learning process not to complicate it (Groundwater-Smith et al., 2007).

3. Assessments are more beneficial when they are designed to be developmental activity that provides diagnostic feedback. Students expect to receive feedback on their learning performance as well as educators need to reflect on own practices and students' achievements (Victorian Curriculum and Assessment Authority, 2007).

4. Assessments are more beneficial when they allow self-assessments as they afford chances to produce work and manage students to reach deeper in their knowledge and accumulate their skills (Board of Studies NSW, 2006).

5. Assessments are more beneficial when they refer to explicit criteria known to students and related to their progress and knowledge they

are seeking which is expected to motivate their performance and guide them to the right way of learning. Teachers are also getting benefits of this principle by setting standards and communicating the requirements of the assessment assisting students to complete the assessment successfully.

6. Assessments are more beneficial when they are considered a teaching tool to facilitate learning and they are submitted to any evaluating, enhancing, and improvement cycles.
7. The progression of quality-oriented assessment in higher education institutions must be more well-thought-out and ceremonial than the learning process if the purpose of the assessment process is the assurance that students have mastered a given discipline. Assessments are more beneficial for teaching–learning process when they are considered critical as indicators of quality because, if structured well, students who pass the disciplinary assessments indicate subject matter mastery of that discipline.

For colleges and universities, considering assessment as a continuous process establishes measurable and clear learning outcomes that identify student's performance at a specific learning extent on the one hand and provision a sufficient amount of learning opportunities to achieve these outcomes on the other hand. The university deliberates assessments for implementing a systematic way of gathering, analyzing, and interpreting evidence to determine how well student learning matches the expectation. The educational system collects assessments' data and information to notify the improvement in student learning, but it is not enough for higher education and quality assurance agencies. Higher education has standards and settings of teaching–learning process; therefore, universities must apply higher education's framework literally to these kinds of assessments so that the teaching–learning process is getting recognized and matching the higher education assessment' settings and criteria.

Some other studies call for detaching teaching from assessment which has challenges and benefits. Separating teaching and instruction from assessments can help to overcome obstacles in terms of applying quality standards. Multi-campus universities usually follow quality assurance management systems for assessments and the educational process they run to get accredited from the mother university and offer a fair chance for students regardless of their campus's locations. Assessments must go through a long procedure separated from tutoring and in different channels of just teaching. Procedures of coordination must be keys of success of designing and applying such assessments. The crucial challenges in such adoptions are more with cultural concepts about educational systems. Old thinking of educational community prefers associating the teaching to assessing of students and taking this as a guaranteed way to guard performance in both teaching and learning. Teachers have always thought that assessments are their own right that cannot be stepped down. For decades of teaching, educators are expected to include assessing as part of their classroom duties, and they like to control the evaluation process. On, students' side, students sometimes fear being abandon from traditional standard which have been used to encourage students to learn harder and compete, which create a mind – resistance to change or switch easily to the new concepts of controlling assessments.

Yet, the evolution of quality-oriented assessment in higher education institutions must be more well-thought-out and ceremonial than the learning process if the purpose of the assessment process is to assure that students have mastered a given discipline. Even well-designed robust assessments that test student mastery of commonly accepted disciplinary knowledge are incapable yet to confirm positively that students have learned what they need to learn. Assessments are critical as indicators of quality because, if controlled well, students who pass the disciplinary assessments indicate subject matter mastery of that discipline. Our next sections will help to understand the importance of assessments in any educational system and propose an educational guide about how a quality-oriented assessment can be.

8.2 Types of Assessments Defined

8.2.1 Definition and Principles

Assessment for learning is any assessment for which the first priority in its design and practice is to oblige the purpose of promoting learners learning. It therefore fluctuates from assessment designed primarily to serve the purposes of accountability, or of classification, or of certifying competence. An educational assessment can support learning if it produces information to be used as feedback by educators, and by their learners in assessing themselves and each other, to amend the teaching and learning process in which they are involved in.

Assessment for learning is the one of the most important purposes of assessments. It is not the only purpose of assessing and is to be distinguished from assessment of learning which is carried out for the purposes of grading and reporting (ARG, 1999). A review into classroom assessment (Black and Wiliam, 1998) has shown that assessment for learning is one of the most influential ways of refining learning and raising principles. A study in 2002 entitled "Assessment for Learning: Research-based principles to guide classroom practice," by Assessment Reform Group has added further evidence that it is important to follow certain guiding principles which reflect the essential topographies of assessment for learning. The principles of assessment for learning that will be presented here have benefited from comments from a wide variety of individuals and associations, whose help is gratefully acknowledged. These 10 principles are actually steps toward changing assessment practice to safeguard the necessary quality of learning experiences needed for achieving the goals of education.

1. Assessments should be part of the effective strategic planning

Strategic planning stage should embrace strategies to ensure that learners understand the goals they are pursuing and the criteria that will be applied in assessing their work. How learners will receive feedback, how they will take part in evaluating their learning, and how they will be assisted to make further improvement should also be premeditated.

University strategic plans should afford effective plans and action procedures supported with expertise and mutual designs that control assessments over the teaching–learning process, opening opportunities for both learners and teachers to acquire and use information about progress toward learning goals. Strategic plans should also include enhancing cycles for these assessments based on feedback and collected information. Therefore, plans have to be elastic to response to developing ideas and required skills.

2. Assessments are central activities of any classroom practice

Much of what teachers and learners do in classrooms can be described as assessment. That is, tasks and questions prompt learners to demonstrate their knowledge, understanding, and skills. What learners say and do is then observed and interpreted, and judgments are made about how learning can be improved. These assessment processes are an essential part of everyday classroom practice and involve both teachers and learners in reflection dialog and decision making.

3. Assessments are to promote understanding of goals and criteria

For effective learning to take place, learners need to understand what it is they are trying to achieve and want to achieve it. Understanding and commitment follow when learners have some part in deciding goals and identifying criteria for assessing progress. Communicating assessment criteria involves discussing them with learners using terms that they can understand, providing examples of how the criteria can be met in practice, and engaging learners in peer- and self-assessment.

4. Assessments are sensitive and constructive

Teachers should be aware of the impact that comments, marks, and grades can have on learners' confidence and enthusiasm and should be as constructive as possible in the feedback that they give. Comments that focus on the work rather than the person are more constructive for both learning and motivation.

5. Assessments foster motivation

Assessment that encourages learning fosters motivation by emphasizing progress and achievement rather than failure. Comparison with others who have been more successful is unlikely to motivate learners. It can also lead to their withdrawing from the earning process in areas where they have been made to feel they are not performing well. Motivation can be enhanced by assessment methods which provide constructive feedback and create opportunities for self-assessments.

6. Assessments recognize all educational achievement

Assessment for learning should be used to enhance all learners' opportunities to learn in all areas of educational activity. It should enable all learners to achieve their best and to have their efforts recognized and awarded.

7. Assessments focus on how pupils learn

The process of learning has to be in the minds of both learners and teachers when assessment is planned and when the evidence is interpreted. Learners should become as aware of the "how" of their learning as they are of the "what."

8. Assessments help learners know how to improve

Learners need information and guidance in order to plan the next steps in their learning. Teachers should: pinpoint the learner's strengths and advice on how to develop them; be clear and constructive about any weaknesses and how they might be addressed; and provide opportunities for learners to improve upon their work.

9. Assessments develop the capacity for peer and self-assessment

Independent learners have the ability to seek out and gain new skills, new knowledge, and new understandings. They are able to engage in self-reflection and to identify the next steps in their learning. Teachers should equip learners with the desire and the capacity to take charge of their learning through developing the skills of self-assessment.

10. Assessments are keys to get professional skills

Teachers require the professional knowledge and skills to: plan for assessment; observe learning; analyze and interpret evidence of learning; give feedback to learners; and support learners in self-assessment. Teachers should be supported in developing these skills through initial and continuing professional development.

With these principles as the framework, some strategies have been developed and are claimed to have a number of constructive effects on students and teachers. A study was conducted by María Teresa Flórez and Pamela Sammons, Oxford University, Department of Education, entitled with "Assessment for learning: effects and impact" (AFL, 2013). Assessment for learning is to be short term as AFL.

The review of AFL focuses on assessment for learning where the first priority is to promote learning. The record commonly mentioned features of AfL in the literature are the better use of questioning, feedback, peer and self-assessment and the formative use of summative tests where these are used in the schools or system concerned. Although most of the literature evaluates the effects of AFL as positive, contextual aspects emerge as possible obstacles for the feasibility of the approach, especially in those contexts in which the ideal conditions observed in research are not given. Further research on this is required as well, but some guidelines can be given on the basis of currently available studies. Policymakers should consider a careful design of dissemination strategies and possible contradictions between different policies. They should give clear messages to the educational community and the public about the view of assessment which is to be promoted or prioritized; the provision of support for dissemination processes through fostering school leadership; enough flexibility to allow some level of appropriation by practitioners; processes of monitoring the progress of dissemination; and commitment to sustain the policy over time (AFL, 2013).

Table 8.1 is re-constructing the previous 10 principles from the view of AFL.

Table 8.1   AFL study 10 principles

Principle	Practice
1. AfL is part of effective planning 2. AfL is central to classroom practice 3. AFL promotes understanding of goals and criteria	These first three principles can be understood as a whole. In brief, they refer to the need to recognize assessment not as a mere accessory to pedagogical practice, but as an integral part of it. Assessment must be intertwined with all the moments of a learning process and, thus, must be considered when planning. For this to happen, teachers must define clear learning goals or criteria and be able to share them with students in an understandable way. Along with this, students should be constantly reminded of these criteria or learning goals; during the learning process, their learning was evaluated and feedback was given to analyze the progress of students and take decisions according to this evidence.
4. AFL is sensitive and constructive 5. AFL fosters motivation 6. AFL recognizes all educational achievement	This second set of principles is related to the impact of assessment in shaping students' motivation, especially in terms of the nature of the feedback they receive. Teachers should be careful in what they say to students and try to give descriptive feedback exclusively centered on the quality and content of each student's work rather than use value-laden terms such as "good" or "poor." They should also suggest ways for students to improve their work. In the context of AFL, there is not only an excellence level which all must achieve in order to have recognition; any learning progress made by the student in relation to his or her previous state deserves recognition and positive feedback.
7. AFL focuses on how pupils learn 8. AfL helps learners know how to improve 9. AfL develops the capacity for peer and self-assessment	The process through which students learn must be a focus of attention in classroom practice, both for teachers and students. This involves developing awareness in the student about his or her learning processes and increasing autonomy through practices of peer and self-assessment in order to support students in developing their own responsibility for their learning. Giving feedback to students on how to improve, and not just on their mistakes, also contributes to the development of autonomous thinking and learning.
10. AFL is a key professional skill	This principle highlights the complexity involved in taking assessment for learning into practice, as it requires teachers to learn how to work from this perspective and to develop the necessary skills for doing so. The ARG recognizes here the need for good quality professional development programs as a fundamental requirement for the successful implementation of assessment for learning in classroom practice.

8.2.2 Educational Assessment Types

Educational systems are structured environments, declared with explicit purpose which is educating students. Most educational systems' terms are designed around a set of values or ideals that govern all educational selections in that system. Such sets include curriculum, organizational models, and design of the physical learning spaces, teaching–learning process and interacting, techniques of assessment, educational accomplishments, and much more. The term assessment is generally used to state to all activities teachers practice to promote students' learning process and to scale students' improvement. Assessment can be divided into the following main categories: Formative and summative forms.

1. Formative assessment: Formative assessment is commonly carried out through a course or project asked during the course period. It can be stated as "educative assessment," as it is used to benefit learning process more than just grading and evaluating. In an educational scenery, formative assessment might be designed by a teacher (or peer) or the learner and more to provide feedback on a student's work and would not essentially be used for grading purposes. Formative assessments can be in the custom of diagnostic, standardized tests, quizzes, oral question, or waft work. Formative assessments are conceded concurrently with instructions that target to understand if the students comprehend the instruction before pursuing a summative assessment.

2. Summative assessment: Summative assessment is generally carried out at the completion of a course or project. In an educational set, summative assessments are characteristically used to consign students a course grade. Summative assessments are evaluative and are made to recapitulate what the students have learned, to determine whether they understand the subject matter well. This type of assessment is typically graded (pass/fail, $0-100$) and can take the form of examinations, tests, or projects. Summative assessments are often used to define whether a student has passed or failed a course. A reproach of summative assessments is that they are reductive, and learners discover how well they have acquired knowledge too late for it to be of use.

Educational researcher Robert Stake explains the difference between formative and summative assessments with the following analogy:

Summative and formative assessments are often referred to in a learning context as *assessment of* learning and assessment for learning, respectively. Assessment of learning is generally summative in nature and intended to measure learning outcomes and report those outcomes to students, parents, and administrators. Assessment of learning generally occurs at the conclusion of a class, course, semester, or academic year. Assessment for learning is generally formative in nature and is used by teachers to consider approaches to teaching and next steps for individual learners and the class.

Assessment (either summative or formative) is frequently deliberated also as either objective or subjective. Objective assessment is a procedure of questioning which has a only accurate solution. Subjective assessment is a system of enquiring which may have more than one correct answer or more than one way of articulating the precise answer. There are innumerable types of objective and subjective queries. Objective question categories include true/false answers, multiple-choice, and multiple-response and identical questions. Subjective questions comprise extended-response questions and essays. Objective assessment is well fit to the increasingly popular high-tech or online assessment format.

8.3 Designing a Quality-oriented Assessment

Higher education institutions can run effectively along with quality assurance agencies. Both can activate effective policies, regulations, and processes which assure that academic contextual and standards for each award of credits or qualification is rigorously conventional at the appropriate level and student's performance is equitably assessed and sustained according to standards. So, and therefore, the quality assurance processes can formulate the assessments to be a key element in setting and maintaining the academic process. The educational process is mainly to focus on the teaching–learning process which is the crucial core of any educational system. All parties (teachers/learners and higher education provider) will be involved and their evaluation is accountable for, in order to get to the ultimate effective operation of different aspects of assessment in all its forms. This kind of integral collaboration between education institution submitting and interacting positively with higher education regulation and quality assurance methods and procedure should at the end benefit the students and academic progress in general.

Assessments as continuous crucial process for a learning system should be given a space and high-skilled expertise as well as open learning resources that can present it to serve the aim of leaning and understanding whether students are coping or the procedure needs alteration in terms of information delivery and teaching–learning interactions. The cycle of quality-oriented educational system itself cannot be running without assuring this inner cycle of collaboration and interaction. Therefore, quality-oriented assessment is a quality-oriented building block that can be added to the learning structural design process that should aim to achieve excellency at all fronts that support the higher education institutions. This should be taken into steps, and in each step, quality assurance system should be the base of processes which ends of a quality-oriented design of assessments that meet all requirements of higher education and quality agencies.

8.3.1 Quality-oriented Assessment Framework

The mainstream of testing sessions ran smoothly as most of higher education institutions had been able to deal with potential problems prior to the start of testing activities. The framework should be with a clear goal of prior to the assessment design. So we are stating a general framework regardless of the strategic being used or the technology being employed. The cycle of a simple assessment should involve a wide range of activities from designing the instruments to validating the quality of the final instruments. Figure 8.1 gives a conceptual model.

Step 1: Assessment frameworks

The first step of the process generally entails unindustrialized a framework so as to establish the purpose of the assessment and outline an agreed upon definition of the domain to be tested. The instruments are then developed based on this agreed upon definition. The framework affords the link between the concept being assessed and the test outcomes. A well-designed framework includes three essential components:

1. A clear definition of what is being assessed;
2. A description of the items to be used and the mix of different types of items;
3. The basis for interpreting the results.

The second stage of the framework development process was to review the draft versions of the frameworks that were developed using existing materials. Experts as educators to design the summative/formative assessments and all technical support team are expected to run at this stage, generating a framework that can be reviewed more than one time for validity purposes. Other teams can be assigned by the academic committees to review the draft framework specifications and content in preparation of a version for wider and targeted consultations. The assessments framework should agree

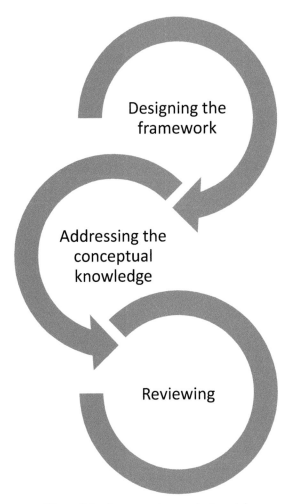

Figure 8.1   Assessment conceptual model.

upon in contrasted higher education and quality assurance procedures that measure and evaluate the design in terms of many point we have referred to before. This space and time allocated for validation and reliability should be measured across diverse of quality assurance standards. The stage may involve demos which increase the evaluation accuracy. Consequently, some decisions may be made to develop "provisional frameworks" not meant to be exhaustive and final, but which should contain most of those elements which the majority of experts in each domain agree are essential.

Step 2: Addressing the conceptual knowledge

This mean kind of metadata can be added to the frame to formalize it and specify kind's skills this framework needs, learning resources, reservations, time of exam, duration of exam, and most specific about tools being used to design this framework. Learners should be aware of how this assessment being conducted for example and should expect to know all details to help him/her to conduct the assessment.

This should also be submitted to assigned committees to match the metadata with the content of assessment that may change some factors to help student to manage the assessment without any kind of hardness. For example, the committee may decide to make the duration longer referring that the acquired knowledge in assessment needs more time than just answering with yes or no. Plus technology-based assessments also affect the conceptually model of assessment that may require more time, more technical support, or maybe more attention and supervision.

Lastly, in implementing instruments, emphasis is placed on ensuring that the resulting measures of learning outcomes would be valid, reliable, and free of bias to the extent possible. Results and analysis should also be taken into consideration how to deal with them in a clear process, how to grade, and how to manipulate the assessment as some assessments need not to give a specific grade for a specific method of answering. Some answers are not expected. A clear procedure should be taken by that. A common form of formative assessment is *diagnostic assessment*. Diagnostic assessment measures a student's current knowledge and skills for the purpose of identifying appropriate program of learning. Self-assessment is a form of diagnostic assessment which involves students assessing themselves. *Forward-looking assessment* asks those being assessed to consider themselves in hypothetical future situations. More to take forms of assessments is the more to go further in addressing the conceptual model. Each classroom teacher should pay attention to the fact that each student is a singular case, and need to be understood and manage the assessment belongs in more particular way, but yet in the frame of a quality assurance that assure equity standards.

More forms usually come with more specifications. For example, performance-based assessment is similar to summative assessment, as it focuses on attainment. It is frequently aligned with the standards-based education change and outcomes-based education movement. Nevertheless, preferably they are significantly different from a traditional multiple-choice test, and they are most ordinarily allied with standards-based assessment which uses free-form retorts to standard questions slashed by human scorers

on a standards-based scale, meeting, falling below, or exceeding a performance standard rather than being ranked on a curve.

A well-defined task is identified and students are asked to create, produce,
or prepare something, regularly in settings that involve real-world application
of knowledge and aids. Proficiency is demonstrated by providing a prolonged
response. Performance setups are further segregated into products and performances. The performance may effect in a product, such as a painting,
portfolio, paper, or exhibition, or it may entail of a performance, such as a
speech, athletic skill, musical recital, or reading.

Some have contended that the peculiarity between objective and subjective assessments is neither useful nor accurate because, in reality, there is
no such thing as "objective" assessment. In fact, all assessments are shaped
with inherent favoritisms built into decisions about relevant subject matter
and content, as well as cultural (class, ethnic, and gender) biases.

Step 3: Review, peer–peer review, and self-assessments

This step is taken as a must for quality assurance purposes. After designing
the conceptual model of an assessment taken into consideration that many
factors have been discussed before, a quality assurance procedure should
be assigned without any influence to evaluate the assessment as one unit.
Usually, in higher education institutions, there is a committee having all
academic coordinators meet to evaluate the assessment as a unit and one goal
to serve the academic teaching–learning process. This review should rely on
the expectations and outcomes of this exam. Feedback, recommendation, and
sometimes clear instruction for altering and modifying the procedure of exam
can be taken and forwarded back to the exam committee who is responsible
about modulating the exam model. This step can hold back the exam till it is
ready to be conducted.

Analyzing the outcome and data after the exam is conducted is very
important. The committee should report that any faults or mistakes were
not taken into consideration in the cycle of exam. Sometimes, unexpected
circumstances can come with new strategic and methods. Documenting these
errors and recommending how to deal with them can save the hassle when the
assessment is conducted next time. The reviewing is not an individual work;
it is a committee and there are lots of coordination between these committees.
The requested coordination is also extended to involve the standards of higher
education and quality assurance procedures.

Guidelines and quality controls are planned to ensure that exam or
assessment with different forms and technology can be a quality-oriented.

All mentioned stages are stated as well for developing and enhancing which goes on while the assessment is being created, designed, and managed to ensure that the instruments in all the various forms are authentic equivalent to all students. In other words, and to ensure the assessment quality, a set of procedures can be structured to cope with quality assurance concepts:

- Quality assurance procedures included defining the design, preparing adaptation guidelines, resolving early on potential equivalence issues, training the academic team and getting them into designing experiments and related assistance, providing consolidated monitoring procedures, and documenting all steps and issues addressed by the team and keeping them followed up and archived.
- Quality control procedures, inspect whether the standards are met and propose corrective action when not. Quality control procedures included a thorough verification of initial versions and models, effective reporting of residual errors and undocumented deviations, expert advice in case corrective action was needed, a final check technique, and standardized quantifiable and qualitative reports on how effectively the conceptual models can be turned into assessment bodies.
- Quality procedures to handle providing data before and after the assessment which means getting students' involvements and matching their expectations to the outcomes. Pre-testing chances such as mock exams, self-assessments, and old versions and materials to train student and give them a close frame of how to be inside the assessment body and what to expect from them are crucial and also an opportunity to collect feedback from students, as well as from teachers or faculty, on the instrument and its items, including their insights on the instrument characteristics such as test length, difficulty, and relevance to the discipline being tested. This huge system of coordination and getting all parties involved and contributed helps a lot in developing cycle and fixing incorrect themes or theories.
- One important quality assurance procedure is sampling. Sampling means that samples of students' outcomes should be documented and reviewed randomly to check for quality standards. Sampling involves all kinds of assessments such as exams, assignments, projects reports, or/and survey. These samples are not limited to students, but also include faculty, coordinators, and administrators. Everyone who is/was involved into designing an assessment is highly appreciated and asked to contribute to feedback the procedure. All these feedback and samples

are reviewed by the individual committee, internally from the institution and sometimes externally for quality assurance purposes. The procedure of quality assurance should be always transparent and accurate to help all to develop the system and come with the high expectations of a teaching–learning process.

The evaluation can be formal or informal. Formal evaluation usually implies a written document, self-evaluation documents, quality assurance agencies' reports, external examiners' feedback, and so on. A formal evaluation is given a numerical score or grade based on the institution's performance, whereas an informal evaluation does not contribute to the final score. An informal assessment usually occurs in a more casual manner and may include observation, inventories, checklists, rating scales, rubrics, performance and portfolio assessments, participation, peer and self-evaluation, and discussion.

8.4 Importance and Impact

In a review of research on assessment and classroom learning, Paul Black and Dylan Wiliam synthesized evidence from over 250 studies (Black and Wiliam, 1998). The research indicates that improving learning through assessment depends on five, deceptively simple, key factors:

- The provision of effective feedback to pupils;
- The active involvement of pupils in their own learning;
- Adjusting teaching to take account of the results of assessment;
- A recognition of the profound influence assessment has on the motivation and self-esteem of pupils, both of which are crucial influences on learning;
- The need for pupils to be able to assess themselves and understand how to improve.

Meanwhile and on the other hand, several constraining influences were acknowledged. Among these are:

- A leaning for teachers to evaluate quantity of work and presentation rather than the quality of learning;
- Greater courtesy given to marking and grading, much of it tending to lower the self-esteem of pupils, rather than providing guidance for improvement;
- A strong emphasis on comparing pupils with each other which demoralizes the less successful learners;

- Teachers' feedback to pupils often serves social and managerial resolves rather than assisting them to acquire more effectively;
- Teachers not being fully aware enough about their pupils' learning requests.

We will try to include some impact of quality-oriented assessments, and how the quality-oriented design is important and mutually impacting the student's performance as well as teaching–learning process.

a. The mutual impact of course coordination system on students' performance

Most of the higher education instructions have adopted a pedagogical model to represent their education systems. Some of these institutions have e-learning, blended learning, or open learning concepts. Others are still following the traditional ways of teaching–learning processes. But although, of sticking to the old teaching methods, technology revolution and student mobility have forced the maximum number to depend on e-learning centers and deliver part of the material using online platforms, these platforms are still submitted to the coordination system which is a supervision architecture to supervise and manage the delivery of knowledge on both cases, online delivery and classroom delivery. So the coordination actually is almost a must system for quality assurance purposes, and actually it is a new procedure to educational systems.

A previous study entitled with: *The mutual Impact of e-Course Coordination System On Students' Performance, by Haifaa Elayyan, 2012,* had discussed the impact of coordination system on students' assessments and performance in general and the difference and added-value by the e-coordination system. The e-course coordination system comes with various layers of perfect e-course coordination and e-monitoring and e-measurement systems that fit with the virtual learning environment (VLE) for quality of assurance applications. The study argued the mutual impact of e-coordination system on students of multi-sections as it presents a comparative analysis depending on Arab Open University (AOU) experience e-coordination system through the VLE and other running courses in regular education without any standardization level.

Using the technology of Internet and the new concept of open learning education along with the advanced e-technologies and innovating learning methods, higher education has been presented in a flexible interface that reduces the need for traditional education procedures while maintaining the assurance of quality that is a vital for any educational institution. Despite of the fact that course coordination system is a recent appreciable system, it has proven efficiency as being one of the highest importance of academic management roles. When teaching process involves multi-sections and mutli-teachers experience, there is a necessitate to have a Course coordination system in deed . A course coordination system is a pre-designed quality-oriented system by coordinators and educators to unify and present the course in one management platform that gurantees equaly oppurtuinties and access for all students in these multi-section to have same teaching and same the delivery of the associated learning subjects, which has always impacted the performances of students in multi-sections positively. It unifies course academic calendar, context covered by the course, syllabus, layout, e-content, assignments, and assessments. Course coordination system plays as a perfect measurement system of the course's reputation, success, and unified average of multi-sections' student's performance that satisfied and aimed at the quality assurance and implementation of the learning process.

In a multi-branch, as well as a multi-campus university like AOU, it is crucial to establish a set of explicit and well-defined academic structures to measure the implementation techniques at all branches. The course coordi-nation system is a measurement technique of most of the teaching–learning processes, particularly the mutual impact of student's performance which clarifies that the educational objectives have been attained and which have not been yet. Having such a measurement system with identified relevant imple-mentations and applied techniques that can be easily applied and tracked in all branches helps for analysis and statistics.

The study presented an example from a university such as King Fahad for Petroleum and Minerals which has a collection of elective course in the computer science program. One of these courses presents the concept of the computer organization and architecture. Usually, this course is offered for second level students. Having multi-section gives a possibility of teaching this course by more than one tutor. Each tutor prepares the content and syllabus and specifies the textbook that has the same concept but it may be a different title and author from the other tutor. Each tutor is responsible about specifying the course work assessments detailing the number of assignments, the weight of each assignment, number of quizzes, and the weight of each quiz. Each tutor separately is responsible about defining the structure and format of these assessments including the final in an unlike way from the other tutor. Let us suppose that we have five sections in the computer organization and

Tutor	Max	Average	Min
1	100	83	70
2	89	75	60
3	83	68	50

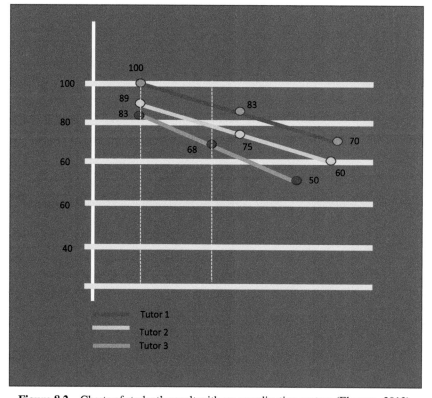

Figure 8.2 Charts of student's result with no-coordination system (Elayyan, 2012).

architecture course. Three tutors had been assigned to teach this course assuming Tutor 1: X: 1 section, Tutor 2: Y: 1 section, and Tutor 3: Z :1 section. Tutors were left to design a course assessment separately. Figure 8.2 shows the deviation results.

Unifying the grade system means that the maximum is 100 for the three sections and minimum is 50 and then the average will be around 75. And grades' distribution can be as follows:

90–100	A
82–90	B+
79–82	B

That means four students in Section 1 will be getting A while only one from Section 2 will get A. None of Section 3 will get the A grade. Such a system will be unfair for students of Section 3 as their assessments and material may be harder than the other sections. Comparing this to the e-coordination system used by AOU for the same course, the system had managed to raise the performance of students having the same assessments' standards and same evaluation guidelines.

Quality assurance systems force actually the coordination as a fair scale for course's assessment designs, regardless of the number of students and their locations, being open learning students or traditional teaching. The coordination system usually has layers of academic coordinators that manage and collaborate for quality assurance purposes.

The study came to conclude that the e-course coordination system is perfect fit mechanism to deal with a huge number of students distributed in multi-locations and multi-sections. This e-course coordination system is supported with features in online platform that makes it possible with high impact on students' performance. The e-course coordination system is also an extension for the quality assurance progress that AOU is committed to. It is worthy to mention that every activity of the-course coordination system is already archived on the VLE as tutors and students' activities which makes more convenient and practical for multi-sections (Elayyan, 2012). Also the study presented that also many studies and research were done and can be found at the main website of OU about the impact of excellent feedback on academic achievements as they have superior aims as the following:

1. Deliver high-quality information about learning.
2. Develop ability to reflect and self-assess.
3. Help clarify what good performance is.
4. Close the gap between current and good performance.
5. Improve motivation and self-esteem.
6. Encourage dialog about learning.
7. Use feedback to improve teaching.

b. The impact and importance of different forms of questioning

The importance of teachers using fitting interrogative methods and techniques to question the knowledge and support student learning is not limited to teacher's skills but also to institution's and program designer's awareness. Some recent researches have observed some problems in the way that teachers were framing and using questions during class interface. They concluded that the questions used by teachers were often insufficiently challenging for the students with many obstacles such as time given to elaborate on an answer was too short, for example. Questioning is a skill that teachers should learn and practice how to do and deliver to students.

Higher education institutions in general have been working toward modifications in questioning interaction in assessments and during classrooms. One previous problem of teacher expecting much from learners to cope and answer in short time was one important to be altered. So many of the quality assurance procedures have recommended to take the assigned time and duration in consideration and give it a priority so that students feel comfortable by receiving the query and thinking to find a proper answer, not in a knowledge race or competition. Specially that individual skills are different from a student to another and actually should not be a factor impacting the student negatively. Some students really need to more time than other to formulate their answers and some with special needs that should be in consideration. Some students are with open character and others are closed which affect the response time. So, assessments which depend on questioning should work to improve student's skills and the quality of questions should be employed by teachers in terms of greater use of aim achieved, of those that are open rather than closed in character. This encourages students to use higher order thinking skills to answer questions.

A chance for student to practice questioning theme and being able to deal with it and pass means that the student is achieving in teaching–learning process and actually is adding values for these assessments. This type of student is able to run and respond positively to a non-threatening climate of dialog, where students feel free to share their ideas and expose their misunderstandings without fear of giving an incorrect answer. The idea is that student learning is enhanced by opportunities for conversation of fallacies and amplification of these in the classroom so that all students can study from one another's mistakes or delusions.

c. The importance of feedback

It is extensively recognized that feedback has a significant impact on the learning cycle. Worthy quality feedback is beheld as a vital aspect in learning process. Many previous studies have shown that students can response positively to sufficient feedback given in turn to their assignments or assessments. Moreover, enlightening and evocative feedback is held to be more proper for a learning-centered perception than the exclusive practice of marking work as right or wrong and giving just an overall mark. Research studies can easily demonstrate that individual reference has more positive effects on students than social reference as assessment norms for teachers.

The effect of feedback according to teachers', students', and other practitioners' perspectives can indicate that formative feedback is generally recognized as a tool for:

1. Improving teaching and learning.
2. Changing old-style classroom practices.
3. Illustrating learning criteria.
4. Raising teachers' and learners' achievements.

It should be mentioned that the positive role of formative feedback is also recognized in other approaches to effective teaching including interactive whole class teaching and direct instruction. To elaborate the concept of feedback in the context of assessment for learning, a crucial factor to consider is the need for feedback to be closely linked with the use of unambiguous and communal learning goals or criteria. It is argued that students can understand only the feedback they receive from teachers or peers if they are first clear about the purposes of learning and what is expected from them. This is a compulsory podium that should strengthen constructive feedback. And to reach this level, we would like to increase the student's involvement and contribution to the feedback process.

There are a number of approaches that can be used to maximize student engagement with the process. One tactic is to design assessment so that students can realize the direct benefits of attending to feedback guidance. This can be done by breaking assignments into phases and providing feedback which is indispensable to the successful navigation of subsequent stages. In addition, students can be compulsory to document how they used feedback to advance to the next stage of the exercise. This strategy has the additional benefits of encouraging students' meta-cognition and making them more energetic participants in the feedback-learning cycle.

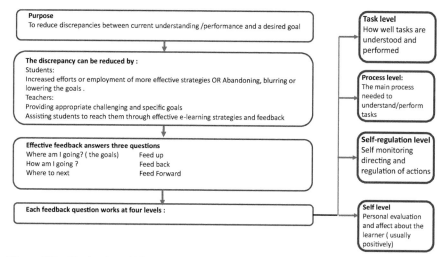

Figure 8.3 Hattie, J., and Timperley, H. (2007). The power of feedback to enhance learning.
Reference: Hattie, J., and Timperley, H. (2007). The Power of feedback. *Review of Educational Research*, 77–87.

The workload for teachers can be offset by the reduction of time needed to give feedback on the final product and by incorporating peer feedback into some of the stages (Nicol, 2008). Another tactic to inspire student consideration on feedback comments is to give a conditional grade, but invite students to talk about their exertion and potentially earn a higher grade. Some reviewers suggest withholding the grade altogether until students have read the comments and indicated this in some way (Taras, 2003).

It is also possible that students do not pay attention to comments because they do not make sense to them (Duncan, 2007) or that they do not understand the purpose of the feedback procedure. Figure 8.3 shows a model of feedback to enhance learning. As feedback should be delivered with a purpose to get the benefits for both teachers and students and should be also supported with a plan or suggest to correct what have been mistaken with.

The model clears all misunderstanding about the feedback and gives steps how to design it and use it for learning enhancement process which actually meets the requirements and standards of a quality assurance procedure. The power of feedback as an educational review (Hattie and Timperley, 2007) 's model presented in Figure 8.3 shows how comments may be connected to these questions on four different levels, task, process, self-regulation, and self-feedback on the task in relative to all three questions usually works best when it is accompanied by explanation of appropriate processes and learning can be excavated if the feedback moreover prompts some degree of self-reflection and supervision.

Good quality of feedback will validate appropriate ways of augmenting the performance of teachers and learners including or assuming that all parties will be invited for more responsibility to improve the procedure. By contrast, feedback "about the self as a person" often has no impact on the learning, because it is not linked to the goals of the task for future learning management or behaviors. According to Hattie and Timperley (2007), "praise addressed to students is unlikely to be effective because it provides little information that provides answers to any of the three questions and too often deflects attention from the task" (p. 96). It should be prominent that this kind of praise should be famed from praise directed to the enactment of the task which can profit learning.

Good feedback can shape some principles that we can evaluate to if feedback does the following:

1. Indorse exchange of ideas around the goals of the assessment.
2. Accentuate the instructional aspects of feedback.
3. Postulate the goals of the assessment task and use feedback to link student performance to the specified assessment goals.
4. Involve students in practical exercises and dialog to understand acquiring criteria.
5. Employ students' conversation to the purposes of feedback.
6. Feedback scheme and comments are expected to invite self-evaluation and future self-learning management.
7. Broaden participation in the feedback conversation-incorporated self- and peer-feedback.

Good assessment and feedback practice should be designed to help clarifying what is the good performance and how to achieve or get closer to the goals and standards. Deliver high-quality feedback information that helps learners self-correct. And provide opportunities to act on feedback to close any gap between current and desired performance. Enhancing the feedback procedure is a step to support the development of teaching–learning in a quality assurance theme.

d. Peer and self-assessment

The curiosity in self- and peer assessment is partially driven by shifting conceptions of teaching and learning. Contemporary tactics emphasize the active engagement of students in their own learning, learner responsibility, metacognitive abilities, and a dialogical, collaborative model of teaching and learning. These are recommended as good strategies for students to develop greater sovereignty in relation to their own life-long learning along with promoting their metacognition and collaborative learning skills.

According to Boud (1995), all assessment including self-assessment comprises two main elements: making decisions about the standards of performance expected and then making judgments about the quality of the performance in relation to these standards. When self-assessment is introduced, it should ideally involve students in both of these aspects. Andrade and Du (2007) provide a helpful definition of self-assessment that focuses on the formative learning that it can promote: Self-assessment is a process of formative assessment during which students reflect on and evaluate the quality of their work and their learning, judge the degree to which they reflect explicitly stated goals or criteria, identify strengths and weaknesses in their work, and revise accordingly (2007, p. 160).

Construction evaluations and self-assessing of learning are integral to the learning process. Self-assessments can give lots of good quality and positive impacts such as building on a natural tendency to check out the progress of one's own learning. Self-assessment is a step to recognize the next step of further.

Self-assessment is not limited to learners. It may also include educators to help them identify basics of teaching–learning progress they are trying to promote which motivates further enhancing and developing of learning. In terms of assessments, self-assessment emphasizes the formative aspects of assessments and encourages a focus on process. It can also accommodate diversity of learners' readiness, experience, and backgrounds and practices can be aligned well with the shift in the higher education literature from a focus on teachers' performance to an emphasis on student learning.

Self-assessment begins to shift the culture from a prevalent one in which students undertake assessment tasks solely in the spirit of pleasing the lecturer (Boud, 1995). Focus shifts away from satisfying the lecturer and more toward the quality of the learning. Boud (1995), talking about the origins of his long interest in self-assessment, invokes a picture of the way in which so many student assessment endeavors are misdirected, when he comments that there was "a slow dawning that it was not others I should be satisfying in my learning endeavors, but myself" (p. 3). Self-assessment with its emphasis on student responsibility and making judgments is "a necessary skill for lifelong learning" (Boud, 1995, p. 11). Additionally, the self-assessment process can help "to prepare students not just to solve the problems we already know the answer to, but to solve problems we cannot at the moment even conceive" (Brew, 1995, p. 57). Also it is about engaging students in the formulation of criteria for self-assessment tasks that help them to deepen their understanding of what constitutes quality outcomes in a specified area.

References and Suggested Literature

[1] QAA, *Quality Enhancement Review Handbook*, 2017, available at: http://www.qaa.ac.uk/en/Publications/Documents/QER-Handbook-17.pdf.

[2] QAA, *Fulfilling our Potential: Teaching Excellence, Social Mobility and Student Choice QAA's Response*, 2016, available at: http://www.qaa.ac.uk/en/Publications/Documents/QAA-Green-Paper-Response-Jan-16.pdf.

[3] QAA, *UK Quality Code for higher education PartB: Assuring and Enhacing Academic Quality*, available at: file:///C:/Users/sharif/Downloads/Chapter%20B6-%20Assessment%20of%20Students%20and%20the%20Recognition%20of%20Prior%20Learning%20(1).pdf.

[4] QAA, *The UK Quality Code for Higher Education, Overview and the Expectations*, 2015, available at: http://www.qaa.ac.uk/en/Publications/Documents/Quality-Code-Overview-2015.pdf.

[5] Available at: https://evolllution.com/opinions/teaching-assessment-quality-assurance-higher-education/.

[6] *Armstrong Collage of education, Educator Preparation Program Quality Assurance & Assessment Manual 2016–2017*, available at: https://www.armstrong.edu/images/uploads/education/16-17_EPPHandbookQuality_AssessmentManual.pdf.

[7] Available at: https://en.wikipedia.org/wiki/Educational_assessment.

[8] Available at: https://www.edutopia.org/assessment-guide-description.

[9] Kuh, G. D., Jankowski, N., Ikenberry, S. O. and Kinzie, J., *Knowing What Students Know and Can Do The Current State of Student Learning Outcomes Assessment in U.S. Colleges and Universities, National Institute for Learning Outcomes Assessment*, 2014, available at: http://www.learningoutcomeassessment.org/documents/2013%20Abridged%20Survey%20Report%20Final.pdf.

[10] Amaral, A., *Quality Assurance and Assessment in Higher Education: recent trends*, 2009, available at: http://www.a3es.pt/sites/default/files/7.%20Oxford%20seminar_0.pdf.

[11] Tremblay, K., *Diane Lalancette Deborah Roseveare, Assessment Of Higher Education Learning Outcomes, Ahelo Feasibility Study Report Volume 1 Design And Implementation*, 2012, ©OECD 2012 available at: http://www.oecd.org/education/skills-beyond-school/AHELOFSReport Volume1.pdf.

[12] QAA, *UK Quality Code for Higher Education – Chapter B6: Assessment of students and the recognition of prior learning*, 2013, available at: http://www.qaa.ac.uk/publications/information-and-guidance/uk-quality-code-for-higher-education-chapter-b6-assessm ent-of-students-and-the-recognition-of-prior-learning1#.WfitX1uCxhF.

[13] Shepard, L. A., *The Role of Assessment in a Learning Culture*, 2000, available at: http://journals.sagepub.com/doi/abs/10.3102/0013189X029 007004?journalCode=edra.

[14] *Reading 13.4 Assessment for learning Assessment Reform Group*, available at: https://set.et-foundation.co.uk/digital-assets/qtlsmap/Resources/ 17/Assessment%20for%20learning-%20assessment%20reform%20gro up.pdf.

[15] *Assessment for Learning:10 principles, Research-based principles to guide classroom practice, Assessment Reform Group 2002*, available at: http://www.hkeaa.edu.hk/DocLibrary/SBA/HKDSE/Eng_DVD/doc/Afl_ principles.pdf.

[16] Flórez, M. T., and Sammons, P., *Oxford University Department of Education, Assessment for learning: effects and impact*, 2013, available at: https://www.educationdevelopmenttrust.com/~/media/EDT/Reports/ Research/2013/r-assessment-for-learning-2013.pdf.

[17] Available at: https://www.edutopia.org/assessment-guide-importance.

[18] Dorothy Spiller Teaching Development, *Wâhanga Whakapakari Ako Assessment: Feedback to promote student learning Teaching Development |Wâhanga Whakapakari Ako*, 2009, available at: http://www.waika to.ac.nz/tdu/pdf/booklets/6_AssessmentFeedback.pdf.

[19] Dorothy Spiller Teaching Development, *Wâhanga Whakapakari Ako, Assessment Matters: Self-Assessment and Peer Assessment Teaching Development |Wâhanga Whakapakari Ako*, 2012, available at: https:// kennslumidstod.hi.is/wp-content/uploads/2016/03/assessment-matters- self-assessment-and-peer-assessment.pdf.

[20] Elayyan, H., *The mutual impact of e-course coordination system on students' performance*, 2012, available at: http://ieeexplore.ieee.org/docum ent/6216685/, *IEEE Xplore*.

[21] Annual Report 2008

9

Quality-oriented Learning Resources

9.1 Introduction

The widespread prevalent axiom "learning resources" does not refer to a specific educational learning resource in specific and actually should not be taken that simply as the term can deceive. It can have many different meanings depending on the design of the educational method employed and the skills of the teacher who end up to create and innovate a learning resource. There are possible ways to build this resource can produce wide variations in quality and in the educational objectives that be aimed to. A taxonomy is needed to facilitate an awareness of these differences in learning resources and to help teachers choose a recourse-based learning method most appropriate for their students and teaching–learning process. The taxonomy we are seeking here is actually build on a proposed topology of quality-oriented networks of learning recourses that presents a mutual dynamic method of building a learning resource and aligning it in an adaptive way for quality assurance fitting and purposes. The topology is supported with educational methods to create and innovate even with limited possessions.

To be able to reach this far, we need to shape the meaning of "learning-resources" term. And it will be easier to break the term into parts:

1. **Learning:** It is a process of knowledge-transporting systems and bringing and adopting new-fangled fluctuations in different aspects, but mainly the way of thinking it and the way of acting for more effectiveness in performance fitting the purpose of this learning. The learner through this process will be exposed to force of acquaintance and expected to accumulate experience. It results in the attainment of knowledge concept and awareness, skill, interests, attitudes, etc.
2. **Resources:** A resource is a static long-term or short-term source of required knowledge relevant to the specific acquiring to fetch all benefits

that can be modulated and produced. It is acquired to accomplish an activity or to achieve desired outcomes.

3. **Learning resources:** It is shaping the resources to fit into the learning purposes and can take any kind of relation and formulation and requires entities to interact, acquire, fetch, and define the suitable outcome needed to accomplish the learning process. It is simply as the case of switching ON/OFF in a computer circuit that requires initiative to acquire the data and fetch it in a processed way to match the purpose. Having the switch ON means that the circuit is ready and all circuits and channels are ready to transport the required data. Having the switch OFF means that the circuit is interrupted and outcome is not expected.

Then it is about linking two terms: Learning and recourses to end up with a cycle or channel that expected to get with outcome fits the proposed inquiry with all required. The learning is accomplished as a mission.

More methodological description we can say that learning resources are any person or kind of devices or procedures that help to make teaching and learning more interesting, more stimulating, more reinforcing, and more effective. This assumption is a successor of selecting the right learning resource to help the process of learning. Which actually makes a big difference in the outcome silhouette. Which makes sense to be precautious about the method of selecting the most fitting learning resource that match our interest, centeredness, and motivation in addition to the educational values that can add to the learning process. Other selection methods and concerns will be discussed later in this chapter.

Educational metadata and brokerage for learning resources are new concepts mapping the use of learning resources in educational systems where users are actually unfamiliar with the content and the way of utilizing the learning resource. So, some educational systems with diversity of users have developed metadata systems to include valuable information and guidance about the embedded learning resources which actually gave more space for updating, re-using, and re-forming in more fitting ways for quality assurance purposes. The aim of this step is to provide a wide obtainability of educational resources and form it to be a more common objective for universities, libraries, archives, and other knowledge-intensive institutions. Although generic metadata specifications (such as Dublin Core) seem to satisfy the need for documenting web-distributed objects, educational resources demand a more specialized treatment and characterization. In this article, we focus on the use of learning object-specific metadata in digital repositories, as they are primarily incarnated in the LOM (learning object metadata) standard.

This kind of re-thinking has added a great value of educational systems and this concept will be briefly described as an important innovation in recent teaching–learning concept related to technology and web-based learning. But as a start, we need to emphasize the importance of learning resources in any educational system, where we cannot ignore nor we can and t complete our representation of quality assurance concept in the of the teaching–learning progress. This chapter is written in the purposes of without defining what we mean by the learning resources and entailing facts and impacts on the teaching–learning process.

9.2 Aligning Learning Resources with Teaching–learning Process for Quality Assurance Purposes

Many recent research works in educational methodologies and strategies have acknowledged that students cannot learn from one type of instructional material alone and many educators being in the field of teaching believe that learning resources are not in hands as before and cannot take the consequences if huge attention is not paid to this important facility that affects and impacts the learning process. Most agree that educators, parents, and administrators who can best determine what content will be effective for learners, the fact that nowadays, learners with the concept of self-learning can make a huge mess if the process is wide open or even limited. On the other hand, while the learning resources are under control, missing the method of selecting the right learning recourse can impact the learning process in diverse ways. Here it comes the urgency to support the teaching–learning process with heterogeneous learning resources that match and cope with the purpose of educational systems.

Delivering an effective education to all students is a shared responsibility. To achieve success, higher education institutions and all levels of government must work together to ensure that every teacher and every student have access to a range of high-quality instructional materials. Engaging, effective, and diverse educational resources that indorse learning, foster skills, and accelerate student triumph are crucial to the core melodies of life-long learning. Each university or college considers that students deserve to be taught with the best of instructional resources, and educators deserve professional-quality publications, development opportunities, and digital materials for use in their classrooms. Educators, parents, and students should expect and demand only the highest quality of standards-based instruction. The learning process does not hold the way to the learning resources being used inside the classrooms, but also supplemental resources help teachers differentiate instruction and engage students who, for whatever reason, need enrichment beyond the core classroom material. No matter which materials are used, though, educators should hold all instructional content providers accountable for the quality of their learning resources.

It comes to reality with the Internet world and advanced technology and mobility; in addition to their teachers as being the main learning resource, students count on a range of resources to help their learning. These vary from physical resources such as libraries or computing facilities to human support in the form of tutors, counsellors, and other advisers to virtual resources that widely open with no limitation of location or time such as e-libraries, virtual labs, virtual teachers, and dynamic/online educational systems, which actually tied to the package of e-learning and blended learning resources in more unconventional approaches than before. Learning resources and other upkeep apparatuses should be actively reachable to students, designed with their needs in mind and receptive to feedback from those who use the amenities provided. Institutions should customarily monitor, review, and improve the effectiveness of the support services offered to their students. Whether the merchandises are print, digital, visual, or web-based, instructors and superintendents hold educational publishers responsible and accountable for accurate, up-to-date content that meets state, district, and curriculum standards and learning goals and addresses the different cultural credentials of today's students. Instructional resources must have developmentally fitting reading levels, language and exercises, differentiated lessons for the various ability levels, comprehensive teacher guides, qualitative and quantitative assessment, and high-level learning goals. In addition, these resources must be evidence-based, objective driven, and premeditated to engage both today's students and teachers.

There is a fact, reader should be conscious of while pursuing this section, which is about segregating between learning resources and teaching materials. "Teaching materials" is a generic expression used to label the resources that educators use to convey teaching. Teaching materials can support student learning and increase student realization. Idyllically, the teaching materials will be personalized to the content in which they are being used and can be recycled into many new ideas, to the students in whose class they are being used, and the teacher. Teaching materials can state to a number of teacher resources; however, the term usually refers to concrete examples, such as worksheets or manipulatives (learning tools or games that students can handle to help them gain and practice facility with new knowledge). Teaching materials are different from teaching "resources," the latter including more theoretical and intangible elements, such as essays or support from other educators, or places to find teaching materials. Attainment of valuable teaching materials is not challenging as it can seem at first. The Internet has many resources for teachers; most of them can significantly increase the contents of teaching materials. Any educator as well can innovate own teaching material that fits specific teaching style and way/or specific science or subject. Every developed learning material will asset the current educator or other educators when they teach a similar unit. An investment of time or money in good teaching materials is an investment in good teaching. And the way of such thinking drives us to the concept of quality-oriented and quality teaching–learning processes as what we are suggesting here is trying to build the concept of learning resources and engage it to the learning process based on quality. Professional standards of quality and consistency should be the reference line for all learning, teaching, and instructional resources. Quality benchmarks include lots of general guidelines that can be applied or taken into consideration while designing a learning resource. First and most important that learning resources are expected to be plainly enunciated learning goals and objectives, appropriate reading levels that enrich students and clearly stated reputable sources.

As main quality process that learning resources are to fit and help learning process, then the learning resources should be engaging, relevant, and up-to-date content. Highly vetted content is accurate, objective, and reliable. The learning resources as well are most likely to cover discriminated learning prospects associated to a learning process along with well design for students, teachers, and other education professionals. These resources to be with more effeteness should be kind of adaptable for individual learning styles and needs. Learners and educators need a comprehensive guide or instructional support materials that come with standards- and evidence-based lessons/learning aligned with high-quality assessments. Keep in mind that learning resources are not only physical or common resources. The concept bears lots of new technologies and can go beyond what is familiar to all.

Well-designed learning resources encourage greater individual engagement by students with information, ideas, and content than is possible with lectures alone. By making such resources an integral part of the teaching and learning process, limited face-to-face teaching time with students can be more effectually used to foster commitment and to nurture discussion, creativity, practical applications, and research activities. In emerging courses and learning resources, teaching staff unsurprisingly use what is obtainable. The cumulative pool of learning resources not only extends their select, but also generates openings for new assets to be adapted to fit the limited context in terms of philosophy and learning needs without demanding protracted copyright dialogs or duplicating content development. Experience shows that, when institutions make respectable quality courses and materials publicly available accessible, they can interest new students, expand their institutional reputation, and advance their public service part.

Such institutions may also further broadcast the research results and thereby attract research funding.

The following can be a general guide for a quality assurance process in terms of designing/managing and updating a learning resource. The process in general can fit and meet the needs of a higher education institution without any complexity, just forward quality assurance process.

9.2.1 Quality Assurance Process for Learning Resource Design and Managing

In order to ensure the quality of their materials, learning resource inventers employ an extensive product development process focused on the needs of the learner.

Step 1: Determine Content

- Consult ceremonial curriculum committees, authors, digital media experts, independent experts/reviewers, national standards' organizations, and national advisory groups.
- Study recognized research base and new research findings.
- Establish plan for customized correlations to state standards.
- Mature preliminary plan for content.

Step 2: Research and Planning

- Classify content experts.
- Survey educators.
- Develop preliminary plans for organization and design.
- Shape out the plan for customized correlations to state standards.
- Cultivate and produce the prototype.
- Review the prototype with authors, digital media experts, and educators.
- Revise the development plan to reflect input from content authors, digital media experts, and educators.
- Develop and test the new prototype.

Step 3: Early Development

- Form development team, including authors, content experts, graphic artists, digital media developers, and other specialists.
- Begin development of customized correlations to state standards.
- Develop detailed outlines and make assignments.
- Establish the project schedule.
- Authors, digital media developers, and content area experts create and evaluate the first draft.
- Design the plan for special features and assign development teams.
- Create the design for main products and all ancillary materials.
- Plan teacher editions/interface and ancillary materials.

Step 4: Editing and Review

- Update as necessary customized correlations to state standards.
- Document all facts from at least two independent sources.
- Edit student and teacher content as well as ancillary materials.
- Review for accuracy (academic reviewers, independent readers, evaluators, and master teachers) and usability (for digital materials).
- Copy edit, fact-check, prove formulas and equations, and proofread.
- Incorporate changes from authors, editors, digital media developers, and reviewers.
- Create pages, produce digital content, develop art, prepare charts and graphs, and choose photographs.
- Check revised content.
- Repeat content checks until all content is correct.
- Check proofs.
- Produce the first version or go to the first printing (intended for use only as marketing samples).
- Distribute the first printing or beta test (digital materials).

Step 5: Quality Reviews of First Version/Printing

- Send student and teacher editions to independent reviewers for complete content read and to test usability (for digital materials).
- Solicit comments from teachers and state review committees.
- Research and verify the accuracy of error reports through authors and independent content authorities.
- Correct errors and create proof of corrected content.
- Correct technical issues.
- Proofread and test corrections.

- Repeat process until all errors are addressed.
- Check proofs of final versions.
- Produce the second digital version or print second printing (which will be sold for classroom use).

Step 6: Continuing Quality Reviews

- Receive and review comments from students, teachers, academics, and review committees.
- Correct text, photographs, charts and graphs, art, and digital interface for errors or clarifications.
- Prepare and distribute errata/updates if errors found.

Step 7: Subsequent Editions/Versions

- Research clarifications, including public comments.
- Hold discussions among authors, developers, and editors.
- Complete entire preparation process – productions, documentation, verification, and editing.
- Reprint (if edition is print) or distribute new versions (for digital materials).

Please note that such a process is for classroom curriculum products. Informal learning products will not need to associate to criteria and have teacher instructions, but they must still make sure that the content is developmentally suitable, employs proven learning strategies, and comprises effective learners in order for the aim to be touched.

Universities as higher education intuitions are the main party we need to highlight and get involved in such quality processes. In this context, it is suggested that higher education institutions to develop institutional strategies for the integration of learning resources with university systems. Support teachers and students and everyone interested with guidelines how to access and manipulate these learning resources. These strategies are to suggest elements and foster initiatives that consider developing corporate strategies for the integration and using resources into a range of activities. Universities are expected to provide incentives to support investment in the development, acquisition, and adaptation of high-quality learning materials. Institutional policies should be reviewed by assigned committees and well-expertise. This kind of committee is to review and state many points associated to well-designed learning recourses and can be the following:

- Reassure sensible selection and adaptation of existing learning resources, as well as development of new materials where necessary;

- Promote the publication of educational materials as learning resources within institutional protocols;
- Promote research on using, reusing, and repurposing learning resources;
- Promote students publishing their work (with the guidance of academic staff and within institutional protocols) under an open license as learning resources;
- Build learning resources into mechanisms for institutional and individual monitoring;
- Promote collaboration both within and beyond the institution in developing materials;
- Provide staff with appropriate incentives and rewards for the development, acquisition, and adaptation of learning materials;
- Ensure that staff workload models allow for curriculum, course, and materials' design and development.

Recognizing the significant role of educational resources within internal quality assurance processes motivates higher education institutions to include establishing and sustaining a rigorous internal process for validating the quality of educational materials prior to their publication as learning resources. Most of the universities consider creating learning resources' center which is a facility within the university staffed by a specialist, containing several information sources, and designated to serve as a supplemental educational tool to balance the concepts that have been used in the teaching–learning process inside and outside the classrooms. Learning resources are profoundly used in schools before but in a limited physical scale but when it comes to higher education level, universities need to expand the definition of this center and promote the scale to cover all educational needs supported with technology.

On the way, to do so, learning resources are undertaken institutional advocacy and capacity building. Ongoing awareness-raising, capacity-building (staff development), and networking/sharing for both women and men can be carried out to develop the full range of competences required to facilitate more effective use of learning resources. Universities' positive response to quality assurance calls to enlighten the role of learning resources in learning progress, and efforts and activities might encourage a shared vision for open educational practices within the organization, which would ideally be aligned to the institution's vision and mission and linked to incentives. And expand it to be a shared network of learning resources among higher educational institutions supported with all web-based technology. Up to a certain level, we can find such scenario of shared learning resources but lots of efforts and work should be organized in terms of flexible copyright policies. Such policies could make it simple for staff to invoke some-rights-reserved copyright or other licensing permutations when this is estimated necessary.

These policies could be part of a wider institutional process to ensure that robust, copyright, and privacy policies are in place and accurately reflected in all legal contracts and conditions of employment. Quality assurance agencies, control and managing are also welcome to contribute here as well as support from the higher education and government levels to ensure for example a stable, steady ICT access for staff, in different locations and multi-campuses as well as students. This means striving to ensure that academic staff and students have ubiquitous access to the necessary ICT infrastructure, software, and connectivity to access the Internet and mature or adapt educational materials of diverse styles. This should include software applications, such as web content editing tools, content management systems, templates, and toolkits that facilitate the creation and use of adaptable, inclusively designed educational resources. It might also entail developing a warehouse of the work of academic staff and students that could oblige as an influential teaching and learning resource, while levitation awareness of the distinction between appropriate sharing/collaboration and plagiarism. Staff and students should also have the right of entry to training/professional development programs and support to use these shared systems.

Huge effort like this gives kind of equality for learners to have access to sort of organized, supervised, and well-designed learning resources. The collaboration work is assumed to develop policies and practices to store and access learning resources using, for example, the latest technology of clouding-based systems and Internet of Things to integrate the different learning resources and deploy them to form network of quality-oriented based of open learning resources system. This includes the capacity to store, manage, and share resources and content, both internally and externally, so that academic endeavors build on a growing base of institutional knowledge but in a broadcasting theme. This might be done most cost effectively as part of a coordinated national strategy or in partnership with emerging global learning resources' networks and repositories based on open standards.

Quality assurance procedures internally inside the institution and from external agencies are evaluating and reviewing.

Learning resources practice periodically. Such reviews will help the institution determine the value of its policies and practices and manage to recover shortages on the spot. They could include reviewing the extent of the use of openly licensed educational materials in the higher education program as well. They could also include assessing the effects of this use on the quality of educational delivery and its impact on the cost of developing/procuring high-quality teaching and learning materials for undergraduate and postgraduate programs. Selecting a learning resource is another layer to communicate here as it may impact the process if the teacher/learner failed to select the right and most fitting learning resource. The selection needs awareness of many factors:

- Fitting to purpose of teaching/learning.
- Suits social levels such as: age, maturity, intellectual level, motives, and social environment of the learner.
- Must capable and meaningful for the proper realization of teaching–learning objectives.
- Simplicity is the key to students to understand and enhance his learning.
- Relevancy and suitability.
- Accuracy and properly maintained in order to provide better learning.
- Financial budget.
- Ability and skills in a teacher to handle aid material.
- Being trained to manipulate the content.
- Utilization to the maximum while using the learning resource.
- Awareness of the nature of the used learning resource.
- Selection of the learning resource should match the teaching/learning conditions: for example, if I am using a web-based learning resource such as e-libraries, I am expected to have stable Internet connectivity.
- Right selection minimizes errors and unexpected conditions or materials while using the learning resource.
- Effective selection and usage of a learning resource is an experience that benefits others.
- Learning resources lose its value if they differ in input and output. Input: finance needed to prepare aid. Output: results in the form of learning.

Working on learning resources and preparing them according to quality assurance concepts that help teachers are still costing the institution to spread the awareness and culture of adopting learning resources as a part of the teaching–learning process that actually cannot be completed without them. As the crucial academic role of universities has evolved toward the academic process, the role of the students has also revolved that with contribution from their side, and desire play a role in shaping the quality of their educational experience, educators and higher education institutions promote the learning resources and enlarge the benefits out of learning environments. In this context, it is suggested that student bodies:

- Realize the issues of learning resources and commence advocacy of learning resources.
- Encourage their members to publish work as learning resources.
- Take a dynamic role in pledging the quality of learning resources through social networks.
- Recognize that ICTs are an increasingly important part of the higher education experience and are often critical for students with special educational needs.
- Encourage student participation in activities to support learning resources' development.

General characteristics of learning resources can simply grip the attention of students and adequate them to use. Therefore, institutions are expected to guide users (teacher/student or any other who are interested in this learning resource), and actually it is a part of a quality assurance procedures as no benefit of accumulating learning resources without informing users how to use them. Guidelines and manuals can be designed for academic staff, and it will be an associated committee in the learning resources center's responsibility. The guidelines should take into consideration challenges facing academic staff being vital agents in ensuring the quality of teaching and learning delivered to students. Teachers face a series of challenges that may include, for example:

1. Time constraints
   Time constraint comes with teaching load and other duties. So academic staff is out of time for preparing curriculum and selecting, adapting, and/or developing teaching and learning materials and assessment tools.
2. Limited access
   Access to high quality, relevant teaching, and learning materials. And sometimes staff are with low ICT level and low Internet connectivity that limits options and selections.
3. Privation of knowledge erection
   Unfamiliarity of how to address the often diverse needs of their learners and demonstrate gender sensitivity.
4. Changing teaching and learning environments
   Teaching environment stability is an important thing to create an innovative teacher. Switching and flipping with new concepts and

technology may impact negatively, for example: Strategies changing from teacher-centered to learner-centered approaches.

5. Academic staff race with technology
   Academic staff cannot 24 hours cope with technology while increased student access to online materials, collaborative networks, and online publishing opportunities may give them merits to be advanced away from their teachers.

Other limitations can be more as legal requirements to broaden access; the need to cover a broad and growing knowledge base; the need to update their ICT skills regularly; high student expectations; and ever-increasing enrolments in many jurisdictions.

Accountability for assuring the quality of any content used in teaching and learning atmospheres will exist in predominantly with the program/course coordinators and individual academic staff members. Whether prescribing core readings/textbooks, suggesting further readings, choosing a video to screen, or using someone else's course plan, they recall absolute responsibility for selecting which materials – open and/or proprietary, digital or hardcopy – to use. For this reason, plenty of the quality of learning resources will be contingent on which resources academic staff decide on to use, how they customize these resources for contextual relevance, and how they assimilate them into various teaching and learning activities.

In this context, it is recommended that academic staff in higher education institutions:

- Develop skills to evaluate learning resources. A good starting point is to increase knowledge of any learning resource which is through exploring present ones in suitable portals/repositories and determining what might be useful in courses and modules. Academic staff may find existing to be useful benchmarks for reflecting on and improving their own curriculum and pedagogy as well as those of others.
- Assemble, adapt, and contextualize existing ones.
- Develop the habit of working in teams.
- Seek institutional support for learning resource design and manipulate skills' development.
- Leverage networks and communities of practice.
- Encourage student participation: Academic staff can be encouraged to use student feedback on learning resources to improve their own materials and encourage students to publish and contribute to learning resources.

Electing digital learning resources such as computer software, interactive media, or online resources, that use technology meritoriously can extant a different set of defies for educators. The main benefit of digital resources is their capability to customize knowledges for students through interactivity, feedback, and constructive engagement. Presentations can be varied to meet the necessities of the learner through diverse sequencing, unconventional material choices, and varied instigation. The practicality of the digital resources such as content, context, and larger learning environment is an enhancement to the learning process.

Digital resources can also combine a variety of media into one involvement through the amalgamation of text, video, sound, and graphics. Where a digital resource affords a mix of media, the assessment benchmarks for the specific media should be painstaking. Digital resources must meet the examinations of usability and functionality. A further consideration and discussion about the digital resources usefulness and design will be in Section 3.

9.3 Quality Ideologies for Digital Learning Resources

The term "learning resources" is deliberately chosen to extricate the artifacts we will study from traditional textbooks. Digital learning resources are different from traditional physical textbook in many ways. One obvious variance is that digital learning resources can be multi-modal, which means that the communication can be made both visually and auditory. On the other hand, and similar to other learning resources, digital learning resources are used in a variety of contexts, by different teachers, and for various purposes. This involves that a given resource can have an educational quality in one particular setting, though not necessarily in another setting. It might be more appropriate to define and clarify exactly what it is that should be assessed, rather than to formulate quality requirements. The concept of digital learning is a kind of jelly and flexible that takes many shapes and fits specific mission for specific time as creating them takes diversity of formats and techniques. For example, visual presentations in digital format can be made not only as still pictures but also as short video sequences or animations. Another metamorphosis is that digital learning resources can be constructed as simulations, where the simulator represents a physical environment in which it is physically nontoxic and not pricey to commit errors.

Sometimes the learning resource can be made into an illustration of the subject matter, like a business. A digital learning resource is mutually an artifact and a semiotic tool with a superior potential than traditional textbooks.

One added dissimilarity is that most textbooks have been industrialized within the outline of the public university system with its specific mores and rules concerning what kinds of objectives students should reach. These quality principles narrate to the strategy of using digital learning resources to support effective learning and teaching. Many broad quality areas are related to all learning resources, nevertheless of whether or not they are digital impacts the learning progress. Elements such as structure, language, use of illustrations, and the scope of the learning assignments are important, autonomous of whether the resources are offered in a printed or digital form. Alike applies to being conscious of gender patterns, discrimination, objectivity, and representativeness. As much as possible, this document will steer clear of general quality criteria for learning resources in favor of the elements that pertain specifically to digital learning resources.

   Quality assurance is more with evaluation procedures and the evaluation consequence of using digital learning resources can be organized in three comprehensive categories:

- User dimension: The interface between the user and the resource.
- The distinctiveness of the digital resource: The possibilities and limitations of the digital resource.
- Subject and education dimension: The educational and evaluation potential.

This section is envisioned as a guide rather than quantifying tool for quality, and it is anticipated that few digital learning resources will themselves incorporate and get integrated with all of the quality philosophies. Whether evaluation is about designing or selecting a resource, the intended educational purpose of the resource will determine which principles are most important for the learning process; the principles are divided into two groups, which are consistent:

1. **Core pedagogic principles:** Which reinforce effective learning and teaching, drawing from learning theory and commonly accepted best practice such as: Digital learning resource design, robustness and support, human–computer interaction, quality of assets, interoperability, accessibility, testing, and verification.
2. **Core design principles:** Covering concerns such as resource design, convenience, and interoperability such as: learner engagement, inclusion and access, effective learning, assessment to support learning, robust summative assessment, innovative approaches, match to the curriculum, and ease of use.

These concepts and principles will be taken in detail; the steps are suggested and properly stated for a quality assurance process to ensure the quality of digital learning resource. But first an introduction of pedagogical framework can help to assemble thoughts.

9.3.1 Pedagogy in Quality Framework

Each higher education institution is highly recommended to adopt a pedagogical framework that is collaboratively developed with the university's expertise and community to ensure "high-quality, evidence-based teaching practices focused on success for every student." Such pedagogical framework actually being used for many quality purposes and one to mention here is presenting a learning environment as an active example for a learning resource where students can acquire and get the knowledge. This requirement acknowledges the impact of quality teaching and the evidence that research validated pedagogy – implemented with consistency across a university setting and supported by instructional leadership – improves students' performance, and develops successful learners.

A. Expectations from this learning environment and resource

Such learning environment and resource implement a research-validated pedagogical framework that:

1. Describes the university morals and philosophies about teaching and learning that respond to the local context and the levels of students' attainment.
2. Sketches procedures for professional learning and instructional leadership to support consistent whole-school pedagogical practices, to monitor and increase the unrelenting impact of those practices on every student's accomplishment.
3. Details procedures, practices, and strategies for teaching, differentiating, monitoring, assessing, and moderating that reflect values and support student improvement.
4. Reflects the following core systemic principles.

B. Core systemic principles

Core systemic principles touch the real value of learning environment and resource as it is about students and surrounding in specific. Student-centered preparation and learning plans which take into consideration techniques:

- Decisions based on knowledge of the students and their prior learning and characteristics.
- Variety of agreed data used to modify learning lanes and target resources.
- Recurrent monitoring and diagnostic assessment to inform differentiation. High expectations such as comprehensive and challenging learning goals for each student based on agreed data sets, deep learning through higher order thinking and authentic contexts, and agreed procedures for ongoing induction, coaching, mentoring, and support in teaching and learning for all staff.

C. Alignment of curriculum, pedagogy, and assessment

Pedagogy aligned with curriculum intent and anxieties of the learning area/subject, general aptitudes, and cross-curriculum priorities. This also includes assessment, with categorical criteria and standards, strategic up front, and aligned with teaching. Course design and delivery, comprising and monitoring data collection practices, are consistent across the university and learning area. Last, moderation practices are to support consistency of teachers' judgment about assessment data.

D. Evidence-based decision making

Teaching and learning processes are to some limits formed by student performance data and validated research deploying the learning environments and resources count on quality evidence of the sustained impact of the agreed teaching methods which is being used to enlighten teaching and learning process. Feedback practices for staff and students can also be employed for further use as a learning resource for others.

9.3.2 Learning Principles to Guide Pedagogy and the Design of Learning Environments and Resources

1. Contiguity Effects: Ideas that need to be associated should be presented contiguously in space and time in the multimedia learning environment and resource. For example, the verbal label for a picture requirement to be placed spatially adjacent the picture on the display, not on the other side of the screen.
2. Perceptual-motor Grounding: Whenever a concept is first designed and introduced as a learning material or resource, it is important to ground it in a concrete perceptual-motor experience. The learner will ideally

visualize a picture of the concept, will be able to manipulate its parts and aspects, and will observe how it functions over time. The teacher and the learner will also gain a common ground (shared knowledge) of the learning material.

3. Dual Code and Multimedia Effects: Digital environments and learning resources with information are expected to be encoded and remembered better when they are delivered in multiple modes (verbal and pictorial), sensory modalities (auditory and visual), or media (computers and lectures) than when delivered in only a single mode, modality, or medium.

4. Testing Effect: There are direct and indirect effects of taking frequent tests. One indirect profit is that regular testing keeps students repetitively engaged in the material. Although students will learn from testing without receiving feedback, there is less forgetting if students receive informative feedback about their performance.

5. Spaced Effects: Spaced timetables of testing produce better long-term retention than a single test.

6. Generation Effect: Learning environments and resources are enhanced when this resource can cope with learners and produce retorts compared to having them recognize this learning material as giving examples according to the generation they live.

7. Organization Effects: Outlining, integrating, and synthesizing information produce a better learning material than revising materials or other more passive strategies.

8. Coherence Effect: The learner needs to get a coherent, well-connected representation of the main ideas to be learned. It is important to remove distracting, irrelevant material from the learning environment and resources.

9. Manageable Cognitive Load: Multimedia learning environments and resources should be compatible with what we know about how people learn. A common error in the design of multimedia learning materials is to "clutter" the learning environment.

10. Segmentation Principle: Information presented in text is necessarily linear because of the constraints of language. When multimedia materials are designed, it is possible to present information simultaneously in multiple modes – auditory, motor, and visual, being the most common.

11. Implications: Learning environments and resources along with teachers should provide challenges that put the learner in cognitive

disequilibrium if the learning objective is to promote deep learning of the material.

12. Discovery learning facility while using the learning resource. Most students have trouble in discovering the important principles on their own, without careful guidance, scaffolding, or materials with well-crafted affordances.

13. Anchored Learning: Designing the learning material with a facility of anchored learning occurs when students work in teams for several hours or days trying to solve a challenging practical problem that matters to the student. The activity is connected to background acquaintance of the learner on a theme that is interesting. Such facility saves time and effort.

9.3.3 Quality Procedures for Pedagogic and Design Principles

1. Core pedagogic principles

Step 1: Effective learning design: The experience of effective learning promotes effective cognitive and behavioral development or change.

Step 2: Assessment to support learning: In order to support learning, teaching and learning should integrate a formative assessment of what has, or has not, been learnt or understood. This comprises providing feedback to the learners on their acquisition of knowledge and skills.

Step3: Robust summative assessment: Summative assessment should be used to provide materials on learner recital that can be used for guidance or selection in relation to future education or work opportunities.

Step 4: Innovative approaches: Digital learning resources may be innovative in their design and use of technology and/or innovative in the approach to teaching and learning that they offer.

Step 5: Ease of use: As well as being clear in their intention, digital learning resources should be as transparent as possible to the user.

Step 6: Match to the curriculum: "Curriculum" refers to any program of learning activity planned by practitioners and/or learners.

2. Core design principles

Step 1: Digital learning resource design: The application of this principle to product design closely reflects many of the elements found in the pedagogic principles.

Step 2: Robustness and support: Digital learning resources should support the user appropriately by having help functions that identify common user problems and their solutions.

Step 3: Human–computer interaction: Digital learning resources should facilitate sound human–computer interaction.

Step 4: Quality of assets: Digital learning resources should ensure that assets are suitable for the context of use.

Step 5: Accessibility: Accessible design of digital learning resources concerns ensuring that no user, practitioner, or learner is unreasonably prevented from benefiting from a resource simply because of their access requirements or preferences.

Step 6: Interoperability: The principle of interoperability has many potential educational benefits for learners and these apply across many aspects of education, for example, the use of learning platforms and e-portfolios, and the transfer of learner data across institutions.

Step 7: Testing and verification: A well-planned development process with effective reviewing and feedback.

Step 8: Effective communication: Effective communication is an underlying principle that is dependent on the implementation of the relevant core pedagogic and design principles. However, it is an important principle that should be considered for all digital learning resources.

The overall objective must be to support the learning consequences of the curriculum. The deliberation of learning resources to help teaching to work further on curriculum must be applied rigorously to all mediums of presentation. These are guidelines that may be useful to department and board/district personnel, administrators, teacher–librarians, teachers, students, and members of the community. These methods and quality assurance suggestions are a collaborative work of scattered researches and recommendations from different studies and centers to help in the selection/evaluation process and can be utilized, adapted, and/or modified to meet specific needs.

9.4 Semantic Connection of Learning Objects

A basic concept in digital learning resources in education is "learning object." A learning object can be created and as a single object and stance on its own with abilities and setting to be interpreted, assimilated into a customized concept to gain meaning by the frame of understanding of that concept for specific purpose. This learning object is designed and can be re-used in different frameworks and concepts holding knowledge to act as a knowledge object by Rob Koper, as a content or information object by Dan Rehak and Robin Mason, and as an asset by Charles Duncan (Reusing online resources, Littlejohn, 2003, p. 10). Object learning is the structuring of framework into chunks of information. Which actually has challenges as how to create, select, and mark up in an appropriate way in terms of size of pedagogical value were those were attempts to address on vicarious learning through dialog (Mayes et al., 2001).

The learning object economy has become possible as we now see:

- Economy: Customers are unwilling to get large monolithic courses while the desired is only a part, or use this part in a different concept; this has become a reality by using and re-using this learning object.
- Expertise: In our days, with technology available for learners and educators, educational experience to deal with such a concept is an easy to get and practiced as well with lots of available demos.
- Technology: Technology in our days is amazingly hiding all ugly details that are irrelative to the learners and educators and may cost training or money. Compatibilities and convenience of Internet connection and applications made the whole world as a plate of knowledge with different learning environments. Clouding and Internet of Things made learning easier to spread and manage despite of the fact that the learning environment has been used.

What we have mentioned above, is most likely to be the first step in the development of a pedagogically-based learning object in addition to the open knowledge initiative, a collaboration between MIT and a number of other institutions, aimed to the development of a modular open-source learning management system (LMS), designed through a rigorous methodology of consultation with university teachers and a detailed analysis of the tasks involved in a campus-based learning, teaching and assessment (Littlejohn).

A. Metaphors galore of learning objects

Metaphor concept is being used to explain and simplify the notion of breaking down a digital learning resource into granular objects holding information and re-assemble them into different learning resource designs for different purposes which include different setting and customization.

When we say breaking down and re-assembling a digital learning resource, we actually refer to the process of disaggregation and aggregation which is a common in a number of fields specially concepts of data structuring and databases. Alsion Littleman has explained the idea in a very simple and interesting example when he stated: From the world of publishing, we can consider the granularity of a book. When visiting a bookshop, the customer is not forced to buy the entire catalog from a published rather the single, selected book. However, it might even better if it were possible to remove some chapters or even pages from the book and replace them with others from a different book. This is a disaggregation that is common among teachers. And some aggregate parts of others and produce new books as well. Another useful way of thinking and understanding of aggregation process is when we talk about Lego bricks which come in different shapes and different sizes but a customization suit that helps to widen the imagination and try to assemble these bricks into a specific form to shape a specific model according to some illustrations on the package showing how far your imagination can be and what can be made with these contents. Yet, user can go with his/her imagination further and comes with illustrations that can be different from what have been specified. The bricks are sufficiently re-usable that the constructions are only limited to the availability of suitable bricks.

Illustrations can be constructed and the planning of these bricks can work simply with well-defined interface. No special skills are required by the user to attain to be able to attach bricks to each other and the way of dispatching is also easy most of the times without affecting the whole illustrations.

B. Levels of granularity

Learning objects' granularity is easy to be defined with the above-presented metaphors, but yet as quality itself with no agreement or describing of this definition. The diversity of definition may be as a result of the aggregation and disaggregation concepts as they have various levels of procedures and actions. The following agreements can be taken into consideration:

1. Educational terms like a course, a module, or a unit.
2. Purpose terms such as asset, re-usable, and learning objects. These terms usually need metadata to describe the content.
3. Size terms: Number of pages, for example, or duration to complete the task. This does not include the comprehensive way to understand how they operate.

The levels of aggregation defined by purpose are the most useful for those who have to perform the aggregation and disaggregation. Three levels of purpose were stated by (Littlejohn,):

1. Raw media assets.
2. Collections of assets that include structure but not educational context.
3. Collections include the educational context and support for educational activities.

Several alternatives approaches to defining granularity are given in Wiley (2000).

C. Aggregation and disaggregation of learning objects

Any learning object such as a file, an image, or a media file, for example, can be used and re-used in many different settings that can be educational setting or commercial settings. The attached metadata that relate extra information about this learning object do not imply the purpose of being used in this setting. The purpose can handle a multitude status. This is why setting and metadata are yet not enough and the aggregation process needs a classification technique to help defining the purpose. This means that one object can be classified into many different contexts. Objects with many different classifications are inherently ripe for customization and re-use being more fundamental building block exactly as a logical gate object might be classified as a computer science/computer architecture but also as electronic engineering/digital principles (Littlejohn). At the level of aggregation, the fundamental building blocks with a classification component are ready to be assembled in this structure and represent a customized learning object in an information object (aggregation structure). The assembly comes with some context yet without a specific according to the classification component that can be assigned later and associated to this learning object. And this classification indicates where this aggregation structure can serve, for example, a video of a previous president is attached to a description and associated to history of events happened during his government time. At this stage, this learning object can fit anywhere and has no specific purpose to be for example educational. But when this assembly is integrated into a course of history, it is classified educational. This is we want to deploy the learning object and classify it to serve specific objective. But we can do both, design an objective and select the best learning objects to fit it and also, design the learning object and import the objective this learning object help to achieve. One learning object as described above can also serve in more than one educational model. It will be convenient to aggregate it, for example, in business course if the industry at that time president was refreshed, for example, or can be aggregated to part of e-library concept helping teachers in their research and training processes. No matter how large this learning object can be as massification of this learning object can lead for a more effective learning purpose. The learning object will be still under modification and alteration to serve different determinations. The highest level of aggregation when learning object needs more defining pedagogical strategies and the services to support them. Which call, emerging standards to describe pedagogy and educational support are often referred to as learning design or educational modeling (IMS-LD, 2002; KOPER, 2001).

Disaggregation concept is no more complex that what we have explained above about aggregation process. Disaggregate a learning object in order to re-build and then incorporate it with other models to serve different purposes and so on.

The discussion above concentrates on the economic and practical advantages of re-use and learning through granular learning objects. Learning objects have added a lot for constructing learning environments and learning resources with the ability of re-using and replacing the defected/undesirable objects. What we will propose in the next is a new way of thinking for this learning object aggregation and disaggregation processes.

D. Semantic connection of learning object

Semantic is the linguistic and philosophical study of meaning, in language, programming languages formal logics and semiotics. It is considered with the relationship between signifiers like words, phrases, signs and symbols and what they stand for, their denotation. It is often used in ordinary language for denoting a problem of understanding that comes down to word selection or connotation. This problem of understanding has been the subject of many formal enquiries.

let's suppose that these learning object can be associated with metadata and a classifying component that be assembled in an aggregation structure to fit into specific purpose. If we can harmonize a semantic connection between these learning objects that happen also to be connecting with a previous setting status of this current learning object to build a new knowledge and history of this learning object that work perfectly in which setting and other learning objects to fit for different purposes. This semantic taxonomy helps imagining a sematic-based network of these learning objects to connect them virtually to help understand the model and the way it can be developed by using the semantic of each learning object and map it with its semantic connections with other learning objects available and used for similar/different purposes. The semantic connection allows these learning objects to act smart in a smart learning environment which also them to communicate and lead the user for better results by connecting them to another learning objects in function but yet can be re-used to complete the aggregation assemble. The question is how this helping to create a useful learning environment and learning resources?

A connection is a semantic connection when it describes the relationship between two entities in a smart environment and focuses on the semantics – or meaning – of the connection between these entities. The semantic connection of these learning objects enables users to manage the connections and information exchanged between these learning objects. The information can be metadata and classification components that were previously associated to these learning objects before being used or assembled. The interaction between these smart learning resources can come with a new pedagogy semantic interoperability platform. Such an interaction can increase the knowledge transported and increase the awareness of the learning this information learning objects hold. Interoperation at a semantic level gives the design another dimension of information and knowledge aiming to help users to some level, users conceptualize the connections and the information these learning objects carry.

References and Suggested Literature

[1] Available at: http://publishers.org/our-markets/prek-12-learning/quality-content-learning-resources

[2] Available at: http://publishers.org/our-markets/prek-12-learning/what-are-learning-resources

[3] Available at: https://www.slideshare.net/MarkFrancisAstom/learning-resource-center

[4] Available at: https://www.slideshare.net/sarishtigarg/learning-resources-46775217

[5] Available at: http://classroom.synonym.com/importance-learning-materials-teaching-6628852.html

[6] Available at: https://www.slideshare.net/sarishtigarg/learning-resources-46775217

[7] Davis, E. A., Krajcik, J. S., Designing Educative Curriculum Materials to Promote Teacher Learning, *ERIC*, 3–14, 2005.

[8] Leacock, T. L., and Nesbit, J. C., A Framework for Evaluating the Quality of Multimedia Learning Resources. *J. Educ. Technol. Soc.* 10, 44–59, 2007.

[9] Anido, L. E., Fernández, M. J., Caeiro, M., Santos, J. M., Rodríguez, J. S., and Llamas, M. (2002). Educational metadata and brokerage for learning resources. *Comput. Educ.* 38, 351–374, 2002.

[10] Koutsomitropoulos, D. A., Alexopoulos, A. D., Solomou, G. D., Papatheodorou, T. S., High Performance Information Systems Laboratory, School of Engineering, Dept. of Computer Engineering and Informatics, 2010.

[11] Vaira, M, Quality assessment in higher education: an overview of institutionalization, practices, problems and conflicts, University of Pavia: Portland Press Ltd., Pavia, 2007.

[12] Available at: https://en.wikipedia.org/wiki/Semantics

[13] van der Vlist, B. J. J., *Semantic Connections: Explorations, Theory and a Framework for Design*, Unpublished Doctoral dissertation, Eindhoven University of Technology, Eindhoven, Netherlands, 2014.

[14] Ramsden, P. (2003). *Learning to Teach in Higher Education.* Abingdon: Routledge. Available at: https://books.google.com.sa/books?hl=en&lr=&id=AsIK_4wAJz4C&oi=fnd&pg=PR1&dq=measurements+and+assessments+in+higher+education&ots=XLTIUJL8Fr&sig=E1t8lY_sN2wlIrrVWjIGhtlvCQs&redir_esc=y#v=onepage&q=measurements%20and%20assessments%20in%20higher%20education&f=false

[15] UNESCO and Commonwealth of Learning, *Guidelines for Open Educational Resources (OER) in Higher Education*, 2015, 2011. Available at: http://unesdoc.unesco.org/images/0021/002136/213605e.pdf

[16] *Learning Resources and Open Access in Higher Education Institutions in Ireland*, Focused Research Report No. 1 2015. Available at: https://www.teachingandlearning.ie/wp-content/uploads/2015/07/Project-1-LearningResourcesandOpenAccess-1607.pdf

[17] *Responsibilities for Quality Assurance in Teaching and Learning*, available at: http://about.unimelb.edu.au/__data/assets/pdf_file/0004/861223/responsibilities.pdf

[18] *Standards and Guidelines for Quality Assurance in the European Higher Education Area, Helsinki, Finland*, 2009, available at: http://www.enqa.eu/wp-content/uploads/2013/06/ESG_3edition-2.pdf

[19] *Textbooks and Learning Resources: A Framework for Policy Development*, UNESCO 2014, available at: http://unesdoc.unesco.org/images/0023/002322/232222e.pdf

[20] *Quality Principles for Digital Learning Resources*, available at: http://39 lu337z5l11zjr1i1ntpio4.wpengine.netdna-cdn.com/wp-content/uploads/ 2015/05/quality_principles.pdf

[21] Harrison, F., *Using Learning Resources to Enhance Teaching and Learning*, available at: http://www.faculty.londondeanery.ac.uk/e-learning/ small-group-teaching/Using_learning_resources_to_enhance_teaching_-_ learning.pdf

[22] *Digital Learning Resources as Systemic Innovation*, 2007, available at: https://www.oecd.org/edu/ceri/38777910.pdf

[23] *Quality Criteria for Digital Learning Resources*, available at: https://ik tsenteret.no/sites/iktsenteret.no/files/attachments/quality_criteria_dlr.pdf

[24] *Quality Principles: For Digital Learning Resources*, available at: http:// 39lu337z5l11zjr1i1ntpio4.wpengine.netdna-cdn.com/wp-content/uploa ds/2015/05/quality_principles.pdf

[25] *Evaluation and Selection of Learning Resources: A Guide*, available at: http://www.gov.pe.ca/photos/original/ed_ESLR_08.pdf

[26] *Learning Principles to Guide Pedagogy and the Design of Learning Environments, Applying the Science of Learning*, available at: http://act ivelearningps.com/wp-content/uploads/2014/07/25-learning-principles-to-guide-pedagogy.pdf

[27] Available at: http://education.qld.gov.au/curriculum/pdfs/pedagogical-framework.pdf

[28] Mayes, J. (2002). Pedagogy, lifelong learning and ICT. Electronic Journal of Instructional Science and Technology, 5(1) Retrieved Feb. 15, 2003 from the WWW at http://www.ipm.ucl.ac.be/ChaireIBM/ Mayes.pdf

[29] Littlejohn, A. and Buckingham Shum, S. (2003). (Eds.) Reusing Online Resources (Special Issue) Journal of Interactive Media in Education, 2003 (1) [www-jime.open.ac.uk/2003/1/], Available online: http://citeseerx.ist.psu.edu/viewdoc/download?doi=10.1.1.593.3121& rep=rep1&type=pdf

10

Crystalize the Bond between University Ranking Systems and University Quality Assurance Systems

10.1 Introduction

Overviewing an educational scenario can be basic and simple supposing many curious learners preparing to alumnus from high school, speculating whether or not they should chase a dream of higher education certificates at any university or any other learning institute is an issue that contemplates heavily on their minds and their parents as well. It is a very difficult verdict that is influenced by voluminous diverse factors, such the social and global awareness of how significant higher education is. Other factors that affect the decision should also be enlightened to understand how the higher education is built and how it is evaluated to grab the thoughtfulness and obsession of students to consider it for their long-life term of learning. Therefore, the front-yard higher education institutions (HEIs) such as colleges and universities are essentially under cataloging systems to classify and categorize the quality on different aspects they present. HE institutions are ordered by various combinations of several factors such as ranking system. The combinations of factors we are conferring might be measures of funding and endowment, research excellence and/or influence, specialization expertise, admissions, student selections, award numbers, internationalization, graduate employment, industrial linkage, historical reputation, and many other criteria that build up a rank correlation to compare two levels for the same set of objects. For example, Spearman's rank correlation coefficient is useful to measure the statistical dependence between the rankings of athletes in two tournaments. Another example is the "Rank–rank hypergeometric overlap" approach, which is designed to compare ranking of the genes that are at the "top" of two ordered lists of differentially expressed genes. Many other examples of ranking can be mentioned here as:

- In politics, rankings focus on the comparison of economic, social, environmental, and governance performance of countries; see the list of international rankings.
- In many sports, individuals or teams are given rankings, generally by the sports' governing body.
- Search engines rank web pages by their expected relevance to a user's query using a combination of query-dependent and query-independent methods. Query-independent methods attempt to measure the estimated importance of a page, independent of any consideration of how well it matches the specific query. Query-independent ranking is usually based on link analysis; examples include the HITS algorithm, Page-Rank, and Trust-Rank. Query-dependent methods attempt to measure the degree to which a page matches a specific query, independent of the importance of the page. Query-dependent ranking is usually based on heuristics that consider the number and locations of matches of the various query words on the page itself, in the URL or in any anchor text referring to the page.
- In Webometrics, it is possible to rank institutions according to their presence in the web (number of webpages) and the impact of these contents (external in-links=site citations), such as the Webometrics Ranking of World Universities.
- In video gaming, players may be given a ranking. To "rank up" is to achieve a higher ranking relative to other players, especially with strategies that do not depend on the player's skill.
- The TrueSkill ranking system is a skill-based ranking system for Xbox Live developed at Microsoft Research.
- A bibliogram ranks common noun phrases in a piece of text.
- In language, the status of an item (usually through what is known as "down-ranking" or "rank-shifting") is in relation to the uppermost rank in a clause; for example, in the sentence "I want to eat the cake you made today," "eat" is on the uppermost rank, but "made" is down-ranked as part of the nominal group "the cake you made today"; this nominal group behaves as though it were a single noun (i.e., I want to eat *it*), and thus the verb within it ("made") is ranked differently from "eat."
- Academic journals are sometimes ranked according to the impact factor, the number of later articles that cite articles in a given journal.

Rankings have most often been conducted by magazines, newspapers, websites, governments, or academics. In addition to ranking entire institutions, organizations perform rankings of specific programs, departments, and schools. Rankings have most often been steered by magazines, newspapers, websites, governments, or academics. In addition to ranking entire institutions, organizations perform rankings of precise programs, departments, and schools. Various rankings consider combinations of processes of funding and endowment, research excellence and/or influence, specialization expertise, admissions, student options, award numbers, internationalization, graduate employment, industrial linkage, historical reputation, and other criteria.

Global university rankings have been a phenomenon in the antiquity of higher education and a contentious indicator of quality in higher education since it first appeared in 2003. Many do not agree that rankings are quantifying or demonstrating quality of higher education in a fair and comprehensive manner. The effortlessness of rankings and the global publicity of the annual rankings outcomes have, however, served a general purpose of putting public and international attention on the role and prominence of higher education – to societies and to individuals. They have put higher education performance on the policy agenda, and fortified the necessity for continuous speculation in higher education. Table 10.1 shows the types and structures of rankings.

The extant chapter seeks to present the ranking systems in a simple way that can be easily understood. The following sections will also compromise and specify how these rankings' results impact the academic institutions and society. The national rankings ARWU and QS will also be presented as well as crystalizing the relationship with quality assurance concepts. The first step is to define and understand ranking itself and specify what is and how ranking correlation works.

Table 10.1 Types of rankings

Types of Rankings	Structure of Rankings
• Unified rankings: Disparate sets of weighted indicators are combined into a single score that reflects the overall quality of a given institution.	• Numerical ranking: Universities are classified with numbers: 1, 2, 3, 4…
• Discipline-based rankings: Institutions are ranked according to the specific programs, specializations, or subjects that are offered.	• Clustering or grouped ranking: Universities are ranked in tires-top, middles, bottom, etc.
• Other: Rankings that are not easily characterized.	• Top-level ranking: Universities are ranked numerically but report only a fixed number at the top.

10.2 Ranking Systems

10.2.1 Definition

A ranking as a word is a rapport concerning a set of items such that, for any two items, the first is either "ranked higher than," "ranked lower than," or "ranked equal to" the second. In mathematics, this is known as a weak order or total preorder of objects. It is not certainly a total order of objects because two reformed objects can have the same position. The rankings themselves are absolutely ordered. For example, constituents are totally preordered by hardness, while degrees of hardness are totally ordered. If two items are equivalent in rank, it is considered a tie. For example, if player 1 has x amount of points and players 2 and 3 of the same x amount of points, then player 1 would be in first and players 2 and 3 would be considered a tie for second. It contemplates a "to be" determined rank, even though three players are involved. By reducing detailed measures to a sequence of ordinal numbers, rankings make it possible to evaluate complex information according to certain criteria. Consequently, for example, an Internet search engine may rank the pages it finds according to an estimation of their significance, making it possible for the manipulator quickly to first-rate the pages they are likely to want to perceive.

Scrutiny of data acquired by ranking universally necessitates non-parametric statistics. In statistics, "ranking" refers to the data transformation in which numerical or ordinal values are replaced by their rank when the data are sorted. For example, the numerical data 3.4, 5.1, 2.6, and 7.3 are observed, and the ranks of these data items would be 2, 3, 1, and 4, respectively. For example, the ordinal data hot, cold, and warm would be replaced by 3, 1, and 2. In these examples, the ranks are allotted to values in ascending order. (In some other cases, descending ranks are used.) Ranks are correlated to the indexed incline of order statistics, which entails of the original data set repositioned into ascending order.

Some kinds of statistical experiments employ shrewdness centered on ranks. Examples include:

- Friedman test
- Kruskal–Wallis test
- Rank products
- Spearman's rank correlation coefficient
- Wilcoxon rank-sum test
- Wilcoxon signed-rank test.

The dissemination of values in declining order of rank is frequently of interest when values fluctuate widely in scale; this is the rank-size distribution (or rank-frequency distribution), for example, for city sizes or expression frequencies. These regularly follow a power law. Roughly, ranks can have non-integer values for tied data values. For example, when there is an even number of reproductions of the same data value, the above-defined fractional statistical rank of the tied data trimmings is 1/2.

A rank correlation is any of several statistics that quantity an ordinal association — the relationship between rankings of different ordinal variables or different rankings of the same variable, where a "ranking" is the assignment of the ordering labels "first," "second," "third," etc., to different observations of a particular variable. A rank correlation coefficient measures the notch of similarity flanked by two rankings, and can be used to evaluate the significance of the relation between them. For example, two collective non-parametric approaches of significance that practice rank correlation are the Mann–Whitney U test and the Wilcoxon signed-rank test.

If, for example, one variable is the identity of a college basketball program and another variable is the identity of a college football program, one could check for a relationship between the poll rankings of the two types of programs: Do colleges with a higher ranked basketball program tend to have a higher ranked football program? A rank correlation coefficient can measure that relationship, and the degree of significance of the rank correlation coefficient can display whether the measured relationship is slight enough to likely be a coincidence. If there is only one variable, the identity of a college football program, but it is subject to two different poll rankings (say, one by coaches and one by sportswriters), and then the similarity of the two different polls' rankings can be measured with a rank correlation coefficient.

As another example, in a contingency table with low income, medium income, and high income in the row variable and educational level — no high school, high school, university — in the column variable), a rank correlation measures the relationship between income and educational level. Some of the more popular rank correlation statistics include

1. Spearman's ρ
2. Kendall's τ
3. Goodman and Kruskal's γ
4. Somers' D.

A cumulative rank correlation coefficient infers increasing agreement between rankings. The coefficient is inside the interval $[-1, 1]$ and assumes the value:

- 1 if the agreement between the two rankings is perfect; the two rankings are the same.
- 0 if the rankings are completely independent.
- −1 if the disagreement between the two rankings is perfect; one ranking is the reverse of the other.

Following Diaconis (1988), a ranking can be seen as a permutation of a set of objects. Thus, we can look at observed rankings as data obtained when the sample space is (identified with) a symmetric group. We can then introduce a metric, making the symmetric group into a metric space. Dissimilar metrics will resemble to different rank correlations.

10.2.2 General Correlation Coefficient

A. Kendall's τ

Kendall (1944) showed that his τ (tau) and Spearman's ρ (rho) are particular cases of a general correlation coefficient. Suppose we have a set of n objects, which are being considered in relation to two properties, represented by x and y, forming the sets of values $\{x_i\}_{i \leq n}$ and $\{y_i\}_{i \leq n}$. To any pair of individuals, say the i-th and the j-th we assign a x-score, denoted by a_{ij}, and a y-score, denoted by b_{ij}. (Note, as these are comparisons, a_{ij} and b_{ij} do not exist for $i = j$.) The only requirement for these functions is that they be anti-symmetric, so $a_{ij} = -a_{ji}$ and $b_{ij} = -b_{ji}$. Then the generalized correlation coefficient Γ is defined as $\Gamma = \dfrac{\sum_{i,j=1}^{n} a_{ij} b_{ij}}{\sqrt{\sum_{i,j=1}^{n} a_{ij}^2 \sum_{i,j=1}^{n} b_{ij}^2}}$ Kendall's τ (tau) as a particular case

If r_i, s_i are the ranks of the i-member according to the x-quality and y-quality, respectively, then we can define

$$a_{ij} = \mathrm{sgn}(r_j - r_i), \quad b_{ij} = \mathrm{sgn}(s_j - s_i).$$

The sum $\sum a_{ij} b_{ij}$ is twice the number of concordant pairs minus the number of discordant pairs (see the Kendall tau rank correlation coefficient). The sum $\sum a_{ij}^2$ is just $n(n-1)$, the number of terms a_{ij}, as is $\sum b_{ij}^2$. Thus, in this case,

$$\Gamma = \frac{2((\text{number of concordant pairs}) - (\text{number of discordant pairs}))}{\sqrt{n(n-1)n(n-1)}}$$

$$= \text{Kendall's } \tau$$

B. Spearman's rank correlation coefficient ρ

Spearman's rank correlation coefficient or Spearman's rho, named after Charles Spearman and often denoted by the Greek letter ρ (rho) or as r_s, is a nonparametric measure of rank correlation (statistical dependence between the ranking of two variables). It assesses how well the relationship between two variables can be described using a monotonic function. The Spearman correlation between two variables is equal to the Pearson correlation between the rank values of those two variables; while Pearson's correlation assesses linear relationships, Spearman's correlation assesses monotonic relationships (whether linear or not). If there are no repeated data values, a perfect Spearman correlation of $+1$ or -1 occurs when each of the variables is a perfect monotone function of the other. Intuitively, the Spearman correlation between two variables will be high when observations have a similar (or identical for a correlation of 1) rank (i.e., relative position label of the observations within the variable: 1st, 2nd, 3rd, etc.) between the two variables, and low when observations have a dissimilar (or fully opposed for a correlation of -1) rank between the two variables. Spearman's coefficient is appropriate for both continuous and discrete ordinal variables [1, 2]. Both Spearman's ρ and Kendall's τ can be formulated as special cases of a more general correlation coefficient.

The Spearman correlation coefficient is defined as the Pearson correlation coefficient between the ranked variables.

For a sample of size n, the n raw scores X_i, Y_i are converted to ranks $\text{rg}X_i, \text{rg}Y_i$, and r_s is computed from:

$$r_s = \rho_{\text{rg}_X, \text{rg}_Y} = \frac{\text{cov}(\text{rg}_X, \text{rg}_Y)}{\sigma_{\text{rg}_X} \sigma_{\text{rg}_Y}}$$

where

- ρ denotes the usual Pearson correlation coefficient, but applied to the rank variables.
- $\text{cov}(\text{rg}_X, \text{rg}_Y)$ is the covariance of the rank variables.
- σ_{rg_X} and σ_{rg_Y} are the standard deviations of the rank variables.

Only if all n ranks are *distinct integers*, it can be computed using the popular formula

$$r_s = 1 - \frac{6 \sum d_i^2}{n(n^2 - 1)}.$$

where

- $d_i = \text{rg}(X_i) - \text{rg}(Y_i)$ is the difference between the two ranks of each observation.
- n is the number of observations.

Identical values are usually [4] each assigned fractional ranks equal to the average of their positions in the ascending order of the values, which is equivalent to averaging over all possible permutations. If ties are present in the data set, this equation yields incorrect results: Only if in both variables all ranks are distinct, then $\sigma_{\text{rg}_X}\sigma_{\text{rg}_Y} = \text{Var rg}_X = \text{Var rg}_Y = n(n^2 - 1)/6$ (cf. tetrahedral number T_{n-1}). The first equation — normalizing by the standard deviation – may even be used even when ranks are normalized to [0;1] ("relative ranks") because it is insensitive both to translation and linear scaling.

This method should not also be used in cases where the data set is truncated; that is, when the Spearman correlation coefficient is desired for the top X records (whether by pre-change rank or post-change rank, or both), the user should use the Pearson correlation coefficient formula given above. The standard error of the coefficient (σ) was determined by Pearson (1907) and Gosset (1920). It is

$$\sigma_{r_s} = \frac{0.6325}{\sqrt{n - 1}}$$

1. Correlated extents

There are several other numerical measures that quantify the extent of statistical dependence between pairs of observations. The most common of these is the Pearson product-moment correlation coefficient, which is a similar correlation method to Spearman's rank that measures the "linear" relationships between the raw numbers rather than between their ranks.

An alternative name for the Spearman rank correlation is the "grade correlation"; in this, the "rank" of an observation is replaced by the "grade." In continuous distributions, the grade of an observation is, by convention, always one half less than the rank, and hence the grade and rank correlations are the same in this case. More generally, the "grade" of an observation is proportional to an estimate of the fraction of a population less than a given value, with the half-observation adjustment at observed values. Thus, this corresponds to one possible treatment of tied ranks. While unusual, the term "grade correlation" is still in use.

A Spearman correlation of 1 results when the two variables being compared are monotonically related, even if their relationship is not linear. This means that all data-points with greater x-values than that of a given data-point will have greater y-values as well. In contrast, this does not give a perfect Pearson correlation as shown in Figure 10.1.

When the data are roughly elliptically distributed and there are no prominent outliers, the Spearman correlation and Pearson correlation give similar values as shown in Figure 10.2.

The Spearman correlation is less sensitive than the Pearson correlation to strong outliers that are in the tails of both samples. That is because Spearman's rho limits the outlier to the value of its rank as shown in Figure 10.3.

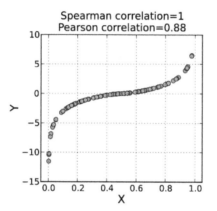

Figure 10.1   A Spearman correlation of 1 results.

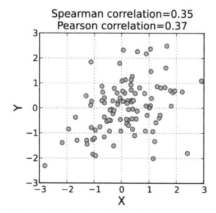

Figure 10.2   Elliptically distributed data.

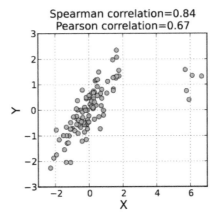

Figure 10.3　Correlation is less sensitive.

2. Interpretation

The sign of the Spearman correlation indicates the direction of association between X (the independent variable) and Y (the dependent variable). If Y tends to increase when X increases, the Spearman correlation coefficient is positive as shown in Figure 10.4, a positive Spearman correlation coefficient corresponds to an increasing monotonic trend between X and Y. If Y tends to decrease when X increases, the Spearman correlation coefficient is negative. A Spearman correlation of 0 indicates that there is no tendency for Y to either increase or decrease when X increases. The Spearman correlation increases in magnitude as X and Y become closer to being perfect monotone functions of each other. When X and Y are perfectly monotonically related, the Spearman correlation coefficient becomes 1. A perfect monotone increasing relationship implies that for any two pairs of data values X_i, Y_i and X_j, Y_j, that $X_i - X_j$ and $Y_i - Y_j$ always have the same sign. A perfect monotone decreasing relationship implies that these differences always have opposite signs. Figure 10.5 presents such relation, negative Spearman correlation coefficient corresponds to a decreasing monotonic trend between X and Y.

The Spearman correlation coefficient is often described as being "non-parametric." This can have two meanings: First, a perfect Spearman correlation results when X and Y are related by any *monotonic function*. Contrast this with the Pearson correlation, which only gives a perfect value when X and Y are related by a *linear* function. The other sense in which the Spearman correlation is non-parametric in that its exact sampling distribution can be obtained without requiring knowledge (knowing the parameters) of the joint probability distribution of X and Y.

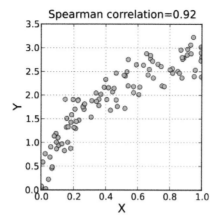

Figure 10.4 A positive Spearman correlation coefficient corresponds to an increasing monotonic trend between X and Y.

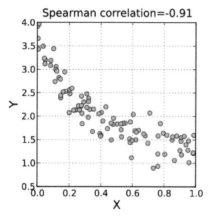

Figure 10.5 A negative Spearman correlation coefficient corresponds to a decreasing monotonic trend between X and Y.

The following example can help to understand more; let us suppose scenario of a necessity to rank two values and find the coefficient. The raw data in Table 10.2 are used to calculate the correlation between the IQS of a person with the sum of hours spent in front of TV per week.

First, evaluate d_i^2. To do so, use the following steps, reflected in Table 10.3.

1. Sort the data by the first column (X_i). Create a new column x_i and assign it the ranked values $1, 2, 3, \ldots, n$.

Table 10.2 IQ per TV watching

IQ, X_i	Hours of TV per Week, Y_i
86	0
112	6
106	7
113	12
110	17
97	20
100	27
99	28
103	29
101	50

Table 10.3 IQ per TV watching and correlation coefficients.

IQ, X_i	Hours of TV per Week, Y_i	Rank x_i	Rank y_i	d_i	d_i^2
86	0	1	1	0	0
97	20	2	6	−4	16
99	28	3	8	−5	25
100	27	4	7	−3	9
101	50	5	10	−5	25
103	29	6	9	−3	9
106	7	7	3	4	16
110	17	8	5	3	9
112	6	9	2	7	49
113	12	10	4	6	36

2. Next, sort the data by the second column (Y_i). Create a fourth column y_i and similarly assign it the ranked values $1, 2, 3, \ldots, n$.
3. Create a fifth column d_i to hold the differences between the two rank columns (x_i and y_i).
4. Create one final column d_i^2 to hold the value of column d_i squared.

With d_i^2 found, add them to find $\sum d_i^2 = 194$. The value of n is 10. These values can now be substituted back into the equation: $\rho = 1 - \frac{6 \sum d_i^2}{n(n^2-1)}$ to give $\rho = 1 - \frac{6 \times 194}{10(10^2 - 1)}$.

which evaluates to $\rho = -29/165 = -0.175757575\ldots$ with a P-value $= 0.627188$ (using the t-distribution). Check the following chart.

This low value shows that the correlation between IQ and hours spent watching TV is very low, although the negative value suggests that the longer the time spent watching television the lower the IQ. In the case of ties in the original values, this formula should not be used; instead, the Pearson correlation coefficient should be calculated on the ranks (where ties are given ranks, as described above).

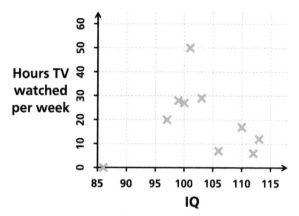

Figure 10.6 Chart of the data presented. It can be seen that there might be a negative correlation, but that the relationship does not appear definitive.

3. Defining significance

One approach to test whether an observed value of ρ is significantly different from 0 (r will always maintain $-1 \leq r \leq 1$) is to calculate the probability that it would be greater than or equal to the observed r, given the null hypothesis, by using a permutation test. An advantage of this approach is that it automatically takes into account the number of tied data values that are in the sample, and the way they are treated in computing the rank correlation. Another approach parallels the use of the Fisher transformation in the case of the Pearson product-moment correlation coefficient. That is, confidence intervals and hypothesis tests relating to the population value ρ can be carried out using the Fisher transformation:

$$F(r) = \frac{1}{2} \ln \frac{1+r}{1-r} = \operatorname{artanh}(r).$$

If $F(r)$ is the Fisher transformation of r, the sample Spearman rank correlation coefficient, and n is the sample size, then

$$z = \sqrt{\frac{n-3}{1.06}} F(r)$$

is a z-score for r which approximately follows a standard normal distribution under the null hypothesis of statistical independence ($\rho = 0$) [7, 8]. One can also test for significance using

$$t = r \sqrt{\frac{n-2}{1-r^2}}$$

which is distributed approximately as Student's t-distribution with $n - 2$ degrees of freedom under the null hypothesis. A justification for this result relies on a permutation argument.

A generalization of the Spearman coefficient is useful in the situation where there are three or more conditions, a number of subjects are all observed in each of them, and it is predicted that the observations will have a particular order. For example, a number of subjects might each be given three trials at the same task, and it is predicted that performance will improve from trial to trial. A test of the significance of the trend between conditions in this situation was developed by E. B. and is usually referred to as Page's trend test for ordered alternatives.

C. Spearman's ρ as a particular case

If r_i, s_i are the ranks of the i-member according to the x and the y-quality, respectively, we can simply define

$$a_{ij} = r_j - r_i$$
$$b_{ij} = s_j - s_i$$

The sums $\sum a_{ij}^2$ and $\sum b_{ij}^2$ are equal, since both r_i and s_i range from 1 to n. Then we have:

$$\Gamma = \frac{\sum (r_j - r_i)(s_j - s_i)}{\sum (r_j - r_i)^2}$$

Now

$$
\begin{aligned}
\sum_{i,j=1}^{n} (r_j - r_i)(s_j - s_i) &= \sum_{i=1}^{n}\sum_{j=1}^{n} r_i s_i + \sum_{i=1}^{n}\sum_{j=1}^{n} r_j s_j \\
&\quad - \sum_{i=1}^{n}\sum_{j=1}^{n} r_i s_j - \sum_{i=1}^{n}\sum_{j=1}^{n} r_j s_i \\
&= 2n \sum_{i=1}^{n} r_i s_i - 2 \sum_{i=1}^{n} r_i \sum_{j=1}^{n} s_j \\
&= 2n \sum_{i=1}^{n} r_i s_i - 2 \left(\frac{1}{2} n(n+1) \right)^2 \\
&= 2n \sum_{i=1}^{n} r_i s_i - \frac{1}{2} n^2 (n+1)^2
\end{aligned}
$$

We also have

$$S = \sum_{i=1}^{n}(r_i - s_i)^2 = 2 \sum r_i^2 - 2 \sum r_i s_i$$

and hence

$$\sum (r_j - r_i)(s_j - s_i) = 2n \sum r_i^2 - \frac{1}{2}n^2(n+1)^2 - nS$$

$\sum r_i^2$ being the sum of squares of the first naturals equals n. Thus, the last equation reduces to $\frac{1}{6}n(n+1)(2n+1)$

Further

$$\sum (r_j - r_i)^2 = 2n \sum r_i^2 - 2 \sum r_i r_j$$
$$= 2n \sum r_i^2 - 2 \left(\sum r_i \right)^2 = \frac{1}{6}n^2(n^2 - 1)$$

and thus, substituting into the original formula these results, we get

$$\Gamma_R = 1 - \frac{6 \sum d_i^2}{n^3 - n}$$

where $d_i = x_i - y_i$ is the difference between ranks, which is exactly the Spearman's rank correlation coefficient ρ.

1. The rank-biserial

Gene Glass (1965) noted that the rank-biserial can be derived from Spearman's ρ. "One can derive a coefficient defined on X, the dichotomous variable, and Y, the ranking variable, which estimates Spearman's rho between X and Y in the same way that biserial r estimates Pearson's r between two normal variables" (p. 91). The rank-biserial correlation had been introduced nine years before by Edward Cureton (1956) as a measure of rank correlation when the ranks are in two groups.

2. Kerby simple difference formula

Dave Kerby (2014) recommended the rank-biserial as the measure to introduce students to rank correlation, because the general logic can be explained at an introductory level. The rank-biserial is the correlation used with the Mann–Whitney U test, a method commonly covered in introductory college courses on statistics. The data for this test consist of two groups; and for each member of the groups, the outcome is ranked for the study as a whole. Kerby showed that this rank correlation can be expressed in terms of two concepts: the percent of data that support a stated hypothesis and the percent of data that do not support it. The Kerby simple difference formula states that the rank correlation can be expressed as the difference between the proportions of favorable evidence (f) minus the proportion of unfavorable evidence (u).

$$r = f - u$$

10.3 Global Academic Ranking Systems

Ranking systems sometimes are called Standings that are lists in forms of guides which compare any category in terms of achievements, abilities, or/and specific factors. Categories like business, sports, entertainments, nations, and institutions either government, academic, or any other meadow. The entities that play roles of factors can be individuals or crew and can be small companies or a fleet. A table or chart (such as a league table, a ladder, or a leaderboard) is designated to visualize and display this ranking or lists. A league table may list several related statistics, but they are generally sorted by the crucial one that determines the rankings. Many diligences and institutions may contend in league tables in order to help fetching new clienteles and patrons. Those tables ranking sports teams are generally used to help determine who may advance to the competitions or another game, who gets endorsed or demoted, or who gets a higher draft pick.

For over, 100 years various organizations have attempted to rank postsecondary institutions. In 1910, James Cattell from Columbia University offered rankings in American Men of Science that assessed the "scientific strength" of elite institutions by looking at the reputations of their science and social science faculty. Most early efforts applied the ranking to the college as a whole, rather than to individual departments. The rankings also tended to be based on what happened to the students after graduation instead of the accomplishments of the school's faculty. Cattell's work is an early exception.

E. Grady Bogue and Robert L. Saunders offered a brief history of graduate school rankings in 1992. They reported that the first graduate school study was conducted in 1925 by Raymond Hughes. He called on his fellow faculty members at Miami University in Ohio to draw up a list of quality universities and to identify national scholars in specific fields of study to serve as raters. Ultimately, in a Study of Graduate Schools of America, Hughes relied on 40 to 60 raters to assess 20 disciplines for graduate study at 36 universities. He followed up this ranking with another in 1934 for the American Council on Education. In this report, he assessed 50 disciplines and increased the number of raters to 100. Graduate programs were not ranked again until 1959, when Hayward Keniston conducted his assessment of them. The list of schools was surprisingly similar to the work done by Hughes in 1925. Two other well-known graduate school studies were done by Allan Cartter in 1966 and Kenneth D. Roose and Charles J. Anderson in 1970.

Since that time, there have been several other notable studies that assessed graduate education. One major schoolwork was conducted in 1982 for the Conference Board of Associated Research Councils. It was far more inclusive than the earlier efforts – covering 32 disciplines at 228 institutions. Then, in 1995, the National Research Council's Committee for the Study of Research-Doctoral Programs assessed 41 disciplines and 274 institutions using over 7500 raters. These 1995 rankings included both reputational ratings based on the opinions of faculty and objective data that focused on student–faculty ratios, the number of programs, and faculty publications and awards. In 1990, *U.S. News and World Report* began to offer their rankings of graduate and professional programs, focusing on business, law, medicine, and engineering.

In overall, the early rankings' exertions were not circulated widely. Most of these endeavors were viewed only by "academic administrators, federal agencies, state legislators, graduate student applicants, and higher education researchers" (Stuart, p. 16). The audience, however, grew substantially when *U.S. News and World Report* began publishing rankings of undergraduate institutions in 1983. By the late 1990s, *U.S. News and World Report, Time* partnering with the *Princeton Review, Newsweek* partnering with Kaplan Testing Service, and *Money* magazine were selling an estimated 6.7 million copies of their special rankings issues annually. As Patricia M. McDonough and her associates illustrated in 1998, rankings have developed immense trade. It should be eminent that there are all varieties of college rankings besides those that look at academic quality. For instance, *Money* magazine determines the "Best College Buys" and the *Princeton Review* names the top party schools.

This chapter is inscribed to help the reader to understand how University and collages are really ranked what ranking process is. For educational systems, various rankings mostly evaluate on institutional productivity by research. Roughly some rankings weigh institutions in the interior of a nation, while others assess institutions worldwide. The subject has twisted much debate about rankings' usefulness and precision. The escalating assortment in rating methodologies and accompanying censures of each indicates the deficiency of consensus in the arena. On the other hand, it seems possible to game the ranking systems through undue self-citations [2]. The variety of academic rankings afford an ample overview and perceptive superintend of different academic institutions on composite capabilities in academia. While the United Nations advocates for the beneficial role that higher education could be the common good of social leverage and educating services to equip everyone participated, yet college ranking is a crystal clear tool for a fair evaluation for the public. Ranking of universities and institutions has attracted wide attention recently, so they are more categorized as strategic network society that keens to cope with internationalization requirements in parallel with internal and regional requirements.

A number of systems have been in running taking the advantage of obsession of globalization and reputation and acquired that attempt to rank academic institutions worldwide. League tables are used to compare the academic achievements of different institutions. College and university rankings order institutions in higher education by combinations of factors. In addition to entire institutions, specific programs, departments, and schools are ranked. These rankings are usually conducted by magazines, newspapers, governments, and academics. For example, league tables of British universities are published annually by The Guardian, The Independent, The Sunday Times, and The Times. The primary aim of these rankings is to inform potential applicants about British universities based on a range of criteria. Similarly, in countries like India, league tables are being developed and a popular magazine, Education World, published them based on data from TheLearningPoint.net.

The two longest established and most influential global rankings are those produced by ShanghaiRanking Consultancy (the Academic Ranking of World Universities; ARWU) and Quacquarelli Symonds (QS). Along with other global rankings, predominantly quantity is the research recital of universities rather than their teaching. They have been criticized for being "largely based on what can be measured rather than what is necessarily relevant and important to the university," and the validity of the data available globally has been questioned. While some rankings attempt to measure teaching using metrics such as staff to student ratio, the Higher Education Policy Institute has pointed out that the metrics used are more meticulously related to research than teaching quality, e.g., "Staff to student ratios are an almost direct measure of research activity," and "The proportion of Ph.D. students is also to a large extent an indication of research activity." Inside Higher Ed similarly states that "these criteria do not actually measure teaching, and none even come close to assessing quality of impact." Many rankings are also deliberated to comprise biases toward the natural sciences and, due to the bibilometric sources used, toward publication in English-language journals. Some rankings, including ARWU, also fail to make any correction for the sizes of institutions, so a large institution is ranked considerably higher than a small institution with the same quality of research. Other compilers, such as Scimago and U.S. News and World Report, use a mix of size-dependent and size-independent metrics.

Some compilers notably QS uses reputational surveys. The validity of these has been criticized: "Most experts are highly critical of the reliability of simply asking a rather unransom clutch of educators and others involved with the academic enterprise for their opinions"; "methodologically (international inspections of reputation) they are flawed, effectively they measure only research performance, and they skew the results in favor of a small number of establishments." However, despite the criticism, much attention is paid to global rankings, predominantly ARWU and QS. Some countries, including Denmark and the Netherlands, use university rankings as part of points-based immigration programs, while others, such as Russia, automatically diagnose degrees from higher ranked universities. India's University Grants Commission entails foreign partners of Indian universities to be ranked in the top most 500 of the ARWU standing, while Brazil's Science without Borders program selected international partner institutions using the QS ranking. The following sub-lists are to help each ranking of ARWU and QS systems. Latter, we include reports of these ranking systems.

10.3.1 Academic Ranking of World Universities (ARWU)

ARWU which is also known as **ShanghaiRanking** is an annual report that is published on periodical time listing universities in order and ranked by ShanghaiRanking Consultancy. Customers and students who are interested can visualize the values by a compiled league table issued by the Center for World-Class Universities (CWCU), Graduate School of Education (formerly the Institute of Higher Education) Shanghai Jiao Tong University. They started the first time in 2003 and the report was the first global ranking with miscellaneous indicators, after which a board of international advisories was established to provide recommendations. The publication currently includes global league tables for institutions and a whole and for a selection of individual subjects, alongside independent regional *Greater China Ranking* and *Macedonian HEIs Ranking*. ARWU is viewed as one of the three most dominant and widely observed university measures. It is praised for its objective methodology but draws some condemnation for narrowly focusing on raw research power, undermining humanities, and quality of instruction.

Since 2009, the Academic Ranking of World Universities (ARWU) has been published and copyrighted by ShanghaiRanking Consultancy. ShanghaiRanking Consultancy is a fully independent organization on higher education intelligence and not legally subordinated to any universities or government agencies. ARWU uses six objective indicators to rank world universities, including the number of alumni and staff winning Nobel Prizes and Fields Medals, the number of highly cited researchers selected by Thomson Reuters, the number of articles published in journals of Nature and Science, the number of articles indexed in Science Citation Index – Expanded and Social Sciences Citation Index, and per capita performance of a university. More than 1200 universities are actually ranked by ARWU every year and the best 500 are published.

Although the initial purpose of ARWU was to find the global standing of top Chinese universities, it has attracted a great deal of attention from universities, governments, and public media worldwide. ARWU has been reported by mainstream media in almost all major countries. Hundreds of universities cited the ranking results in their campus news, annual reports, or promotional brochures. A survey on higher education published by The Economist in 2005 commented ARWU as "the most widely used annual ranking of the world's research universities." Burton Bollag, a reporter at Chronicle of Higher Education, wrote that ARWU "is considered the most influential international ranking."

One of the factors for the significant influence of ARWU is that its methodology is scientifically sound, stable, and transparent. The EU Research Headlines reported ARWU work on 31st December 2003: "The universities were carefully evaluated using several indicators of research performance." The Chancellor of Oxford University, Chris Patten, said that "it looks like a pretty good stab at a fair comparison." Professor Simon Margison of Institute of Education, University of London, commented that one of the strengths of "the academically rigorous and globally inclusive Jiao Tong approach" is "constantly tuning its rankings and invites open collaboration in that."

ARWU and its content have been widely cited and employed as a starting point for identifying national strengths and weaknesses as well as facilitating reform and setting new initiatives. Figure 10.7 shows a historical diagram of ARWU of the consistency of ranking methodology since 2005. Bill Destler, the President of the Rochester Institute of Technology, drew reference to ARWU to analyze the comparative advantages that the Western Europe

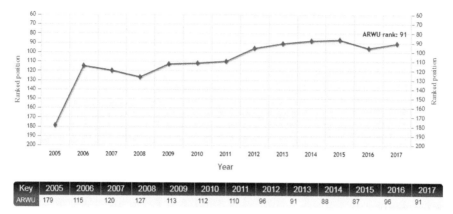

Key	2005	2006	2007	2008	2009	2010	2011	2012	2013	2014	2015	2016	2017
ARWU	179	115	120	127	113	112	110	96	91	88	87	96	91

Figure 10.7 The ARWU first appeared in 2003. Although there have been some changes in the criteria used to rank universities, the ranking methodology has been fairly consistent since 2005.

and United States have in terms of intellectual talent and creativity in his publication in the journal Nature. Martin Enserink referred to ARWU and argued in his paper published in Science that "France's poor showing in the ShanghaiRanking ... helped trigger a national debate about higher education that resulted in a new law... giving universities more freedom."

1. Performance indicators

Table 10.4 lists the arguments that are used in ARWU ranking system, and Table 10.5 presents the ARWU subject area of ranking.

Table 10.4 ARWU indicators

Value	Indicator
Alumni winning Nobel Prizes and Fields Medals, with the value declining over time since the winner's graduation (100-year time limit)	10%
Staff winning Nobel Prizes and Fields Medals with the value declining over time since the award was granted (100-year time limit)	20%
Highly cited researchers affiliated to the institution on the Thomson Reuter's Highly Cited website in 21 broad subject categories	20%
Articles published in Nature and Science over a 5-year window, weighted for the position of the institution in list of affiliations	20%
The number of articles and published proceedings papers listed in the Science Citation Index and Social Sciences Citation Index for the previous calendar year	20%
The per capita academic performance (on the indicators above) of an institution	10%

Table 10.5   ARWU subject area

ARWU Broad Subject Area	UWA's 2016 Rank
Life and agricultural sciences	24
Clinical medicine and pharmacy	51–75
Social sciences	76–100
Natural sciences and mathematics	201+
Engineering/technology and computer sciences	201+

2. Broad subject area rankings

In recent years, along with many other ranking systems, the ARWU has branched out and now produces a number of broad subject area rankings and more refined subject field rankings. However, only the top 200 universities are listed, and individual ranked positions are limited to the top 50.

The methodology for subject rankings also departs somewhat from the ARWU. Rather than a Nature and Science score, the "Percentage of papers published in top 20% journals" in the field/discipline is used. Also, the per capita academic performance indictor has been removed, and the value of staff winning Nobel Prizes and Fields Medals has been reduced to 15%.

For the engineering field, a new indicator, "Total engineering-related research expenditures," has also been introduced, which replaces the Award and Alumni criterion for Nobel Prizes and the Fields Medal.

3. Subject rankings

In 2017, a new subject ranking was introduced, which expanded the subjects ranked from 5 to 52 and changed the methodology used to determine the rankings. Indicators for ARWU subject ranking are presented in Table 10.6.

Table 10.6   Indicators for ARWU subject ranking

Indicators	Definition
PUB	PUB is the number of papers authored by an institution in an academic subject during the period of 2011–2015. Only papers of "Article" type are considered. Data are collected from InCites database. Papers in different Web of Science categories are grouped into relevant academic subjects (Classification of Web of Science Categories into Academic Subjects).
CNCI	Category Normalized Citation Impact (CNCI) is the ratio of citation of papers published by an institution in an academic subject during the period of 2011–2015 to the average citation of papers in the same category, of the same year and same type. A CNCI value of 1 represents world-average performance while a value above 1 represents performance above the world average. Only papers of

(Continued)

Table 10.6 Continued

Indicators	Definition
	"Article" type are considered. Data are collected from the InCites database.
IC	International collaboration (IC) is the number of publications that have been found with at least two different countries in addresses of the authors divided by the total number of publications in an academic subject for an institution during the period of 2011–2015. Only papers of "Article" type are considered. Data are collected from the InCites database.
TOP	TOP is the number of papers published in top journals in an academic subject for an institution during the period of 2011–2015. Top journals are identified through ShanghaiRanking's Academic Excellence Survey or by journal impact factor. In 2017, 94 top journals selected by the survey are used in rankings of 33 academic subjects. The list of the top journals can be found here. For academic subjects that do not have journals identified by the survey, the JCR top 20% journals are used. Top 20% journals are defined as their journal impact factors in the top 20% of each Web of Science category according to Journal Citation Report (JCR) 2015, and then aggregated into different academic subjects. Only papers of "Article" type are considered. In the Academic Subject of Public Health, although a top journal is selected by the survey, there is no "Article" type paper on it, and therefore we use JCR top 20% journals instead.
AWARD	AWARD refers to the total number of the staff of an institution wining a significant award in an academic subject. Staff is defined as those who work full time at an institution at the time of winning the prize. If a researcher was retired at the time of winning the award, we count the institution where the researcher's last full-time academic position was at. The significant awards in each subject are identified through *ShanghaiRanking's Academic Excellence Survey*. The list of the significant awards in each subject can be viewed here. If a winner is affiliated with more than one institution at the time of winning the award, each institution is assigned the reciprocal of the number of institutions. If the award is awarded to more than one winner in 1 year, weights are set for winners according to their proportion of the prize. Different weights are set according to the periods of winning the prizes. The weight is 100% for winners in 2011–2015, 75% for winners in 2001–2010, 50% for winners in 1991–2000, and 25% for winners in 1981–1990.

(Continued)

Table 10.6 Continued

Subject	Weight (%)			UWA 2017 Rank	
	PUB	CNCI	IC	TOP	AWARD
Engineering					
Biomedical engineering	23.8 23.8 4.8 23.8 23.8			201–300	
Biotechnology	31.3 31.3 6.3 31.3			201–300	
Chemical engineering	23.8 23.8 4.8 23.8 23.8			201–300	
Civil engineering	23.8 23.8 4.8 23.8 23.8			201–300	
Computer science and engineering	23.8 23.8 4.8 4.8 42.9			151–200	
Electrical and electronic engineering	23.8 23.8 4.8 23.8 23.8			151–200	
Energy science and engineering	23.8 23.8 4.8 47.6			201–300	

(*Continued*)

Table 10.6 Continued

Subject	Weight (%)			UWA 2017 Rank	
	PUB	CNCI	IC	TOP	AWARD
Environmental science and engineering	23.8 23.8 4.8 23.8 23.8			16	
Marine/ocean engineering	31.3 31.3 6.3 31.3			9	
Mechanical engineering	23.8 23.8 4.8 23.8 23.8			101–150	
Mining and mineral engineering	31.3 31.3 6.3 31.3			7	
Water resources	31.3 31.3 6.3 31.3			76–100	
Life sciences					
Agricultural sciences	31.3 31.3 6.3 31.3			14	
Biological sciences	23.8 23.8 4.8 23.8 23.8			30	
Human biological sciences	31.3 31.3 6.3 31.3			151–200	

(*Continued*)

Table 10.6   Continued

Subject	Weight (%)		UWA 2017 Rank		
	PUB	CNCI	IC	TOP	AWARD
Veterinary sciences	23.8			151–200	
	23.8				
	4.8				
	47.6				
Medical sciences					
Clinical medicine	23.8			42	
	23.8				
	4.8				
	23.8				
	23.8				
Medical technology	31.3			151–200	
	31.3				
	6.3				
	31.3				
Pharmacy and pharmaceutical sciences	23.8			51–75	
	23.8				
	4.8				
	23.8				
	23.8				
Public health	31.3			51–75	
	31.3				
	6.3				
	31.3				
Natural sciences					
Earth sciences	23.8			51–75	
	23.8				
	4.8				
	23.8				
	23.8				
Ecology	31.3			22	
	31.3				
	6.3				
	31.3				

(*Continued*)

Table 10.6 Continued

Subject	Weight (%)		UWA 2017 Rank		
	PUB	CNCI	IC	TOP	AWARD
Geography	31.3 31.3 6.3 31.3			76–100	
Mathematics	23.8 23.8 4.8 23.8 23.8			201–300	
Physics	23.8 23.8 4.8 23.8 23.8			401–500	
Social sciences					
Economics	36.6 12.2 2.4 24.4 24.4			201–300	
Education	41.7 13.9 2.8 41.7			76–100	
Finance	48.4 16.1 3.2 32.3			201–300	
Hospitality and tourism management	48.4 16.1 3.2 32.3			51–75	

(*Continued*)

Table 10.6   Continued

Subject	Weight (%)			UWA 2017 Rank	
	PUB	CNCI	IC	TOP	AWARD
Management	41.7			101–150	
	13.9				
	2.8				
	41.7				
Psychology	41.7			101–150	
	13.9				
	2.8				
	41.7				

Source: http://www.research.uwa.edu.au/staff/about/uwa-ranking/arwu-ranking-methodology

10.3.2 QS World University Rankings

1. Definition

QS World University Rankings is an annual publication of university rankings by Quacquarelli Symonds (QS). Previously known as *Times Higher Education–QS World University Rankings*, the publisher had collaborated with *Times* Higher Education magazine (THE) to publish its international league tables from 2004 to 2009 before both started to announce their own versions. QS then chose to still use the pre-existing methodology while Times Higher Education adopted a new methodology. The QS system now comprises the global overall and subject rankings (which name the world's top universities for the study of 46 different subjects and five composite faculty areas), alongside five independent regional tables (Asia, Latin America, Emerging Europe and Central Asia, the Arab Region, and BRICS). It is the only international ranking to have received International Ranking Expert Group (IREG) approval [2], and is viewed as one of the most widely read of its kind, along with Academic Ranking of World Universities and Times Higher Education World University Ranking. However, allocating undue weight to subjective indicators and having highly fluctuating results are its major criticisms [7–9].

A perceived need for an international ranking of universities for the United Kingdom purposes was highlighted in December 2003 in Richard Lambert's review of university–industry collaboration in Britain for HM Treasury, the finance ministry of the United Kingdom. Among its recommendations were world university rankings, which Lambert said would help the United Kingdom to gauge the global standing of its universities.

The idea for the rankings was credited in Ben Wildavsky's book, *The Great Brain Race: How Global Universities are Reshaping the World*, to then-editor of *Times Higher Education* (*THE*), John O'Leary. *THE* chose to partner with educational and careers advice company Quacquarelli Symonds (QS) to supply the data, appointing Martin Ince, formerly deputy editor and later a contractor to *THE*, to manage the project. Between 2004 and 2009, QS produced the rankings in partnership with *THE*. In 2009, *THE* announced that they would produce their own rankings, the Times Higher Education World University Rankings, in partnership with Thomson Reuters. *THE* cited an asserted weakness in the methodology of the original rankings, as well as a perceived favoritism in the existing methodology for science over the humanities, as two of the key reasons for the decision to split with QS.

QS retained intellectual property in the prior rankings and the methodology used to compile them and continues to produce rankings based on that methodology, which are now called the QS World University Rankings.

THE created a new methodology with Thomson Reuters, and published the first Times Higher Education World University Rankings in September 2010. QS Publishes the rankings results in the world's media and has entered into partnerships with a number of outlets, including The Guardian in the United Kingdom and Chosun Ilbo in Korea. The first rankings produced by QS independently of *THE*, and using QS's consistent and original methodology, were released on September 8, 2010, with the second appearing on September 6, 2011.

QS designed its rankings in order to assess performance according to what it believes to be key aspects of a university's mission: teaching, research, nurturing employability, and internationalization.

2. Academic peer review

This is the most controversial part of the methodology. Using a combination of purchased mailing lists and applications and suggestions, this survey asks active academicians across the world about the top universities in their specialist fields. QS has published the job titles and geographical distribution of the participants.

The 2016/17 rankings made use of responses from 74,651 people from over 140 nations for its Academic Reputation indicator, including votes from the previous 5 years rolled forward provided that there was no more recent information available from the same individual. Participants can nominate up to 30 universities but are not able to vote for their own. They tend to nominate a median of about 20, which means that this survey includes over 500,000 data points. The average respondent possesses 20.4 years of academic

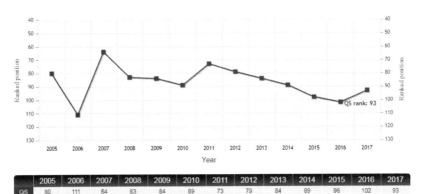

Figure 10.8 Example of university ranking changing over years.

experience, while 81% of respondents have over a decade of experience in the academic world. Figure 10.8 presents an example of QS university ranking changing over years.

In 2004, when the rankings first appeared, academic peer review accounted for half of a university's possible score. In 2005, its share was cut to 40% because of the introduction of the Employer Reputation Survey.

3. Methodology
The QS World University Rankings focuses on four broad areas which they feel are of particular interest to prospective students: research, teaching, employability, and international outlook.

These four key areas are assessed using six indicators, with each given a different percentage weighting. A strong emphasis is placed on "Academic reputation," the "Faculty/student ratio," and the "Citations per faculty." The other indicators are "Employer reputation," "international student ratio," and "international staff ratio."

The QS World University Rankings places a strong emphasis on reputation. The "academic reputation" is assessed through a global survey completed by academics and the "employer reputation" is assessed through a survey of employers worldwide. While the rankings also score research citations, smaller and specialist institutions can be included in this ranking if they have a strong global reputation.

4. Performance indicators
The QS ranking is based on six indicators as shown in Table 10.7, with 50% of the score determined by academics and business employers' responses to reputation surveys.

Table 10.7 QS ranking indicators

Indicator	Percentage	
Academic reputation	40%	Based on an internal global academic survey
Employer reputation	10%	Based on a survey on graduate employers
Faculty student ratio	20%	A measurement of teaching commitment
Citations per academic (Scopus)	20%	A measurement of research impact
International academic ratio	5%	A measurement of the diversity of the academic staff
International student ratio	5%	A measurement of the diversity of the student community

QS has excluded self-citations since 2011. In 2015, QS introduced a hybrid model where normalization is applied to citation totals at the faculty area level. Given that around 49% of global citations are in the life sciences and medicine area, applying this adjustment made a substantial difference to the indicator and overall performance of certain institutions, including UWA. In 2016, QS excluded papers with more than 10 affiliated institutions.

5. Subject rankings

QS employs a different set of measures for their subject rankings such as indicators' data shown and presented in Table 10.8, which list the top 200 universities in a given field. Each of the 46 subjects (across five broad areas) has different weightings applied to four factors.

Table 10.8 QS ranking indicators' data

Indicator
Academic reputation: Data are pulled on a per-subject basis from the survey employed in the main ranking.
Employer reputation: Data are pulled on a per-subject basis from the survey employed in the main ranking.
Citations per paper: Data are extracted from Scopus, and are calculated across a 5-year window (2010–2014 for the 2016 ranking).
H-Index: Two h-indices are calculated: One for all the papers that are attributable to the given subject (h1) and one to the papers that are only attributable to that subject (h2). These are aggregated with double weight given to h2.

(Continued)

Table 10.8   Continued

Subject	Weight (%) Academic Rep	Employer Rep	Citations/ Paper	H-index	UWA 2016 Rank
Arts and humanities					108
Archeology	70	10	10	10	51–100
Architecture/built environment	70	10	10	10	101–150
Art and design	90	10			101–150
English language and literature	80	10	10		51–100
History	60	10	15	15	51–100
Languages	70	30			151–200
Linguistics	80	10	5	5	201–250
Performing arts	90	10			33
Philosophy	75	5	10	10	151–200
Engineering and technology					124
Chemical engineering	40	30	15	15	151–200

(Continued)

Table 10.8   Continued

Subject	Weight (%) Academic Rep	UWA 2016 Rank Employer Rep	Citations/ Paper	H-index
Civil and structural engineering	40		48	
	30			
	15			
	15			
Computer science and information systems	40		151–200	
	30			
	15			
	15			
Electrical engineering	40		101–150	
	30			
	15			
	15			
Mechanical engineering	40		101–150	
	30			
	15			
	15			
Mineral and mining engineering	50		11	
	20			
	15			
	15			
Life sciences and medicine			55	
Agriculture and forestry	50		31	
	10			
	20			
	20			
Anatomy and physiology	40		13	
	10			
	25			
	25			
Biological sciences	40		51–100	
	10			
	25			
	25			
Medicine	40		51–100	
	10			
	25			
	25			

(Continued)

	Weight (%)	UWA 2016 Rank		
Subject	Academic Rep	Employer Rep	Citations/ Paper	H-index
Pharmacy and pharmacology	40		201–250	
	10			
	25			
	25			
Psychology	40		41	
	20			
	20			
	20			
Natural sciences			143	
Chemistry	40		201–250	
	20			
	20			
	20			
Earth and marine sciences	40		43	
	10			
	25			
	25			
Environmental sciences	40		51–100	
	10			
	25			
	25			
Geography	60		51–100	
	10			
	15			
	15			
Mathematics	40		151–200	
	20			
	20			
	20			
Physics and astronomy	40		201–250	
	20			
	20			
	20			
Social sciences and management			78	
Accounting and finance	50		51–100	
	30			
	10			
	10			

Table 10.8   Continued

(Continued)

Table 10.8 Continued

Subject	Weight (%) Academic Rep	Employer Rep	Citations/ Paper	H-index	UWA 2016 Rank
Business and management studies	50 / 30 / 10 / 10				51–100
Communication and media studies	50 / 10 / 20 / 20				101–150
Economics and econometrics	40 / 20 / 20 / 20				101–150
Education	50 / 10 / 20 / 20				51–100
Law	50 / 30 / 5 / 15				101–150
Politics and international studies	50 / 30 / 10 / 10				151–200
Sociology	70 / 10 / 5 / 15				101–150
Sports-related subjects	60 / 10 / 15 / 15				20
Statistics and operational research	50 / 10 / 20 / 20				151–200

Certain commentators have expressed concern about the use or misapplication of review data. However, QS's Intelligence Unit, responsible for compiling the rankings, states that the extent of the sample size used for their surveys means that they are now "almost unbearable to manipulate and very challenging for institutions to game." They also state that over 62,000 academic respondents contributed to their 2013 academic results, four times more than in 2010.

Indicators used by global and national rankings clear that differences between global rankings and national rankings are observed in the lists of ranking indicators as well as their corresponding weightings. Global rankings have an inclination to focus on research/staff-related indicators, in particular publication volume and citations, and allocate extensive weighting to these indicators. Teaching indicators, when used, are limited to student staff numbers/ratios and the teaching/learning environment. On the contrary, national rankings (Guardian, USNWR Best Colleges and CHE University Ranking) tend to focus on student/study related indicators, with little or no place at all for research indicators. Ranking systems have made a remarkable differences in the Higher Education insinuations' academic progression that can not be easily summarized nor presented in few sections. Our trial to present the importance of such systems is a only step to make them easier to understand and motivate readers to learn more about their methodologists. We do highly recommend to pay a visit to their websites as follows:

1. THE RANKING SYSTEM: https://www.timeshighereducation.com/
2. QS ranking system: https://www.topuniversities.com/qs-world-university-rankings
3. ARWU ranking system: http://www.shanghairanking.com/

Also we would like to bring reader's attention that recently well-known academic trials and researches were published to present graphical comparisons of world ranking systems that we also recommend to read and get benefits of such as the one entitled "Graphical Comparison of World University Rankings" by idal, Philippe & Filliatreau, Ghislaine. (2014) Higher Education Evaluation and Development. 1–14. Also a publication entitled with "TRANSPARENCY OF UNIVERSITY RANKINGS IN THE EFFECTIVE MANAGEMENT OF UNIVERSITY" by Marta JAROCKA (2015) – Business, Management and Education. A third publication titled with "Collecting University Rankings for Comparison Using Web Extraction and Entity Linking Techniques" by Nick Bassiliades, Springer International Publishing Switzerland 2014. References are mentioned in the references section of this chapter.

In the following section, this book continues to lay out the positives and negatives of ranking systems as well as mentioning some impacts on higher education insinuations (HEI) as main interest of our book for quality assurance purposes.

10.4 Positives, Negatives, and Impact of University Rankings

10.4.1 Positives and Negatives of University Ranking

The evolution of global rankings overlaps with the advance of globalization, the new protagonist of higher education as a beacon for mobile capital and talents, marketization of higher education, and the rapid development of online digital media. Along with the massification of higher education, the first independent QA agencies emerged in the early 1990s. In general, reasons for ranking universities can be simply the vast growth in higher education across the world in recent spans. Universities are better-quality than in the past. University without quality-oriented concept of running, performing, and teaching is no longer considered a dream-university. As the number of students has grown and rapid process of technology increasing the number of universities and specializations, students and parents need guidance and information to be able to decide the right one to follow. And also, the sheer cost of attending a university is not any more affordable for all and should be taken into consideration while ranking at least for student's adjacent.

As mentioned, the courses and programs offered by universities are increasingly diverse. Observers such as Michael Porter of Harvard Business School point out that advanced economies produce increasingly varied and specialized employment. This means that innovative forms of university education are appearing to the surface and, in turn mean a cumulative need for information and statistics. This issue is not limited to the global ranking, but also any national university ranking is mainly driven to the need graduate students to be employed coping with the need of society and giving positive impact to community. Ranking systems work on both sides, universities and students: universities to grant information and guide to get the right category of students to enroll and get admitted according to their spectra and students to choose the right university that allow them to catch with dream of right program. Measures such as class size, staff/student ratio, completion rates, the achievement of high honors degrees, and your likelihood of employment after you leave are the sort of thing that gets counted for all parties.

In a Niche survey of more than 800 users, 93% reported a college's reputation being somewhat to extremely important.

Repeatedly and naturally, students and learners want to know where a specific university of their choice stacks up against other universities, what it is known for, and whether attending for 4 years will be advantageous to their career. Attending to a university is a speculation, and ranking gives some awareness to whether making this investment a potential and profitable. Growing prestige of a HEI rankings counterparts inflating the fees and required tuitions. Figure 10.9 records the satisfaction and the feedback of how the ranking system is important and reflect the awareness of using such ranking systems.

Universities worldwide are increasingly being confronted with ranking and classification initiatives both at national and international levels. While many institutions have reservations about the methodologies used by the ranking compilers, there is a growing recognition that rankings and classifications are here to stay and many feel the need to respond. University leadership and policy wonks excitedly await the results of yearly university rankings. Part of the excitement is waning, however, due to the same (old) institutions being listed every year. In the EAIE Barometer study, 35% of practitioners indicated that improving international reputation or position in rankings is one of the top three reasons for internationalizing. Rankings are used by some governments in their higher education policy, by institutions looking for international partners, and by prospective students searching for a

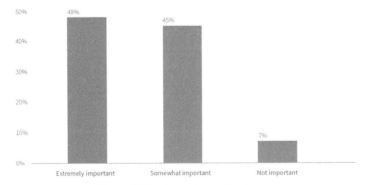

Figure 10.9   Importance of ranking.

place to study – due, often, to the lack of other widespread metrics. But how powerful are rankings in the higher education world?

Many proposals and projects have been floating, but most worthwhile one was a collaboration work among a few of valued academic institutions. For 2 years from 2012 to 2015, The European University Association (EUA) worked on a project entitled "Rankings in Institutional Strategies and Processes" (RISP), in partnership with the Dublin Institute of Technology (DIT), the French Rectors' Conference (CPU), and the Academic Information Centre (AIC) in Latvia. The research on institutional responses further shows that despite the widely acknowledged methodological shortcomings, rankings and classifications have a growing impact on institutional decision-making and actions. Detaching light on the ways in which universities react to rankings and other transparency tools – consciously or unconsciously – would be the earliest step toward identifying opportunities for using rankings in constructive ways for institutional development. The objectives of the project were:

- To gain a deep understanding of the impact and influence of rankings and similar transparency tools on European higher education and institutional strategic decision-making;
- To identify how HEIs can use rankings and similar transparency tools as a strategic tool to promote institutional development;
- To provide input to the HE policy community on the potential effects of rankings on HE systems;
- To enhance European cooperation and sharing of good practices in the field.

The RISP project was the first pan-European study of the impact and influence of rankings on European universities. The project aimed to build understanding on how rankings and similar transparency tools impact the development of institutional strategies and processes, and to propose recommendations on how they can be used to promote institutional development while also identifying potential pitfalls that the universities should avoid.

The project objectives were achieved through the following activities:

- An online survey (March–July 2013) among European universities and HEIs, on the influence of rankings on institutional strategic decision-making, policies, and organization;
- A series of site visits, in October–December 2013;
- A roundtable in June 2014, with senior university managers and stakeholders selected through an open call for participation to provide comments and input as well as to benefit from peer-learning through sharing experiences;

A final publication, including policy recommendations, was released at the end of 2014. It is available in English and French and was launched at two events that also provided an opportunity for discussion, in Brussels (for the English version) and in Paris (for the French version).

Further responses to this collaborative work (RISP) have declared in points why ranking is important to any academic institutions, universities in specific.

1. National policy

The study Rankings in institutional strategies and processes (RISP): Impact or illusion? upholds that many governments use rankings to provide funding for selected institutions deemed as capable of becoming world class, whereas others make use of rankings for classification purposes. In recent years, countries as diverse as India, the Russian Federation, and the Netherlands have made use of rankings in their partnership schemes, recognition and immigration policies.

2. Institutional decision making

According to the majority of the respondents of the same RISP survey, rankings affect institutional decision making. In it, 27% of respondents report that policies have been revised, 26% report that focus has shifted to new features of existing procedures, while 23% indicate that changes have taken place in the research areas prioritized and 21% that the criteria for recruitment and promotion have been affected.

3. Monitoring and benchmarking

The vast majority of RISP survey respondents report that they monitor their institutional performance in rankings and that the senior institutional leadership is involved in this process. Some even have dedicated staff or units for this purpose. Many use rankings not only to monitor their own performance, but also that of their partners and competitors for benchmarking purposes.

4. Data collection

A briefing paper from the Institute of Higher Education Policy states that, in practice, rankings impact discussions about, and collection of, comparative data on both a national and institutional level. Rankings encourage not only the collection but also the publication of education data – according to the *Global* university rankings and their impact study.

5. Partnering

Most research in the field indicates that rankings impact institutional partnering. This applies particularly to international partnering, as knowledge of (prospective) partner institutions is often not sufficiently available. Having a highly ranked partner can also be used for reputational purposes.

6. Branding

The Trends in Higher Education Marketing, Recruitment, and Technology *study* shows that university branding requires constant effort and data to support a desired image. A university's position in rankings serves this purpose well. Ranking outcomes are often mentioned on institutional websites, on social media, and on institutional presentations in order to increase institutional visibility and credibility.

7. Student choice

The Effects of Rankings on Student Choices and Institutional Selection study outlines how rankings affect the choice of study destination, particularly when studying internationally. Especially as, in such cases, information about institutions and education system in the direct network of prospective students is often limited.

8. Quality of enrolled students

The same study purports that the quality of the students enrolled correlates with how well a university performs in rankings. That is, students with good academic records prefer to enroll at highly ranked institutions perceived as offering better education – or, at the very least, a more impressive diploma.

9. Attracting researchers

Researchers tend to seek to employment at institutions that are perceived as prestigious in their field. Respondents of the RISP survey believed that rankings influence prospective researchers. This was particularly perceived to be the case for internationally ranked institutions.

10. Research choices

Due the indicators used, rankings reportedly affect research choices in three different ways: the language of publication, with an increased preference for English; publishing on matters of international interest; and prioritizing publishing in specific journals.

In sum, rankings influence our view of academia and prestige and therefore affect institutional and personal decision-making – as well as policies and practices. Many argue that this is not due to their merit, but rather due to the lack of other comparative international data on the same scale. The influence rankings have varies, however. It is dependent on the type of HEI, the higher education system it is located in, and the availability of other information.

To summarize positives, we can say that ranking is done for a multiplicity of motives, but mainly for the following points:

- To provide the public with information (whatever the specifics of the ranking format) on the standing of HEIs for individual or group decision-making (potential students, parents, politicians, foundations, funding agencies, research councils, employers, international organizations, etc.);
- to foster healthy competition among HEIs;
- To provide additional evidence about performance of particular HEIs and/or study programs;
- To stimulate the evolution of centers of excellence;
- To provide additional rationale for allocation of funds.

There are, generally speaking, four sources of data and information that are collected for the conduct of rankings:

1. Surveys that allow the receipt of opinions from various stakeholders (there is a clear preference for peer-review) in order to obtain a set of comparable data on quality and prestige for different institutions, study programs, and/or other activities generally available (public domain);
2. Information collected by government agencies and other various agencies involved in higher education and research;
3. Information collected by institutions of higher education, which can be of two types – data on governance and management that is usually collected by institutions, and/or data and information exclusively requested and provided to those drawing up the rankings;
4. Bibliometric/scientometric databases, such as those run by Thomson Reuters (ISI Web of Knowledge) and Elsevier (Scopus), which facilitate a multi-disciplinary and research performance assessment.

On the other hand, business wise, rankings "enjoy a high level of acceptance among stakeholders and the wider public because of their simplicity and consumer type information" (AUBR, 2010). This remains factual, to a large scope, in the past 5 years. However erudite the ranking methodology and data collection process may be, the presentation of rankings boils down to a simplified league table, sometimes allowing users to select a few of their choices for detailed comparisons. In most cases, the universities are ranked in ordinal numbers based on the scores they earned according to the criteria defined by ranking providers.

All the previous facts drive us to the question: Have ranking systems developed to match our expectations? Since 1983, U.S. News and World Report has released its annual list of the best colleges and graduate schools, but since then, the face of college rankings has changed drastically. Despite their increasing sophistication, the previous explained ranking systems are basically unidimensional: they provide rankings based on research criteria. Although they claim to take account of other dimensions of universities' performance, it is essentially research that they measure. Beyond the international rankings, national rankings also exist in an increasing number of countries, generally hoarded by commercial entities. In the United Kingdom, many newspapers circulate rankings of what they entitlement to be the "best" universities. These are of varying validity but, because they are mostly based on more energetic data, are more highly regarded than international rankings.

Lots of opponents argue against any positives of ranking. Four major negative factors principally affect the international rankings:

1. The impact of rankings – The extent to which they stimulus the decisions of governments and universities themselves;
2. The data on which the rankings are based – If the data are not robust, then the rankings that follow will not be robust either;
3. The way rankings are vacant – The results are generally presented as an ordinal list from best to worst, according to the number of points scored in the rankings scheme;
4. Choice of appropriate weights.

Opponents argue that rankings can divert universities' attention away from teaching and social responsibility toward the nature of scientific research valued by indicators used for ranking aerobics. There have also been anxieties that by applying an imperfect set of criteria to world universities, and given the resilient craving to feature in the top 200 universities, rankings actually inspire the homogenization of HEIs, making them less receptive and less pertinent to their instantaneous contexts. The fact that rankings are also said to favor the advantage relished by the 200 best-ranked institutions has significant implications for evenhandedness [215].

Along to go, with regard to international rankings is that they only quantity research performance. Although they claim to measure other things, fundamentally the measures used are measures of research performance and activity. Specifically, reputation surveys, which account for half of the contributions to the QS ranking and one-third of those to the Times Higher education ranking (*THE*), can be nothing much more than appraisals of research reputation. How, other than concluded knowledge of research articles, conference presentations, historical prestige, and so on, is an academic in, say, Belgium likely to be aware of a university in Australia? They are certainly most unlikely to distinguish anything about the quality of the teaching or outreach, which may be outstanding. And the survey of employers is even less likely to provide meaningful evidence about the proportional quality of universities in different countries. It gives the impression far-fetched to expect an employer in Belgium to provide a view about the quality of graduates of an Australian university.

In an article titled The Globalization of College and University Rankings and appearing in the January/February 2012 issue of *Change* magazine, Philip Altbach, a professor of higher education at Boston College and also a member of the *THE* editorial board, said that "The QS World University Rankings are the most problematical. From the beginning, the QS has relied on reputational indicators for half of its analysis it probably accounts for the significant variability in the QS rankings over the years. In addition, QS queries employers, introducing even more variability and unreliability into the mix. Whether the QS rankings should be taken seriously by the higher education community is questionable" [46].

Simon Marginson, a professor of higher education at University of Melbourne and a member of the *THE* editorial board, in the article "Improving Latin American universities' global ranking" for University World News on 10 June 2012, said: "I will not discuss the QS ranking because the methodology is not sufficiently robust to provide data valid as social science" [47]. QS's Intelligence Unit counters these criticisms by stating that "independent academic reviews have confirmed these results to be more than 99% reliable."

Academic staff to student ratio is a measure claimed QS to be an indicator of teaching quality. But there is no attempt to separate out the research effort of the staff concerned: and the more research a university does, the more staff it will have. Staff to student ratios are an almost direct portion of research activity. The recruitment of international staff is much more likely to be related to research expertise than teaching ability. The proportion of Ph.D.

students is also to a large extent an indication of research activity. It says little about the quality of the education.

The ARWU's use of the number of Nobel Prize winners among a university's staff as a measure of teaching as distinct from research excellence appears astonishing. Only the ratio of international to domestic students can reasonably be claimed to be a factor independent of research but that partly arises from a country's migration policies. Moreover, it accounts for only a tiny percentage of the value in any ranking.

Calculated properly over 85% of the measures attached to the QS rankings — and 100% of those of ARWU — are in one way or another research related. So, the only way of improving performance in the international rankings is to improve research performance. This drives universities around the world, at the outlay of a focus on teaching, on broadening contribution, and on outreach. Such a focus on research is appropriate only for a small number of universities. One significant point to elaborate is the function of universities is to mature the human wealth of a country and enable personalities to accomplish their latent. Most institutions should be converging on their students, and a ranking scheme that takes no account of that cannot, as all do, prerogative to identify the "best" universities. In their apprehension to escalation up the international rankings, universities have ranked research over other activities.

Governments too have reformed their strategies explicitly in response to the performance of their universities in international rankings. The governments of France, Germany, Russia, and China, among others, have put in place policies aimed unambiguously at enlightening the position of their universities in international rankings. The consequence is that very large amounts of money are being delivered to a small number of selected universities in order to enable them to improve their research — money that could have been used elsewhere in the system to improve other aspects of university recital. Data issues' valid comparisons require confidence that the data on which the rankings are based have been gathered to comparable standards and using the same definitions. No such confidence exists, other than in relation to research publication data where the arrangements for journal publication, peer review, and citations are well established and internationally recycled. But for no other data, there are such internationally comparable arrangements. There is not even a mutually used classification of a student: the notion of a full-time equivalent student does not exist everywhere, and in some schemes, a graduate student is not distinguishable from an undergraduate.

The description of a full-time fellow of academic staff also contrasts from country to country. QS produces own definitions, but other than for research publications, they cannot rely on international databases for the data. Universities quantity their own data and the compilers of the rankings consent the data as complete robotically without any solid quality procedures to manage or to build enough confidence that the thousands of universities around the world that are assessed in the rankings are using these definitions in a consistent way. It is worse that there is no operational effort by the compilers of the rankings to review or assure the quality of the data that are submitted. The data are the most important factor in this ranking system. While QS ranking have had aspects of their rankings audited — their internal processes and calculations — these audits do not extend to the accuracy of the data submitted by institutions.

Data integrity and reliability for the most part are a matter left to the institutions themselves. A case in point is provided by the most recent Times Higher education (*THE*) rankings, where before publication Trinity College Dublin concerned that its position had deteriorated, investigated, and discovered that on a key measure it had misplaced the decimal point in its returns — an error of a factor of 10! There was no effective data verification route other than that conducted subsequently by the University. Trinity College Dublin is a reputable institution and, although what it revealed was a data error that worked against its interests, we can be poised that it would have drawn attention to errors that worked in its favor.

But there may be other universities that submit inaccurate data — whether deliberately or accidentally — and there can be no confidence in the accuracy of their data nor whether the data are twisted to common classifications and standards. Although the ranking bodies do have automated checks which are intended to guarantee, for example, that the data reimbursed are within credible bounds, these fall far short of an audit of the data returns themselves. While relying on universities to supply their own data stretches that rise to problems, those problems are compounded by the run-through engaged in by QS of "data scraping." This involves the seeking out of data from a variety of data sources where a university does not itself offer data to the ranking organization, and where there is unquestionably no control over the data that are being garnered.

However, Martin Ince, the chair of the Advisory Board for the Rankings, points out that their volatility has been reduced since 2007 by the introduction of the Z-score calculation method and that over time, the quality of QS's data gathering has improved to reduce anomalies. In addition, the academic and employer review are now so big that even modestly ranked universities receive a statistically valid number of votes. QS has published extensive data [43] on who the respondents are, where they are, and the subjects and industries to which the academicians and employers respectively belong.

League tables of different Ranking systems are mostly different from each others and most of cases ended that one university has different ranks. Many academic surveys have enlightened this point as the one was published in 2007 and was kind of interesting with title: *College and University Ranking Systems GLOBAL PERSPECTIVES AND AMERICAN CHALLENGES,* INSTITUTE FOR HIGHER EDUCATION POLICY (EDITOR), APRIL, 2007. The survey came with a number of things. Perhaps most important, there were vast differences between university league tables in terms of what they measure, how they measure it, and how they implicitly define "quality." Some of these differences appear to be geographic or cultural in nature. There was notable clustering of certain types of indicators and certain types of data sources. It was indistinct whether this reflects genuine differences in opinion about the definition of what constitutes "quality" in universities or cross-national differences in the collection and availability of data. The lack of common indicators across countries explains why the large international league tables (Shanghai Jiao Tong and *THE*'s) are so reliant on measures of publication outputs and on reputational surveys (respectively), as they are the only indicators that do not rely on governments or institutions to first collect and process the data. At the same time, despite major inconsistencies in the methodologies used to rank universities, there is a surprising level of agreement between ranking systems as to which universities in a given country are "the best." To the extent that different methodologies give differing opinions about the quality of an institution, the variance between observations grows as one moves down the ordinal rankings. After more than 11 years of this survey, we have a question: how far the difference is reduced by now? and we are concerned and motivated to have a ranking system's leagues table in terms of ranking universities which are more closer and more convenient to each other.

10.4.2 Impact of Internal and Global University Ranking Systems

Logically when we start talking about the impact of a university ranking system, we should start reviewing the impact on the higher education and state it as a challenge. Contemporary higher education is multifaceted that should cope with the transition of knowledge and technology while still guarding the quality-oriented concepts which is a permit for globalization. Yet, it is inflated and but vital for learners according to UNSCO surveys. HE as a structure of many relations internally and externally could grab the attention of everyone's courtesy. Politicians and employers, not to mention potential students and their families, are in need of quantified evidence about quality and performance without going too much into the workings of a particular institution or for the no-less-valid argument of not having sufficient capacity to undertake a fully fledged analysis of performance [3]. The reason and need for simplified expressions of quality and performance are evident. Of no less importance, and associated with ranking, is a quality debate, which is one of the key policy areas of higher education. This is particularly evident in the case of the Bologna Process, a major pan-European initiative to harmonize the structures and functioning of national systems of higher education among 46 participating European countries. The objective is the functional creation of the European Higher Education Area by 2010.

In this context of discussion about ranking and classification, it should be mentioned that criticism of the so-called one-dimensional rankings is based on their conventional of indicators providing summative ordering of various HEIs and, as such, not taking into account institutional diversity in higher education. In other words, such criticism is arguing that "classification is a prerequisite for sensible ranking" and should be combined with "multidimensional" ranking (van der Wende and Westerheijden, 2009).

Despite of globalization, higher education is a local institution based on government affiliation and support, in turn to mainly support the internal society and understand what the employer needs to perform efficiently. Partners of higher education want higher education institution to provide the society with graduate that can be easily get the business owner's faith and eager to hire depending that the knowledge-based system the graduate had been building is just a fit for the his/her business. In other academic words, the primary focus of higher education should be to produce generalists because the world changes so fast that specific knowledge is soon purified obsolete. And the other perspective in which HE should focus on developing professional knowledge because this delivers the basis for professional expertise. The Hungarian Higher Educational System has to be strengthen its connections with the members of the labor market. As Malcolm Frazer wrote, "higher education has been seen by many as "a secret garden," better infrastructures, nationally and internationally, and more openness in many other fields of activity have meant that universities can no longer fleece behind the defense of academic sovereignty. In this section and roughly, we have chosen specific impacts to be enlightened more, and at the end, we propose a new theory of scholarship impact of university ranking that is still a hot field of research and building solid conclusions around.

10.4.3 Impact of Global Rankings on Higher Education Research and the Production of Knowledge

In recent years, ranking systems have emerged as powerful yet controversial instruments which exert considerable influence in the higher education policy-making arena. Ranking systems are imperative indicators for research universities pursuing to brand themselves as world-class entities in this field. At the same time, they have attracted the attention of governments seeking to build higher education systems which assure quality provision in both research and teaching. Given that higher education rankings have become an established phenomenon to be found around the world, it is not surprising to see that the ranking scene is evolving – first and foremost in a good direction. Those who produce and publish rankings are increasingly aware that they put their reputation on the line if ranking tables are not free of material errors or if they are not carried out with due attention to basic deontological procedures. In this context, an important initiative was undertaken by a group of experts and producers of university rankings known as the International Ranking Expert Group (IREG), which in May 2006 came up with a set of guidelines and goals for bodies involved in the practice of ranking – as the working paper was known, "The Berlin Principles on Ranking of Higher Education Institutions."

It is very true that the provinces of activities accomplished by HEIs that are then imitated in rankings are directly related to quality and enactment (as well as excellence and reputation). They can be defined, understood, and interpreted differently by different stakeholders, thus representing a notch of fuzziness. However, there is ample evidence that these drawbacks do not totally prevent such exercises from being undertaken, nor from being taken into consideration in case of individual as well as group decisions. It should be revealed that higher education is not the only domain in which the decision-making process has to deal with woolliness and with trying to turn subjective concepts into more or less eloquent statistical evidence – other sectors, for example, are health services and banking. In addition, even if rankings are irregular and ready and a discreet reflection of performance and quality, they can give the opportunity of determining whether a particular policy is effective. In the world of today, it is growing difficult for leaders of

HEIs to contest peripheral impost using "a shield of exceptionality" for each study program or institution.

Equally important, the higher education community and other stake-holders are better informed about what rankings can and cannot present. There is a growing indulgent among leaders of HEIs that using "a shield of exceptionality" for each study program or institution has its limits and that externally carried-out assessment is part of the quality culture. It would not be far-fetched to vice the proliferation in rankings and league tables with the "massification," or unprecedented increase in enrolments, in higher educa-tion around the world. In addition, the torrent of cross-border sequestered and distance providers, the tendency toward internationalization of tertiary education, and the related increase in stakeholders' stresses for greater responsibility, clearness, and efficiency have all contributed to increased incentives for quantifying quality.

It is not surprising, then, that, in the present tertiary education world characterized by increased global competition for students, the number of rankings has grown as governments and the municipal at large are ever more inattentive with the relative performance of tertiary education institutions and with getting the superlative superficial assessment as consumers of education. In such a context, "quality" and "prestige" become important distinctions in brand battles and one of the crucial rudiments in a competition for scarce resources and talents. Moreover, whereas in the past, the system of higher education was assuring the virtuous quality of a prearranged university, nowa-days, it is the institution or group of institutions that do well in international rankings that is raising prestige of the wider system.

That impact is consistent with previous research findings on the topic and with other studies, such as that one carried out by the Higher Education Fund-ing Council for England (HEFCE). Despite criticisms of rankings such as too much importance being given to them over other institutional assessments and practices, or that rankings can tilt institutional resources to favor research over teaching, the study found overall that rankings could also attend as power to make higher education more effective and innovative (HEFCE, 2008).

Ranking systems can help higher education to upgrade data-based decision-making, as HEIs increasingly use data to inform their decision-making and to document student and institutional success. Ranking systems awaken universities to prompt institutional discussions about what constitutes success and how the institution could better document and report that success. Participation in broader discussions about measuring institutional success encourages institutions to move beyond their interior conversations to contribute in broader national and international considerations about new ways broadcasting academic success and excellence. Despite the fact that most ranking institutions prefer research as a criteria over teaching, institutional use of rankings to hasty changes that directly improve student learning experiences, representing that rankings can also lead to positive change in teaching and learning practices. Contribution to documentation and replication of model programs in the framework of the peer benchmarking function of the ranking of HEIs. And contributing to an increase of institutional collaboration.

The last finding of the HEFCE is particularly motivating in view of the criticism of rankings as an instigator of struggle among institutions to the detriment of collaboration and cohesion. The IHEP study, too, recommended that rankings foster collaboration, such as research partnerships, student and faculty exchange programs, and alliances. More specifically (IHEP, 2009): "Rankings can be important starting points to identify institutions with which to collaborate and partner." In highlighting ways that ranking systems can positively impact institutional decision making, this issue brief also underscores the continued need for attention to potential negative effects of rankings.

These comprise the degree to which rankings – and an importance on developing world-class universities – threaten college access for disadvantaged student populations; an unbalanced emphasis on research over teaching; the ratio between full-time and adjunct faculty; the perfection of key ranking variables as a substitute for ample, institution generated strategic planning; and the funding of world-class institutions at the outflow of institutions that further other nationwide goals. Institutions should deliberate the concerns upstretched about the possessions of rankings as catalysts for undeviating policy actions to moderate potential negative impacts.

Despite methodological flaws, global rankings do more than benchmark presentation. They have become a paradigm of the marketization of higher education and the global combat for world-class excellence. By ranking higher education, they afford a framework through which national and institutional desire and competitiveness can be measured as the number of knowledge-producing dimensions and talent-catching higher education institutions (HEIs) in the top 20, 50, or 100. By benefitting particular chastisements and fields of investigation, outputs, and achievements, rankings – like comparable research valuation exercises – support to reiterate an outmoded consideration of knowledge fabrication and research, and its international division of labor.

10.4.3.1 How Rankings Impact Research

Rankings compare HEIs using a range of different indicators, which are weighed differently according to each ranking system. Information is generally drawn from three different sources: (1) independent third party sources, e.g., government databases; (2) HEI sources; or (3) survey data of students, employers, or other stakeholders. Given the absence of reliable publicly available cross-national comparative data, global rankings (are forced to) measure research in broad brush strokes, rather than the full range of HE activity. As such, they rely heavily on traditional research outputs as captured in the bibliometric and citations databases developed by either Thompson-ISI or Elsevier-Scopus. Research productivity is measured by the number of publications in peer-reviewed journals, and research excellence and impact are measured by the number of citations. Essentially, peer publications and citations attempt to measure the extent to which research impacts on and influences the global science community. SJT takes this argument one step further by specifically focusing on publications in Nature and Science, and the number of Nobel or other major prizes winners employed by an individual HEI, as a proxy for scientific excellence. Because the outcome is a derivative of institutional size, SJT does attempt to control for this by assigning 10% of its score to this while the Taiwan system accounts for institutional age by assigning a special weighting for publications in the current year. Research capacity (or potential) is measured by faculty output, which is also the reasoning behind prizes.

10.4.3.2 Institutional Responses to Rankings

Most significantly, rankings appear to be influencing priorities, including curriculum. However, the biggest changes are apparent in rebalancing teaching/research and undergraduate/postgraduate activity, and re-focusing resource allocation toward those fields which are likely to be more productive, better performers, and indicator sensitive/responsive. Regardless of what kind of HEI, the message is clear: "research matters more now, not more than teaching necessarily but it matters more right now at this point in time."

Rankings are also supporting national strategic objectives, attitudes toward the higher education system, and the protagonist of individual institutions. Government speeches impulse HEIs to be more competitive and receptive to the marketplace and customers, define a distinctive mission, are more efficient and productive, and become world-class. In turn, governments are enquiring if research and research preparation investment should be concentrated "through much more focused funding of research infrastructure in [one or two] high accomplishment institutions" or "support for an unspecified number of high performing research intensive universities" or "support for excellent performance, wherever its institutional setting" (Review of HE, 2008). Reviewing the various "excellence" and policy initiatives internationally (Salmi, 2008), two policy positions are discernable – reflecting the fact that policies reflect choices.

1. The neo-liberal model aims to create greater reputational (vertical) differentiation using rankings as a free market mechanism to drive the concentration of "excellence" in a small number of research-intensive universities in order to compete globally. China, France, Germany, Japan, Korea, and Russia prefer to create a small number of world-class universities, focusing on research performance via competitions for centers of excellence (CE) and graduate schools. This model has two main forms: Model A which jettisons traditional equity values (e.g., Germany) and Model B (e.g., Japan) which upholds outdated status/hierarchical standards. The United Kingdom attempted another variation of this prototypical by formally distinguishing between teaching and research institutions, but abandoned this relying on the impact of performance measurement, e.g., the UK Research Assessment Exercise (RAE). Table 10.9 presents how academic institutions map the actions against rankings assuring the importance of these indicators when it comes to evaluate the university's performance.

2. The social-democratic model aims to build a system of horizontally differentiated high performing, globally focused institutions and student experiences. In contrast to an emphasis on antagonism as a driver of excellence, Australia, Ireland, and Norway aim to support "excellence wherever it occurs" by supporting "good quality universities" across the country, using institutional compacts to drive clearer mission differentiation. Rather than elevating a minor number of exclusive institutions to world-class status, the fresh Australian Review of Higher Education seeks to figure a world-class system so that "wherever students are in this country, whatever institution they're at, they're getting a world-class education" (Gillard, 2008; Review of Australian Higher Education, 2008).

As shown above, different rankings have different purposes and target users. The impacts, of rankings, intended or unintended, should also be

Table 10.9   Mapping institutions actions against rankings

Criteria	Response
Research	Increase output, quality, and citations
	• Reward faculty for publications in highly cited journals
	• Publish in English-language journals
	• Set individual targets for faculty and departments
Organization	Merge with another institution, or bring together discipline complementary departments
	• Incorporate autonomous institutes into host HEI
	• Establish centers of excellence and graduate schools
	• Develop/expand English-language facilities, international student facilities, laboratories, and dormitories
	• Establish institutional research capability
Curriculum	Harmonize with EU/US models
	• Favor science/bio-science disciplines
	• Discontinue programs/activities which negatively affect performance
	• Grow postgraduate activity relative to undergraduate
	• Positively affect student/staff ratio (SSR)
	• Improve teaching quality
Students	Target recruitment of high-achieving students, esp. Ph.D.
	• Offer attractive merit scholarships and other benefits
	• More international activities and exchange programs
	• Open international office and professionalized recruitment
Faculty	Recruit/head-hunt international high-achieving/HiCi scholars
	• Create new contract/tenure arrangements
	• Set market-based or performance/merit-based salaries
	• Reward high achievers
	• Identify weak performers
	• Enable best researchers to concentrate on research/relieve them of teaching
Public image/ marketing	Professionalize admissions, marketing, and public relations
	• Ensure common brand used on all publications
	• Advertisements in Nature and Science and other high focus journals
	• Expand internationalization alliances and membership of global networks

Source: Adapted from Hazelkorn (2009).

assessed against the stated resolves and mark users of the rankings concerned. In the case of ARWU, for example, its impacts on China's higher education policy are evident in its goal-setting role for the "211 and 985 projects" (China's HE "excellence initiative"). Beyond that, it also has unintended impacts on other higher education systems and HEIs, some of which later become ARWU's new target clients – governments and universities.

One of the concrete examples is the 2013–2014 Ranking of Macedonian Higher Education Institutions commissioned by the Ministry of Education and Science of Republic of Macedonia in 2013.

In contrast to the high-profile ARWU, another research-only ranking, the Leiden ranking has not caught as much global courtesy. The two rankings have very similar backgrounds in that they are both run by university researchers and are now both spin-off companies of public universities. However, the effect of the Leiden ranking has largely been confined to what it intends to have – spin-off consultancy services for "universities, academic hospitals, research institutes, funding bodies, government/European Union, industry, and network university quality indicators: a critical assessment of organizations." Among them, Leiden University, University of Amsterdam, University of Manchester, University College London, University of Southampton, Heidelberg University, Uppsala University, ETH Zürich, and EPFL, are said to be the users of CWTS' bibliometric analysis, which also forms the backbone of the Leiden ranking (CWTS B.V., 2015).

10.4.4 Impact on Graduates

Universities as Higher Education Institutions are required to expose and to elucidate to society what they are about and in what way they are doing it. Immense intercontinental companies and partnerships need up-to-date and competitive knowledge, which can be used in the professional area too. Ranking systems have to integrate the opinion of the labor market too, to inspect the different problems, and to compile work toward being precautious how to overcome any obstacles. Globalization of higher education has brought different kinds of unexpected impacts. But mainly the employment of a graduate is a worry and concern. Owners have a point of view which is fresh graduates and knowledge they learned yet cannot be used in the "real life," their knowledge is not precious for multinational companies. Unemployed graduates after their catastrophes will try to establish the value of their knowledge, partly with changing their education which is a waste of resources and human values. Studies of the impacts of rankings on student employment are mostly related to national rankings (e.g., USNWR college rankings) and business school rankings (FT Global MBA rankings). This originates as no astonishment considering that national and business school rankings aim mainly at guiding training choices. These rankings are found to have a significant impact on a school's aptitude to attract more new applicants (Standifird, 2005; Peters, 2007 cited in Wilkins and Huisman,

2012), in particular, "high achievers" including the most able students as well as top scholars (Clarke, 2007; Hazelkorn, 2008; Wilkins and Huisman, 2012). Conversely, poor rankings are said to have impacted negatively on staff morale, making it difficult to retain good staff (Hazelkorn, 2008).

Ranking systems as QS had released report about Employment of graduates as a part of estimating and evaluating the university in terms of output of their academic programs. The report details will be mentioned in the next section.

10.4.4.1 QS Graduate Employability Rankings 2018 Methodology
https://www.topuniversities.com/employability-rankings/methodology

The QS Graduate Employability Rankings is an innovative exercise designed to provide the world's students with a unique tool by which they can compare university performance in terms of graduate employability outcomes and prospects.

For the current edition, they aimed to significantly increase this ranking's scope. Whereas in previous years they acquired and analyzed data pertaining to 300 institutions and published a list of 200, this year, they have taken a bold step forward by doubling the number of evaluated institutions and publishing the top 500 universities that we have ranked. We introduced a minimal but still-Signiant recalibration of the weightings we use, aiming to both reduce the reliance on self-reported provide an enhanced normalization mechanism for the results — necessary, given their global scale.

The alumni outcomes' indicator now carries a weighting of 25%, while the employer–student connections' ratio has a reduced weight of 10%. Additionally, the employer reputation index reflects the changes recently introduced in the 2018 QS World University Rankings, with the domestic component of the indicator receiving increased weight. As expected, the extended coverage and the methodologies have had an impact on the results, introducing greater volatility into the rankings table compared to previous editions. We fully expect this volatility to be conned to this instalment. Each institution's score is composed of carefully chosen indicators. Employer reputation expected that all metrics used are, currently, unique to the QS Graduate Employability Rankings. These indicators and the main methodological enhancements introduced this year are described below: Employer reputation (30%).

QS conventionally includes the employer reputation as a key performance area in all its ranking exercises. Of course, this metric adopts a leading role in a ranking focused solely on employability. The employer reputation metric is based on over 30,000 responses to the QS Employer Survey, and asks employers to identify those institutions from which they source the most competent, innovative, and elective graduates. The QS Employer Survey is also the world's largest of its kind. Previously, international responses were weighted at 70%, with inland responses contributing 30% of the total score for this metric. As was the case in the 2018 QS World University Rankings, this has been changed this year: international and domestic responses now each contribute 50% to an institution's score. Alumni outcomes (25%) a university that values the careers of its graduates tend to produce success-ful alumni. Here, QS has recognized the alma maters of those individuals featuring in over 100 high achievers' lists, each measuring desirable out-comes in a particular walk of life. In total, QS has analyzed more than 30,000 of the world's most innovative, creative, wealthy, entrepreneurial, and/or philanthropic individuals to establish which universities are producing world-changing individuals.

This represents a data set approximately 40% larger than that used in the previous edition. A higher weighting is applied to those individuals featured in lists focused on younger profiles, to ensure a high level of contemporary relevance. Likewise, undergraduate degrees have a higher weighting than post-graduate degrees, as it is assumed that the early stages of the higher education learning process are more formative in establishing an individual's employability. Considering the size of the data set and the robustness of the results, the weighting of this indicator has been increased to 25% (versus 20% in previous editions).

Partnerships with Employers per Faculty (25%): This indicator comprises two parts. First, it uses Elsevier's Scopus database to establish which uni-versities are collaborating successfully with global companies to produce citable, transformative research. Only distinct companies producing three or more collaborative papers in a 5-year period (2011–2015) are included in the count. This year's ranking accounts for university collaborations with 2000 top global companies, as listed by Fortune and Forbes.

Employer/Student Connections (10%): This indicator involves summing the number of individual employers who have been actively present on a university's campus over the past 12 months, providing motivated students with an opportunity to network and acquire information. Employer presence also increases the opportunities that students have to participate in career-launching internships and research opportunities. This "active presence" may take the form of participating in careers fairs, organizing company presentations, or any other self-promoting activities. This count is adjusted by the number of students, accounting for the size of each institution. To compensate for the increased significance of the alumni outcomes' indicator, the weight of this metric has been set at 10%, *falling from 15%* in previous editions.

Graduate Employment Rate (10%): This indicator is the simplest, but essential for any understanding of how successful universities are at nurturing employability. It involves measuring the proportion of graduates (excluding those opting to pursue further study or unavailable to work) in full or part time employment within 12 months of graduation. To calculate the scores, we consider the difference between each institution's rate and the average in the country in which they are based. To preclude significant anomalies, the results are adjusted by the range between the maximum and minimum values recorded in each country or region. This accounts for the fact that a university's ability to foster employability will be affected by the economic performance of the country in which they are situated.

Estimated Scores: Whenever QS has not been able to collect data directly from institutions or reliable sources, a conservative estimate is used for missing records. This calculation is based on the records available from institutions based in the relevant country or region. QS has recently released the league table in terms of employment as shown in Figure 10.10.

10.4.5 Proposing a New Impact: Impact on University Selection for Higher Education Affordable Scholarships — The Passive Motivator

The reputation and legitimacy of universities obtained through rankings have material impact on universities' acquisition of resources, such as tuition fees, sponsorship, government funding, as well as talented students and faculty (Zhao, 2007; Marginson, 2009; Hazelkorn, 2008; Wilkins and Huisman, 2012). It is common sense that old established universities are more likely to compete for the highest rankings and to be influenced by the mainstreaming pressure of rankings (Harley and Lee, 1997). Public institutions, rather than private institutions, are more likely to be impacted by rankings, probably because private universities have more flexibility to respond to rankings (Meredith, 2004). Ranked societies took more actions, either to make use of rankings or to control damage, than unranked institutions which were likely to dispute the validity of rankings and did nothing to improve their positions (Hazelkorn, 2008; Wilkins and Huisman, 2012). So, their higher education as a government institution is not a premature in this field. Higher education

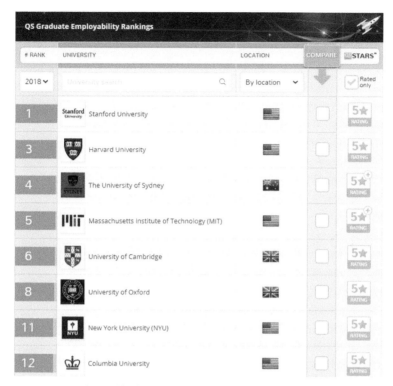

Figure 10.10 QS Graduate Employment ranking.
https://www.topuniversities.com/employability-rankings/methodology

responses and responsibilities can affect other's decision and contributions. Government and funding agencies tend to support highly ranked universities (Marginson, 2009). Some use rankings to allocate funding to universities or outbound mobility scholarships for students. Two recent examples are the Russia's Global Education Program, which selected eligible schools and programs based on their inclusion in ARWU, *THE*, and QS ratings (Global Education Program, n.d.), and The Hong Kong Scholarship for Excellence Scheme which draws reference to the top 100 institutions ranked by QS, ARWU, *THE*, and BGU; and the top 30 colleges in the National Liberal Arts Colleges Rankings published by USNWR (Hong Kong Education Bureau, 2014). An earlier example of the Mongolian government's scholarship scheme for outbound mobility, drawing reference to world rankings, was also documented (Clarke, 2007).

So, as long as, not all the impacts of rankings can be directly qualified to individual rankings, most of the impacts of rankings are indirect or unintended impacts that are difficult to form causal interactions with rankings. We are proposing a new indirect impulsive impact which is about how higher education scholarships and partnerships' selection list being affected with the unintentionally recommendations and motivations of other senior and old higher education's institutions of another country in terms of foreign universities for a scholar. This mutual influence yet lacks of empirical studies to show this relationship. The scenario is more conventional in developing countries where the HEIs are still premature and facing problems with quality and ranking systems. Old and established higher education and HEIs in developed countries usually pay effort and space for marketing, reputation, and legitimacy. Governance practice usually supports and elevates auxiliary for its HEIs. Logically thinking but yet a theoretically that higher educations of countries with low ranking internally seek partnerships and affiliate scholars to random universities but yet supported and recommended by the old and well-established HE. Scholarships and partnerships are real business on another front. And they are actually very costly, especially when they guarantee high standards of quality.

To simplify the theory, we would like to present our model in Figure 10.11 and say the following:

Global University Ranking is an important issue these days for more than one party. Educational institutions are the direct impacted with the league table and responsible to react and response promoting education process to the level of real race irrespective of the criteria of ranking. One side of this response is also related to the investment in education process to promote the process from low levels to a competitive level and consequently aim to be build a well-established knowledge-based learning process. Partnerships with higher education both internally and externally are an important way and method to work on developing side. As well, higher education itself as a government institution in most of world countries has lots of contributions and investments with students' scholarships and erudite students internally in their country and externally. The higher education process of making decision to decide which partnership is more profiting and which university is selected for scholars is not easy and requires lots of authentic data about the partner and/or the university welling to receive the scholar. Most of cases, higher education institution follows a guarantee way which be a follower and get impacted and positively influenced by selections of other senior agents in same field. The motivation is born with high recommendation and

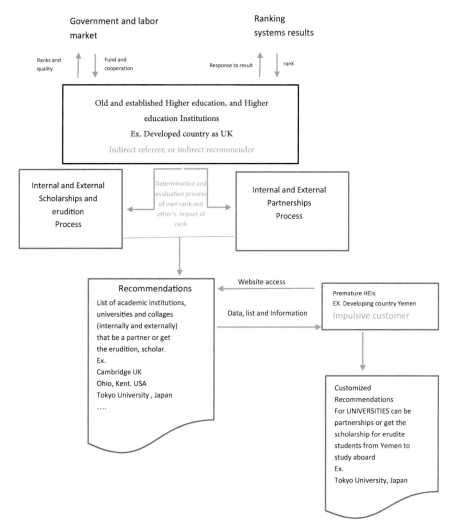

Figure 10.11 Scholarship and partnerships passive and influential impact.

feedback from the senior institution. One important assumption is that the recommendations are accessed unintentionally, and therefore the impact is indirect. We conclude that the ranking can be an active ingredient in higher education institutional strategic planning, it is vital that each university has an articulate mission and strategic plan, and stays true to that mission.

At the same time, we suppose that HEIs should understand the limitations of rankings and not let themselves become preoccupied or captivated by them.

10.5 Data-mine-based Data Pool of Students' Learning Experience on Social Network for Ranking Purposes

10.5.1 Universities Create their Own Social Networks for Students

Online social networking has developed rapidly in the past decade (Boyd and Ellison, 2007; Rainie and Wellman, 2012), including a number of social networking sites (SNSs) aimed specifically at academics (Nentwich and König, 2012). A study was released with title: @Academic and their networks: Exploring the role of academic social networking sites @ by Katy Jordan (2014). In this study, one criterion had been discussed, as What do academics perceive to be beneficial by being part of the network? The study presented sampled statistics for some websites, as summarized in Table 10.10.

In addition to collecting information about profiles' connections (to form the basis of network graphs), data were also collected about academic position and discipline. Disciplines were coded using the Higher Education Statistics Agency (HESA) subject area codes (HESA, 2013), while positions of academics were coded into the following categories: Undergraduate students; graduate students; alumni; tutor; or consultant (a catch-all term used by the university for a range of freelance-type posts); academic support (including librarians, learning technologists, and administrators, for example); researcher (research assistants, postdoctoral researchers, and research fellows); lecturer (lecturers, associate lecturers, and senior lecturers); and professor (professors and readers). Data about nodes (academics) and edges (links between academics) were entered into Gephi, in order to visualize the network graphs and conduct basic SNA.

Survey sought to explore the reasons behind the trends observed in network structure and gain insight into the network participants' perspective. A sample was constructed within the OU-affiliated academics at Academia.edu to reflect differences in seniority and discipline (focusing upon academics in the graduate student, lecturer, researcher, professor and biological sciences, computer science, social studies, historical and philosophical studies, and education categories, respectively). On this basis, 162 academics were invited to take part in the survey. The sample was drawn from the Academia.edu population as it is the largest network sampled, and sampling within this alone rather than the combined Mendeley and Academia.edu data sets removed the possibility of duplication. In answering Part 2 of the survey, participants were asked to focus upon their use of Academia.edu in particular, as it is the only site common to all included in the sample. The survey comprised

Table 10.10 Summary of data collection from academic SNS

Site	Date of Data Collection	Number of OU Academics	Number of First Degree Contacts	Total Nodes	Total Edges	Graph Type
Academia.edu	8 February 2013–15 February 2013	1045	10,283	11,328	22,342	Directed
Mendeley	15 November 2012	70	289	359	412	Undirected
Zotero	14 December 2012	115	14	129	18	Directed

two parts: demographic information, and questions relating to their use of academic SNS.

Part 1 collected brief demographic information, including: department; position; and SNS they have a profile on and level of use. This provided sub-groups to draw comparisons between responses in the second section. The second part addressed how participants view their participation in the network and potential benefits. An inventory of Likert scale items was included (Bell, 2005; Likert, 1932), with statements drawing upon the existing literature (Table 10.11). A five-point scale was used, ranging from "strongly disagree" to "strongly agree," allowing a neutral ("neither agree nor disagree") mid-point.

The above study shows the possibility of collecting random data form specific category in social networks to classify trends of researches and specialization. Some other institutions that have decided engaging with students through social media require more than just a Facebook page and a Twitter account. These universities are creating their own social networks centered on the student experience, with many of them at least building toward covering the span from prospective student to accepted student, enrolled student, and alumnus. We are intending to simulate a classifier to classify data afforded by students voluntarily by interpersonal communication about their learning experiences in any institutions. Classy the recommendation they are forwarding to each other unintentionally by commenting on a joke or formal post asking for recommending a university or a specific program. The purity and honesty responses and replies in such communication are closer to real experiences they faced during their study. And this of data can be a pool of verified information (especially if they are collected from the university's social network) for ranking purposes and analysis in bibliometric researches.

10.6 Quality-oriented Crystallization Process of University Ranking Systems

Crystallization process in chemistry is the course of bringing into being a crystalline material from a liquid, gas, or unstructured solid. The crystals thus formed have highly regular internal structure, the basis of which is called the crystal lattice. Since ranking systems are a vital process for any university either internally and externally as well as having a building block of quality-oriented structure of education, we would like to present a new

Table 10.11   Inventory of Likert scale items and the theme each draws upon

Item	Rationale and Basis
Being part of an academic social networking site is useful	Initial warm-up question
I see my profile as an online business card	How academics view the role of profiles – Bukvova, 2012; Veletsianos and Kimmons, 2013
I use my profile as a research journal	
My online academic and personal identities are separated	
I actively interact with others via the site	
I only follow people who I know personally	Exploring trends in network structure identified in network analysis
I follow people as a way of staying in touch with people I used to work with	
I follow people who I would like to work with in the future	
If someone follows me, I follow them back	
Being able to ask questions of the online community is important	Following up on differences in terms of question posing and answering identified in previous studies – Almousa, 2011; Menendez, *et al.*, 2012
Academic social networking sites allow me to draw upon a wider community of expertise when I need help	
Academic social networking sites are a useful way to support working in collaboration with other researchers	Collaborative aspects of academic social networking – Jeng, *et al.*, 2012; Oh and Jeng, 2011
Having a profile will enhance my future career prospects	
Academic social networking sites are a good way of promoting my own academic publications	Exploring importance of academic publications as part of an online academic identity
Academic social networking sites are a good way of finding out about new publications of interest	

formation of such a highly ordered education structure that prohibits foreign disorders and irregular actions from being incorporated into the lattice, to produce a solid product that – oriented on quality and purity of education is obtained, and proudly has no fears of Globalization and University Ranking systems. The simultaneous formation and purification takes us to the makes crystallized relationship between quality Assurance and University Ranking.

All crystallization processes are aimed at creating a supersaturated building block to drive force under whose influence and indicators to help the quality-oriented blocks to grow and illustrate a new proposal of the relationship in between.

QA ranking has fundamentally different purposes other than ranking. The (quantified) persistence of most rankings sampled for the study is to identify "excellence," in terms of the best HEIs. In addition, rankings have often (unstated) commercial purposes. In contrast, the main purposes of external QA are to guarantee compliance with (minimum) standards and to support quality enhancement. By providing sovereign information, QA is to help building trust in higher education, which is expected to provide a better basis for recognition. The different purposes of QA and rankings are reflected in their legal and institutional framework.

The basic procedure that is created has these two features at its core, and the results obtained by each institution are weighted and summed to calculate their relative positions. There may be variations on this. For example, the institution with the highest score on an indicator may be given a value of 100 and others awarded marks depending on their scores relative to the top scoring institution – but the fundamental principles of dimensions, indicators, and weights remain the same. Although the basic workings of rankings are common between the compilers of rankings, the specific dimensions, indicators, and weights vary. In large part, the variations are pragmatic, constrained by the data that are available; but in part the dimensions and indicators themselves, and the weights in particular, follow from value judgments on the part of the compilers about what is important in judging the quality of universities and, more critically, on the data that are available or that can be constructed.

Among the benefits claimed for rankings are that they provide information to potential students, particularly those from backgrounds where there is little knowledge of higher education or of individual providers, and they similarly provide information to policymakers and other stakeholders. A further claim made for rankings is that they provide benchmarking information for universities themselves and are a stimulus to improvement both as a result of such benchmark information and because of the incentives to improvement that competition provides. For example, Jan Konvalinka, Vice-Rector of Charles University, Prague, has said that the availability of international comparisons might enable him to shake his staff out of their complacency by showing that, although they are first in their own country, they are way down in international terms – a sentiment often repeated by institutional leaders.

How Data Are Collected? A key issue in the preparation of league tables and rankings is the method by which data are collected. The most complete and most detailed sources of data on universities are of course universities themselves, and they are thus potentially a very rich source of data. Quality-oriented procedures and assurance can help a lot in accomplishing this task with high standards and specific routines, simply because the university should be built and cope with internal quality assurance requirements and standards that require lots of steps and quality of education to be internally recognized and potential for external recognition. The pool of data presented to quality assurance is not a blend of information. As huge as they are, yet very well classified to cover and maintain specific task. This solves a problem that ranking systems always face, which usually data collected from the university not mainly for the purpose of compiling league tables but rather as an administrative by-product of ordinary business. As a result, over-reliance on this source of data can lead to a situation where indicators are chosen simply on the clear data with the fully contribute to a sensible definition of quality.

The benefit of this approach is that one can — in theory — answer a number of questions about quality that cannot otherwise be answered. The main drawback of the current ranking is that there is absolutely no guarantee that institutions will actually report the data to the ranker on a consistent basis, as all have a clear incentive to manipulate data in a manner which will benefit them. Indeed, at some institutions in the United States, there are staff positions within institutional research offices that require the incumbent to do nothing but provide institutional data to US News in a favorable light.

What League Tables Measure — A Look at Indicators and Weightings

It should come as no surprise to learn that different ranking systems use very different indicators in order to obtain a picture of "quality."

In their proposal for a system of measuring quality in post-secondary education, a conceptual framework for quality measurement was developed based on the following four elements:

1. Beginning characteristics represent the characteristics, attributes, and abilities of incoming students as they start their programs.
2. Learning inputs come in two main types: (i) resources, both financial and material, available to students and faculty for educational ends, and (ii) staff, both in terms of the numbers but also the way in which they are

deployed to teach and the learning environment they create, as measured by the amount of contact time students have with their teachers, the kinds of exams they face, etc.

3. Learning outputs represent the "skill sets" or other attributes of graduates that culminate from their educational experiences, such as critical thinking, analytic reasoning, and technical knowledge. They also include records relating to retention and completion.

4. Final outcomes represent the ultimate ends to which the educational system may contribute, including not only such traditional measures as employment rates and incomes but also any other outcome deemed to be important to individuals and society, such as job satisfaction, being a "good citizen," etc. As it turns out, these four elements or categories actually encompass the majority of indicators used by the ranking systems covered by this study.

It is also noteworthy that some of the international rankers have created the Observatory on Academic Ranking and Excellence IREG, which works along QA lines and awards a label of compliant ranking providers. In other words, IREG uses QA methodology to assure the quality of rankings. QA and rankings have fundamentally different purposes. The main purpose of rankings is to pick the best. The main purposes of external QA is to guarantee compliance with (minimum) standards and support quality enhancement. Thus, QA covers all HEIs, not only the top segment.

- Rankings primarily measure an institution's research quality; QA tends to focus on the quality of teaching and learning.
- The main target groups of QA are public authorities and the HEIs concerned. The target groups of international rankings are less clear. In the case of ARWU, it was initially the Chinese government which needed data on the country's progress toward building world-class universities. *THE* and QS are predominantly driven by commercial motives.
- The main functions of QA are securing compliance with minimum standards and quality enhancement. Rankings are viewed as creating a whole set of (mostly unstated and predominantly undesirable) effects. Hard evidence is in short supply concerning the impacts of both QA and rankings.
- QA reports are hard to understand for non-experts, while international ranking results appear to be easily readable. This is, however, a "fake simplicity."

10.6.1 How Ranking Helps Quality Assurance and Vice Versa

To utilize the two well-functioning systems: Quality assurance and ranking, an integration in between is desirable to get maximum benefits of enhancing and developing both systems. The integral connection in between, can promote quality assurance approaches responding to the rapidly changing higher education landscape. The European and national quality assurance frameworks and systems need to react to the challenges and trends in higher education, such as life-long learning, massive open online courses, e-learning, the increasing focus on learning outcomes, and cross-border education, to name a few. To respond to those changes, it is advisable to further include representatives of HEIs in the political dialog on quality assurance. Furthermore, the existing flexible "soft regulation" laid down in the ESG is much appreciated since it allows member states to opt for the external quality assurance system most suitable for their context while complying with the shared framework.

Encourage QA agencies to put more emphasis on accessibility and comprehensibility of quality assurance outcomes. The ESG expect reports to be easy to find and to be written in an understandable way. This is not yet always the case, which limits the transparency of quality assurance. We therefore recommend producing, in addition to the comprehensive technical reports, short and comparable summaries allowing all readers to immediately recognize the main outcomes of any quality assessment. A further step might be to collect such summaries on a European web-based platform, thus providing better access to the results of quality assessments undertake.

When talking about ranking, we call and encouragem to create a universal compatible independent quality assurance mechanism for any higher education institution ranking. There is a real need for QA of rankings and ranking providers. The center of such an organization exists already, in the guise of the IREG Observatory on Academic Ranking and Excellence. It offers an audit which leads to the label "IREG approved." Even though we have no indication at all of a lack of quality of IREG's work, we would like to point out that its structure entails a potential conflict of interest. IREG is a membership organization, and the providers of the main international rankings are members of IREG. We would recommend the creation of an independent European ranking watchdog mechanism, which would provide soft regulation and develop, in parallel to the ESG in QA, a set of minimum European standards for ranking methodologies. This mechanism should involve the relevant HE stakeholders and it should also be entrusted with the information tasks.

References and Suggested Literature

[1] Lehman, A., *Jmp for Basic Univariate and Multivariate Statistics: A Step-by-step Guide*. Cary, NC: SAS Press, 123, 2005.

[2] Myers, J. L., Well, A. D., (2003). *Research Design and Statistical Analysis*, 2nd edn. Mahwah, NJ: Lawrence Erlbaum, 508.

[3] Dodge, Y., *The Concise Encyclopedia of Statistics*. New York: Springer-Verlag, 502, 2010.

[4] Yule, G. U., Kendall, M. G., *An Introduction to the Theory of Statistics*, 14th edn. London: Charles Griffin and Co, 268, 1968 [1950].

[5] Piantadosi, J., Howlett, P., Boland, J., Matching the grade correlation coefficient using a copula with maximum disorder. *J. Indus. Manage. Optim.* **3**, 305–312, 2007.

[6] Choi, S. C., Tests of Equality of Dependent Correlation Coefficients. *Biometrika*. **64**, 645–647, 1977.

[7] Fieller, E. C., Hartley, H. O., Pearson, E. S., Tests for rank correlation coefficients. I. *Biometrika*. **44**, 470–481, 1957.

[8] Press, W. H., Vettering, W., Teukolsky, S. A., Flannery, B. P., *Numerical Recipes in C: The Art of Scientific Computing*, 2nd edn., 640, 1992.

[9] Kendall, M. G., Stuart, A., *The Advanced Theory of Statistics, Volume 2: Inference and Relationship*. Griffin, 1973 (Sections 31.19, 31.21).

[10] Page, E. B., Ordered hypotheses for multiple treatments: A significance test for linear ranks. *J. Am. Stat. Assoc.* **58**, 216–230, 1963.

[11] Kowalczyk, T., Pleszczyńska, E., Ruland, F., (eds) *Grade Models and Methods for Data Analysis with Applications for the Analysis of Data Populations. Studies in Fuzziness and Soft Computing*. Berlin: Springer Verlag, 151, 2004.

[12] Kruskal, W. H., Ordinal Measures of Association. *J. Am. Stat. Assoc.* **53**, 814–861, 1958.

[13] Available at: http://firstmonday.org/article/view/4937/4159

[14] Available at: http://www.studylondon.ac.uk/application-advice/university-rankings-explained

[15] *International university rankings: For good or ill? Bahram Bekhradnia*, Available at: http://www.hepi.ac.uk/wp-content/uploads/2016/12/Hepi_International-university-rankings-For-good-or-for-ill-REPORT-89-10_12_16_Screen.pdf

[16] Available at: http://www.ihep.org/sites/default/files/uploads/docs/pubs/collegerankingsystems.pdf

[17] Sadlak, J., *Ranking in Higher Education: Its Place and Impact.* This essay was first published in The Europa World of Learning 2010. For further information see the final page of this PDF or visit www.worldoflearning.com. ©Routledge 2011, all rights reserved.

[18] ABET [Accreditation Board for Engineering and Technology], *Criteria for Accrediting Computing Programs, 2015–2016*, 2015a. Available at: http://www.abet.org/caccriteria-2015-2016/ [Accessed 12 January 2015].

[19] ABET [Accreditation Board for Engineering and Technology], *During the Accreditation Process*, 2015b. Available at: http://www.abet.org/duri ng-accreditation-process/ [Accessed 12 January 2015].

[20] ABET [Accreditation Board for Engineering and Technology], *Decision and Notification*, 2015c. Available at: http://www.abet.org/decision-notification/ [Accessed 12 January 2015].

[21] Accreditation Council, *Procedure of the Accreditation Council for the monitoring of accreditations undertaken by the agencies, Resolution of the Accreditation Council adopted by circulation procedure on 21.09.2006*, 2009. Available at: http://www.akkreditierungsrat.de/filead min/Seiteninhalte/AR/Beschluesse/en/AR_Ueber pruefung_Agenturen_ en.pdf [Accessed 21 November 2014].

[22] Accreditation Council, *Rules of the Accreditation Council for the Accreditation of Agencies, Resolution of the Accreditation Council of 08.12.20091 as amended on 10 December 2010*. Available at: http://www.akkreditierungsrat.de/fileadmin/Seiteninhalte/AR/Beschlue sse/en/AR_Regeln_Agenturen_en_aktuell.pdf [Accessed 21 November 2014].

[23] Accreditation Council, Rechtsgrundlagen für die Akkreditierung und die Einrichtung von Studiengängen mit den Abschlüssen Bachelor und Master in den einzelnen Bundesländern, 2011. Available at: http://www.akkreditierungsrat.de/fileadmin/Seiteninhalte/AR/Sonstige/ AR_Uebersicht_Rechtsgrundlagen_Akkreditierung_Laender_aktuell.pdf [Accessed 20 November 2014].

[24] Accreditation Council, *Report of the Accreditation Council on the evaluation of first experiences with system accreditation*, 2012. Available at: http://www.akkreditierungsrat.de/fileadmin/Seiteninhalte/ AR/Veroeffen tlichungen/Berichte/en/AR_Bericht_Auswertung_Systema kkreditierung_en.pdf [Accessed 27 November 2014].

[25] Accreditation Council, *Rules for the Accreditation of Study Pro-grammes and for System Accreditation, Resolution of the Accreditation*

Council of 08.12.2009, last amended on 20.02.2013, 2013. Available at: http://www.akkreditierungsrat.de/fileadmin/Seiteninhalte/ AR/Beschluesse/en/AR_Regeln_Studiengaenge_en_aktuell.pdf [Accessed 20 November 2014].

[26] Accreditation Council, *Members of the Accreditation Council*, 2014a. Available at: http://www.akkreditierungsrat.de/index.php?id=18&L=1 [Accessed 28 November 2014].

[27] Accreditation Council, *Accredited Degree Programmes — Central Database - Query Form*, 2014b. Available at: http://www.hskompass2.de/ kompass/xml/akkr/maske_en.html [Accessed 19 December 2014].

[28] Accreditation Council, *Verfahren des Akkreditierungsrates zur Überwa chung der seitens der Agenturen durchgeführten Akkreditierungen, Beschluss des Akkreditierungsrates vom 21.09.2006, zuletzt geändert am 25.02.2014*, 2014c. Available at: http://www.akkreditierungsrat.de/file admin/Seiteninhalte/AR/Beschluesse/AR_Ueberpru efung_Agenturen. pdf [Accessed 28 January 2015].

[29] *Act of 27 July 2005 Law on Higher Education. Dziennik Ustaw – Official Journal of Laws of 2005, No.164, item 1365*, as amended.

[30] Available at: http://www.eua.be/activities-services/projects/past-projec ts/quality-assurance-and-transaparency/Rankings-in-Institutional-Strate gies-and-Processes.aspx

[31] Available at: http://www.eua.be/Libraries/publications-homepage-list/ EUA_RISP_Publication

Further Reading

[1] Corder, G. W., Foreman, D. I., *Nonparametric Statistics: A Step-by-Step Approach*. New York: Wiley, 2014.

[2] Daniel, W. W., *"Spearman rank correlation coefficient."* Applied Non-parametric Statistics, 2nd edn. Boston: PWS-Kent, 358–365, 1990.

[3] Spearman, C., The proof and measurement of association between two things. *Am. J. Psychol.* **15**, 72–101, 1904.

[4] Bonett, D. G., Wright, T. A., Sample size requirements for Pearson, Kendall, and Spearman correlations. *Psychometrika.* **65**, 23–28, 2000.

[5] Kendall, M. G., Rank correlation methods, 4th edn. London: Griffin, 1970.

[6] Hollander, M., Wolfe, D. A., *Nonparametric Statistical Methods*. New York: Wiley, 1973.

[7] Caruso, J. C., Cliff, N., Empirical size, coverage, and power of confidence intervals for Spearman's Rho. *Educ. Psychol. Measure.* **57**, 637–654, 1997.

[8] Cureton, E. E., Rank-biserial correlation. *Psychometrika.* 21, 287–290, 1956.

[9] Everitt, B. S., *The Cambridge Dictionary of Statistics.* Cambridge: Cambridge University Press, 2002.

[10] Diaconis, P., *Group Representations in Probability and Statistics*, Lecture Notes-Monograph Series, Hayward, CA: Institute of Mathematical Statistics, 1988.

[11] Glass, G. V., A ranking variable analogue of biserial correlation: implications for short-cut item analysis. *J. Educ. Measure.* **2**, 91–951965.

[12] Kendall, M. G., *Rank Correlation Methods*, London: Griffin, 1970.

[13] Kerby, D. S., The Simple Difference Formula: An Approach to Teaching Nonparametric Correlation. *Comp. Psychol.* **3**, 1, 2014.

[14] *External links.*

[15] Weunsch, K. L., *Nonparametric effect sizes*, 2015.

[16] Available at: www.wikipedia.com

[17] Vidal, Philippe & Filliatreau, Ghislaine (2014). Graphical Comparison of World University Rankings. Higher Education Evaluation and Development. 1–14.

[18] Marta JAROCKA, Business, Management and Education ISSN 2029-7491 / eISSN 2029-6169 2015, 13(1): 64–75. doi:10.3846/bme.2015.260 Available online: file:///C:/Users/maste/Downloads/2168-Article%20Text-4676-1-10-20180530.pdf

[19] Nick Bassiliades, Collecting University Rankings for Comparison Using Web Extraction and Entity Linking Techniques, Springer International Publishing Switzerland 2014, V. Ermolayev *et al.* (Eds.): ICTERI 2014, CCIS 469, pp. 23–46, 2014. DOI: 10.1007/978-3-319-13206-8_2 Available Online: file:///C:/Users/maste/Downloads/9783319132051-c2%20(1).pdf

Special Thanks

[1] THE RANKING SYSTEM: https://www.timeshighereducation.com/

[2] QS ranking system: https://www.topuniversities.com/qs-world-university-rankings

[3] ARWU ranking system: http://www.shanghairanking.com/

11

Toward Excellence of Higher Education

11.1 Introduction

a. What is excellence?

Excellence as one word is the state, quality, or condition of excelling superiority. It is a quality-oriented aptitude which is unusually virtuous and so surpasses ordinary standards. It is also used as a standard of performance as measured when excellence means excelling in input and output, and emphasizing on the "level" of input and output, is an absolutist measure of quality. Strictly in significant disputes, excellence can have the primarily definition as exhibiting physiognomies that are excellent. In the educational descriptive context, excellence preserves one aspect of quality and according to the traditional view, it links quality with the exceptional performance. From this point of view, excellence of quality is a measure of roughly special that is not always being attained. Excellence is not a quality to be judged against a set of criteria, but a quality which is isolated and unattainable for most people.

More technically, excellence is a designed process of which proficiency is the earliest stage. For this reason, it cannot be defined as a simple outcome. Excellence, which popularly appears in the mission statements of many universities, has not been well defined in the academic researches and studies which make the estimation and appreciation of excellence vague and difficult. The remarkable notion of quality sees it as something special. According to Harvey and Green, there are three variations:

1. Traditional notion of quality

The concept of quality has been linked with the notion of singularity, of something special or "high class." The traditional notion of quality infers exclusiveness where quality is not determined through an assessment of delivered data, but it is based on an assumption that there is uniqueness and solitariness of is itself as a word "quality."

351

2. Exceeding high standards

Superiority and excellence are often used interchangeably with quality and perceived as "high" standards with competence base while at the same time ensuring that these are almost unattainable. It is a selective approach in as the sense that it sees quality as only possibly attainable in limited circumstances.

3. Checking standards

A "quality" product is one that has conceded a set of quality criteria. The criteria are based on realistic performance indicators that are designed to reject "defective" items. Therefore, "quality" is attributed to all the items fulfilling the bottom of standards set by the monitoring system implying that quality can be measured as the result of "scientific quality control."

Describing excellence is not a simple task as defining a quality itself, but few steps can be initiative on how we should think about excellence. Superiority includes insight and being out of regular limitations, in other words thinking out of the box and having creativity to generate new and powerful ideas. Critical Skills and analytical thinking are necessary at this stage to ensure that the presented ideas are good ones. Practice thinking to implement the ideas and persuade others of their value. And also, being wise in wise ensures that the ideas help build a common good concept of quality-oriented achievements.

11.2 Excellence in Higher Education

11.2.1 Introduction

Higher education is shifting, and many institutions are in a state of significant transition. It is both exhilarating and unsettling that new educational models and means of conveying educational programs and services are developing at all levels of higher education. Therefore, academic calls have increased for more definition, specifications, accountability, and review of student learning. Higher education coping with globalization and international race has development revised accreditation principles and more calls for standards. The new definitions have brought a common understanding of the term to a mark of distinction, describing something that is exceptional, meritocratic, outstanding, and exceeding normal expectations. Excellence in higher education institution is a form of commendation generally linked to the reputation of institutions and to the achievements of students. If some provision is recognized as excellent, it implies that the majority of other benefactors are simply sustaining standards. The concept has no meaning if all are excellent and there is no way of distinguishing the performance of individual institutions and subdivisions. However, not all would subscribe to this elite view of excellence. It can be perceived as both a comparative and a complete perfect concept.

Investing in excellence:

The University's vision is "(...) to create a stimulating, challenging, and rewarding university experience in a world-class learning community, through sharing a unique fusion of education, research, and professional practice that inspires our students and staff to enrich the world." Times Higher Education, Appointments, 5 September 2013.

Vision:

To achieve excellence in higher education through quality assurance

Quality Assurance and Accreditation Council, Sri Lanka

Miscellaneous system of higher education (as now exists within the most of developed countries such as United Kingdom, United States) deliberations about notions of excellence in teaching and learning acme underlying strains between notions of excellence as a positional good with attendant concerns for reputation and prestige hierarchies on one theoretical hand and excellence relating to higher education's role in broader collective terms on the practical hand. The role of excellence should specify some standards and assure these standards can be measured and evaluated, which actually comes to the conclusion that the way to achieve excellence or at least to start the journey to the excellence, quality assurance procedures, controls, and managements are a fundamental element to link the context and the value of excellence.

11.2.1.1 Definitions

Describing excellence in higher education is not a tranquil assignment. There are many possible definitions and excellence can be with a complicated selection of universal meanings, because it is simultaneously linked to the social and cultural environment values and principles of that culture and also linked to the dogmatic and profitable contexts in that country. But yet, excellence in higher education is not a premature concept. One classic definition of excellence refers to the fulfillment of a certain standard that had been agreed on. This could be inferred as fitness for purpose taking the quality as a base. Quality-oriented excellence concept can be a measure

Table 11.1   Elements that constitute the definition of excellence in higher education

Ownership and Aims	Range	Awarding Boundaries	Diversity in the Use of Values	Use of Indicators
Who is defining/ measuring excellence?	Excellence is a reality	Excellence is exclusive	There is a universal definition of excellence for each reality	Preference in the use of objective indicators
Why is a definition of excellence promoted?	Excellence is a developmental process Excellence is a horizon	Excellence is inclusive	Different definitions of excellence can be used for the same reality	Preference in the use of subjective indicators

Source: ENQA Excellence WG Report the Concept of Excellence in Higher education

of performance according to a predefined set of standards that meet at the end the goals and aims of the educational system. A different elucidation of excellence could lead one to an entirely new scenery if the concept is linked with unforeseen outcomes those which demonstrate better than anticipated. Inspiration of expectations and priorities from different elements and factors can be easily spotted when statements about excellence in higher education are conscripted.

In general, definition of excellence should take into account the use of quantitative and qualitative parameters or the custom of objective and subjective indicators. The consideration of all these arguments will facilitate the identification of a tolerable definition of what is excellence in higher education. Table 11.1 presents elements that constitute the definition of excellence in higher education.

Higher education institutions can been seen as a community dedicated to the pursuit and dissemination of knowledge, to the education and clarification of values, and to the advancement of the society it serves. To support these objectives, institutions of higher education working in a well-designed theme can help a lot to adopt the excellence in the journey of education without facing hassles or obstacles that may abort the excellence achieving process. Regardless of possibility to encounter multiple and conflicting views of excellence in higher education, higher education institutions can adopt the core standards of evaluation areas that are used to outline excellence in higher education theme which actually should also get along with the main missions and vision of that institution. The European Foundation for Quality Management (EFQM) "Excellence Model" is a self-assessment framework for measuring the strengths and areas for improvement of an organization across all of its activities. The EFQM Excellence Model establishes broad criteria, which any organization can use to assess the progress toward excellence. These nine criteria are divided between enablers and results:

1. **Leadership:** Excellence process needs excellent leaders to mature and enable the achievement of the mission and vision.
2. **Policy and strategy:** Institution should be able to apply strategic plans to instruct their mission and vision by developing a stakeholder-focused strategy that takes account of the market and sector in which it operates. These policies should be based on excellence standards to help the ultimate achievement.
3. **People management:** Smart management contributes to develop and release the full potential of their people at an individual, team-based, and organizational level. They promote fairness and equality and involve and empower their people.
4. **Partnerships and resources:** Excellent plans to manage external partnerships, suppliers, and internal resources in order to support policy and strategy and the operative procedure of processes.
5. **Process management:** Excellent organizations enterprise, manage, and improve progresses in order to placate and generate collective value for customers and other stakeholders.

Criteria 6, 7, 8, and 9 are to represent the organization's results and whether to consider the organization has achieved or not. They represent encompass of satisfaction among the organization's employees and customers, and how it impacts on the community and key performance indicators:

6. **Customer Satisfaction:** The organization can't confirm achievement if the services have not been accepted by customers/client. Continuous communication with clients will add huge advantage and make the process easier for the organization to measure the stratification of its customers and how to achieve this. Customers' feedback can make a difference in the organization's performance.
7. **People results:** This examines the workforce the process of the organization produces with solid backgrounds of relative marketplace's requirements and manages to cope with challenges.
8. **Society results:** Organization has ties with it's surrounding environment and community. The process of the organization should plan well to involve the community and sense it's expectations from the organization.

9. **Business results:** These results are expected to examine the organization's performance and improvement in its dominate business factors: Customer satisfaction, financial and marketplace performance as well as the operational performance. Business results also evaluate how the organization performs to its competitors.

In a study by Mike Pupius, Director, Centre for Integral Excellence, Sheffield Hallam University, entitled with "Quality and Excellence in Higher Education" presented the impact of these standards and in workshop, authors presented a conceptual model for the impact. Figure 11.1 presents the fundamental concepts of excellence detailing a continuous learning and improvement cycles. Figure 11.2 Presents the EFQM excellence model.

The Baldrige model is another model that has been developed as well and has been widely adopted. It covers many of the same areas as EFQM but is more wide ranging in scope and is more directly appropriate to educational institutions. Baldrige's Education Criteria pressure student learning whereas they identify education organizations' varying missions, roles, and programs. The criteria sight students as key patrons. In the Education Criteria, the concept of excellence includes three components:

1. A well-apprehended and well-executed assessment approach;

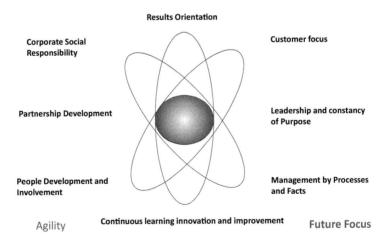

Figure 11.1   Fundamental concepts of excellence. © Sheffield Hallam University. http://slideplayer.com/slide/5170517/

Source: https://www.kodolanyi.hu/images/tartalom/File/hefop/quality2_mike_pupius.pdf

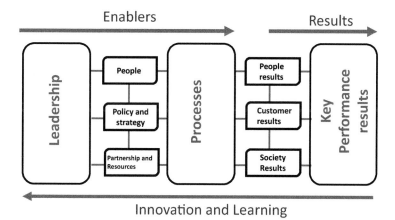

Figure 11.2 EFQM Excellence Model®, The EFQM Excellence Model is a registered trademark © Sheffield Hallam University.

Source: https://www.kodolanyi.hu/images/tartalom/File/hefop/quality2_mike_pupius.pdf
http://www.efqm.org/sites/default/files/modelteaser.pdf

2. Year-to-year improvement in key measures and indicators of performance, especially student learning;
3. A verified performance evaluation and control, and concert improvement relative to analogous organizations and appropriate benchmarks.

Since "managing for innovation" is one of the fundamental values of the Baldrige Criteria for Performance Excellence, it is dignified as an operative tool to deliver a methodical process for dynamic change. The model permits institutions to associate their contemporary practices against reputable standards in other institutions and other economic sectors. The Education Criteria for Performance Excellence, Criteria for Performance Excellence (business/non-profit criteria), and Health Care Criteria for Performance Excellence are all built on the same seven-part framework. The outline is adaptable to the necessities of all associations. Using a common context for all sectors of the economy fosters cross-sector collaboration and the sharing of best practices.

Knowing that education organizations may report these requirements differently from organizations in other sectors, the Education Criteria interpret the language and basic concepts of business and organizational excellence into correspondingly important notions in education excellence. The core tenets and concepts of the Education Criteria are personified in the following seven categories:

Leadership; strategic planning; student, stakeholder, and market focus; measurement, analysis, and knowledge management; faculty and staff focus; process management; and organizational performance results. Figure 11.3 shows the Baldrige model.

Excellence in the United States Brent Ruben's book on excellence in American higher education concentrates on essentials to establish and maintain an outstanding educational society, department, or program. The outline is framed around the incorporation of tactics to assessment, planning, and improvement. It allures elements from management audits, disciplinary reviews, and strategic planning to afford a simple and basic but at the same with a quality-oriented model that can be broadly applicable across all roles and levels of an institution. The following estimation extents are used to define excellence in higher education in United States (ENQA, 2014):

1. Leadership
2. Purposes and plans

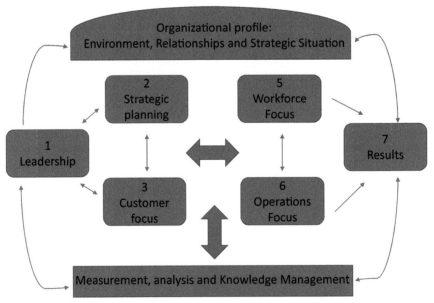

Figure 11.3 Baldrige framework.

Source: http://www.baldrige21.com/Baldrige_Framework.html

3. Beneficiaries and constituencies
4. Programs and services
5. Faculty/staff and workplace
6. Assessment and information use
7. Outcomes and achievements

Excellence in higher education provides a structured guide for reviewing each of these areas as they work contained by a particular institution, department, or program. Evaluation areas 1–5 are important building blocks in any operative organization. Evaluation area 6 emphasizes on systems and procedures in place to evaluate quality and effectiveness in each of these five areas. Evaluation area 7 deliberates the consequences and achievements that are predictable through the assessment process. The model can be used by an intact college or university and also by discrete administrative units, for example. It can also be recycled at the level of academic departments and among programs within the institution for more nested procedures that check, validate, and evaluate. Figure 11.4 presents a general model of excellence for higher education.

11.2.2 Characteristics of Excellence for Professional Practice in Higher Education

There are a few quantifiable factors or performance indicators that adequately express excellent practice, or simple "quality metrics" which encapsulate the

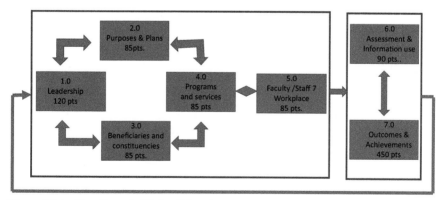

Figure 11.4 Excellence in Higher Education: Framework, Categories, and Point Structure.

Source: Lead Your College or University to Higher Performance. EXCELLENCE IN HIGHER EDUCATION WORKBOOK, AND SCORING GUIDE, developed by Brent D. Ruben, Ph.D.
https://files.eric.ed.gov/fulltext/ED524348.pdf

various manifestations of excellence in the range of institutional activities. The characteristics of excellent institutions are scrutinized and deliberated in an effort to specify the scopes of excellence in relation to what institutions do, what they stand for, and how they conduct their business.

A criterion-referenced approach to excellence suggests that there are values which are commonly acknowledged and which are fitting for all establishments. Some foundations may have strengths in some precise extents and may have courses and programs which spectacle the characteristics of excellence, but to achieve the eminence of excellence, institutions, in general, would be probable to illustrate a good practice in a range of diverse areas. The following are proposed as a guide to the recognition of excellence rather than a "check list" of compulsory settings. Suggested criteria should imitate the characteristics of an excellent institution. It is not a comprehensive list. There are many further ways in which organizations and academic disciplines may state their drive and persona. Features assume a philosophy and practice of life-long learning and professional development shared by individual practitioners and their institutions. Characteristics are grouped into general knowledge and skills, interactive competencies, and self-mastery. These features are optimistically presenting a frame and a sense of what an excellent institution might be [Council for the Advancement of Standards. (2006). *CAS professional standards for higher education,* 6th edn.].

11.2.2.1 General Knowledge and Skills

General Knowledge

1. Understands and supports the broad responsibility of the institution for enhancing the collegiate experience for all students
2. Possesses appropriate knowledge of relevant theories, literature, and philosophies on which to base informed professional practice
3. Knows values, historical context, and current issues of one's profession
4. Has developed, can articulate, and acts consistently with a sound educational philosophy consistent with the institution's mission
5. Understands and respects similarities and differences of people in the institutional environment
6. Understands relevant legal issues.

General Skills

7. Manages and influences campus environments that promote student success

8. Works to create campus and related educational environments that are safe and secure
9. Effectively utilizes language through speaking, writing, and other means of communication
10. Engages disparate audiences effectively
11. Teaches effectively directly or through examples
12. Thinks critically about complex issues
13. Works collaboratively
14. Is trustworthy and maintains confidentiality
15. Exercises responsible stewardship of resources
16. Engages in evaluation and assessment to determine outcomes and identify areas for improvement
17. Uses technology effectively for educational and institutional purposes
18. Bases decisions on appropriate data
19. Models effective leadership.

11.2.2.2 Interactive Competencies

With Students

20. Counsels, advises, supervises, and leads individuals and groups effectively
21. Knows the developmental effects of college on students
22. Knows characteristics of students attending institutions of higher education
23. Knows students who attend the institution, use services, and participate in programs
24. Interacts effectively with a diverse range of students
25. Provides fair treatment to all students and works to change aspects of the environment that do not promote fair treatment
26. Values differences among groups of students and between individuals; helps students understand the interdependence among people both locally and globally
27. Actively and continually pursues insight into the cultural heritage of students
28. Encourages student learning through successful experiences as well as failures.

With Colleagues and the Institution

29. Supervises others effectively

30. Manages fiscal, physical, and human resources responsibly and effectively
31. Judges the performance of self and others fairly
32. Contributes productively in partnerships and team efforts
33. Demonstrates loyalty and support of the institution where employed
34. Behaves in ways that reflect integrity, responsibility, honesty, and with accurate representation of self, others, and program
35. Creates and maintains campus relationships characterized by integrity and responsibility
36. Effectively creates and maintains networks among colleagues locally, regionally, nationally, and internationally
37. Contributes to campus life and supports activities that promote campus community.

11.2.2.3 Self-mastery

38. Commits to excellence in all work
39. Intentionally employs self-reflection to improve practice and gain insight
40. Responds to the duties of one's role and also to the spirit of one's responsibilities
41. Views his or her professional life as an important element of personal identity
42. Strives to maintain personal wellness and a healthy lifestyle
43. Maintains position-appropriate appearance
44. Stays professionally current by reading literature, building skills, attending conferences, enhancing technological literacy, and engaging in other professional development activities
45. Manages personal life so that overall professional effectiveness is maintained
46. Belongs to and contributes to activities of relevant professional associations
47. Assumes proper accountability for individual and organizational mistakes
48. Espouses and follows a written code of professional ethical standards
49. Abides by laws and institutional policies and works to change policies that are incongruent with personal and professional principles
50. Re-evaluates continued employment when personal, professional, and institutional goals and values are incompatible and inhibit the pursuit of excellence

51. Robust and progressive strategic governance and management. Institutions may be expected to demonstrate a strong commitment to excellence in institutional mission and purpose
52. High standards of academic achievement
53. A strong track record in student destinations
54. An exceptional student experience
55. Positive stakeholder satisfaction
56. High levels of student satisfaction
57. Commitment to research and academic development
58. Support for social, economic, and cultural development
59. Recognition of the social benefit of education
60. Commitment to internationalization
61. Promotion of equity and academic freedom.

11.2.3 Quality Assurance and Excellence of Higher Education

Quality as defined by the British Standards Institution (BSI) states to the totality of features and characteristics of a product or service that bear on its ability to satisfy stated or implied needs (BSI, 1991). The International Organization of Standardizations (ISO) terms quality as something that can be determined by comparing a set of inherent characteristics with a set of requirements. If those inherent characteristics meet all requirements, high or excellent quality is achieved. If those physiognomies do not light all requirements, a truncated or deprived level of quality is attained.

The quality of something depends on a set of inherent characteristics and a set of requirements and how well the former complies with the latter (ISO, 2010). Meanwhile, Greene and Harvey (1991) identified five different approaches to defining quality as follows:

1. In terms of exceptional or exceeding high standards and passing a required standard;
2. In terms of consistency or exhibited through "zero defects" and "getting right the first time," making a quality culture;
3. As fitness for purpose – meaning that the product or service meets the stated purpose, customer specifications, and satisfaction;
4. As value for money through efficiency and effectiveness;
5. As transformative in terms of qualitative change.

Lastly, quality assurance as defined by ISO is a set of activities intended to establish confidence that quality requirements will be met.

"Excellence" is linked to the idea of social responsibility and activity directed to the improvement of conditions for individuals. It is based on an understanding of the social, economic, and cultural contribution of higher education. At a personal level, it is enriching and encourages the realization of human potential. There should be a link between excellence and quality and how it affects quality assurance procedures and improvement of quality. Universities may be expected to integrate the concept of excellence in their internal quality systems and culture. The concept and approach apply both to external quality assurance and internal quality procedures, but essentially institutions have more control over their own procedures and can focus on internal processes to secure expectations around quality. Excellence is derived more from external perceptions and can be established through benchmarking one university with another. One of the main orientations of QA procedures is to check the current status of excellence in terms of external quality and also to foster excellence. This section will consider how QA procedures as presented in Figure 11.5 (Modified form of Mike Pupius), can help to define excellence in a particular university and, thus, improve the quality culture. Excellence will be described here as a result of QA procedures.

The World Declaration on Higher Education for the Twenty-First Century provides in its vision to action statement that: quality in higher education is a multidimensional concept, which must encirclement all its functions and activities: teaching and academic programs, research and scholarship, staffing, students, buildings, facilities, equipment, services to

Figure 11.5   The evolution of quality and excellence (Modified form of Mike Pupius).

Source: https://www.kodolanyi.hu/images/tartalom/File/hefop/quality2_mike_pupius.pdf

the community, and the academic environment. Internal self-evaluation and external review, conducted openly by independent specialists, if possible with international expertise, are crucial for enhancing quality. At a purposeful level, excellence of knowledge might be seen as linked to a higher education institution's research mission, whereas access to (excellent) knowledge can be seen as linked to the institution's teaching mission (Calhoun, 2006). A teaching mission necessarily embraces both a concern for teaching and a concern for the end product of the teaching progression, that is, the student learning experience.

Independent national bodies should be established and comparative standards of excellence, recognized at international level, should be defined. Due courtesy should be paid to particular institutional, national, and regional settings in order to take into account diversity and to avoid consistency which means in other words that these institutions are still not adopting the quality concepts. And because this can harm not only the educational systems in these institutions but also, the higher education as a foster of this organization, We do encourage stakeholders to participate and an important part of the institutional evaluation process which will get them a close view of how process is going on and what the outcome is expected given that stakeholders are committed to the value of higher education which is providing quality of education.

There should be no exception or tolerating any more with quality procedures. Lots of academic calls encouraged to support the excellence them with "centers of excellence," defined as centers of advanced research with visible results in the national and international context toward excellence of higher education. Effective teaching and learning are essential if we are to promote excellence and opportunity in higher education. High-quality teaching must be recognized and rewarded, and best practice shared (DfES, 2003, p. 11).

... recognize excellent teaching as a university mission in its own right, University title will be made dependent on teaching degree awarding powers – from 2004–05 it will no longer be necessary to have research degree awarding powers to become a university (DfES, 2003, p. 51).

A number of specific standards, criteria, and indicators can be developed to evaluate these centers nationally, and some quality assurance agencies may get in link and use them and cooperate with to define the excellence and have standards and procedure. QA agencies need to show readiness to include excellence in their mission and statements of intent. Also, it is necessary to clarify the meaning of the word "quality method" as used by QA agencies. The method contains a set of standards, criteria, performance indicators, procedures, or any other type of activity aiming at evaluating an institution, a study program, or a department, with the purpose of making judgments on the quality of activities such as teaching, research, and student support.

Standards for excellence can be found in the literature. A survey in Romania, under the framework of an EU funded program, showed that in the United States, one group of standards refers to the institutional context and another group of standards to informative efficiency. The first group of standards looks at: mission and objectives of the institution; planning methods, allocation of resources, and innovation; leadership and governance; administration; integrity; self-evaluation; and external evaluation. If appropriately met, this group of standards would set the "obligatory (but not sufficient)" circumstances under which the institution would be "eligible" to be reflected as "excellent" (see Chapter 1 section on the Baldrige model).

The second group of standards examines the educational efficiency and is thus directly related to the educational activities and other services that the institution offers its students: admission of students and their performance, including graduation rates; support services for students; quality of curricula and faculty; quality of instructive offerings, including academic content, coherence, learning objectives, expected learning outcomes, and expected skills of graduates; general education learning outcomes, including communication skills, critical thinking, technological skills, etc.; evaluation of achieved learning outcomes of students; and other standards related to study programs. A recent study of the World Bank, led by Jamil Salmi, lists, in a given order of importance, the external and internal conditions that, if met, lead to proper functioning of an educational system: equity, teaching, achieving the expected learning outcomes, research, knowledge or technology transfer to society, and the acceptance of a set of values.

While excellence has been used to be a concept closely linked with an individual virtue or quality (a result of the outstanding quality of an academic's work) over the last 30 years (due to the global changes referenced above), it has become an organizational characteristic, a result of the high level of quality that distinguishes the best universities from the others. Consequently, there is an increasing emphasis on vertical stratification, which promotes an "aura of exceptionality." Thus, within the vertically differentiated systems of higher education, excellence is being equated to "being better" which could mean, according to Altbach's list of characteristics of leading international universities, the following:

- Excellence in research;
- Top quality professors;

- Favorable working conditions;
- Job security and good salary and benefits;
- Adequate facilities;
- Adequate funding, including predictability year to year;
- Academic freedom and an atmosphere of intellectual excitement;
- Faculty self-governance.

Governments and QA agencies are focusing on excellence as a means for enhancing the quality of university teaching and research, and for disseminating good practices. This approach also encourages competition between institutions for recognition as centers of excellence, which may enhance their profile and standing. There is a perceptible shift in thinking away from utilitarian notions of equity and the view of higher education as a "social good" toward the promotion of a more competitive market for institutions in the belief that competition will improve standards and quality. By recognizing the "best" providers, it is expected that standards will be established for the sector as a whole. Not all may achieve excellence but all can benefit from the recognition of best practices and the pursuit of enhancement. Categorization of providers as "excellent" is an alternative to league tables for establishing the status and reputation of institutions. It avoids the relative positioning of institutions in rank order and establishes goals to which all providers strive for. Excellent institutions may also gain recognition in an international context and demonstrate the capacity to compete with other countries.

11.2.4 Accreditation and Excellence of Higher Education

Excellence in accreditation procedures has become a noticeable trend in quality assessment, and this trend is bound to become especially relevant after two or three rounds of accreditation. After the accumulated experience and practice, it is desirable to review and adjust the existing approaches used by any agency. There are a number of reasons for the focus on excellence in the academic community. When a HEI applies for re-accreditation, especially for the third or fourth round, agencies and experts anticipate something more than what they experienced previously and search for improvement and development. Thus, it is important to see the progress in a HEI's performance, not stability but the movement forward (not to prove but to improve). Additionally, the accreditation system itself has to evolve too. In most cases, the first evaluation round establishes minimum requirements, so re-accreditation implies progress in evaluation. It means that not only do experts expect high quality and development from educational institutions, but a HEI itself expects more profound goals, requirements, and standards from the accreditation procedure. The practice of excellence is not just a case of using good practice, but the philosophy of outstanding and distinguished practice. Unlike the practice of excellence, accreditation agencies set minimum and sufficient requirements for the quality of education. It is important that these requirements could be achieved if not by the majority of HEIs but at least by half of them. Accreditation, being a social norm, is supposed to set initially achievable results. On the contrary, the practice of excellence becomes vitally necessary for efficient, well-run HEIs which easily meet accreditation standards and for which threshold standards cannot be considered as the vector of development. Strong universities search for other approaches and methods. If accreditation agencies could offer such procedures, they would be much in demand. Such approaches already exist, and the practice of excellence can serve this purpose perfectly.

11.3 The Never Ending Journey in Pursuit of Excellence of Higher Education

11.3.1 Journey Just Started

So many things have been written about quality assurance and excellence in higher education and so many efforts have been made to instill quality and excellence among HEIs and make it a way of life. Governments have played roles and have been doing their fair share to regulate HEI providers and to assure that provision of education is one that is based on the principle of quality and excellence as evidenced by its enumeration of what it intends to accomplish in higher education.

Excellence is a highly qualified notion. It indicates a verdict that evaluates if and to what extent something or someone retains definite intrinsic physiognomies to be deliberated as excellent. The convinced characteristics can follow a comprehensive scheme or a more all-purpose framework. We can consequently make the next statements about the way excellence is attributed to someone or something:

- Excellence is closely linked to an evaluation.
- Someone or something is always declared excellent "in context," not in itself but in relation to a reference framework that entails criteria.
- The judgment of excellence depends on the criteria that have been chosen.
- Those criteria evolve through time, space, and perspective but are not arbitrary.

- The judgment of excellence thus depends on the person or the society that makes it.
- Moreover, this judgment can be emotionally distorted (personal interests, ignorance of certain realities, bias, inconsistency, and passion).

Linking excellence to quality needs to distinguish between norm-referenced and criterion-referenced conceptions of excellence. A norm-referenced conception will define excellence relative to the performance of others. Thus, people will be in competition for excellence and, as a matter of logic, not everyone can attain it. A criterion-referenced view, however, will define excellence in relation to a standard such that people are not in competition for it and, in principle, if not necessarily in fact, everyone can attain it (quality). Logically, and step by step in an excellence way, we assume quality and superiority at the earliest stages. Then, conception of excellence will tend to be defined by purposes. If our purposes are to develop human capital, we are going to have a norm referenced conception of excellence at the top of the pyramid. What seems to be relevant from Strike's argument is that excellence should be a concept available for every HEI and not only for a few of them. A criterion-referenced conception of excellence would be more impartial and the description of those criteria should also be more closely related to the purposes, the missions, and the values of different HEIs.

Therefore, instead of perceiving excellence as a univocal concept, one might conceptualize "excellences," not as a polysomic or ambiguous concept, but as a concept that incorporates different modalities according to the types of institutions, their different contexts, purposes, and missions.

On the international recognition of excellence, it is defined as "exclusivity" reinforces merit and positions an institution in some real or imaginary ranking. This description of excellence is comparative to the performance of others, which suggests that organizations will be in antagonism for excellence and that not everyone can achieve it. Within this scenario, the term "excellence" is used not only in the sense of claiming a position within a hierarchy but also as a way of emphasizing precise initiatives concerned with enhancing global competiveness. The term is also used to reinforce the merit of some higher education aspects not traditionally related with excellence. The analysis of national policies for excellence allowed Rostan and Vaira to identify some of its dimensions:

- The international dimension in which excellence is embedded.
- The emphasis on research and its dominant role in defining excellence.

- The parallel retrenchment of the traditional link between excellence and elite education.
- The recognition of excellence through the activity of external evaluation.

Many educational systems create a distinction system between internal quality assurance which is intra-institutional practices in view of monitoring and refining the quality of higher education on the one hand and external quality assurance which is interior supra-institutional schemes of assuring the quality of higher education institutions and programs on the other hand. Quality assurance activities depend on the existence of the necessary institutional mechanisms preferably sustained by a solid quality culture. Quality management, quality enhancement, quality control, and quality assessment are means through which quality assurance is ensured. The scope of quality assurance is firm by the shape and size of the higher education system.

According to this operating cycle, we can conclude that even if excellence is an all-encompassing concept, the judgment of excellence is highly relative to the reference framework on which it is based. Each moral system or theory has its corresponding ideals, its own notion of what makes someone or something excellent in a given field. Quality assurance is an all-embracing term covering all the policies, processes, and actions through which the quality of higher education is sustained and developed (Campbell and Rozsnyai, 2002, p. 32). The UNESCO definition enlarges on the context of quality assurance when it states that it is: An all-embracing tenure referring to an ongoing, incessant process of evaluating (assessing, monitoring, guaranteeing, maintaining, and improving) the quality of a higher education system, institutions, or programs. As a regulatory mechanism, quality assurance focuses on both accountability and improvement, providing information and judgments (not ranking) through a settled upon and consistent process and well-established criteria. A new model can be proposed here depending on four main juncture arguments:

1. Quality assurance
2. Educational system
3. Strategic planning
4. Globalization

Our new concept of excellence pyramid can be simply sketched as Figure 11.6 presents. The main four factors and building blocks are bonded as the following.

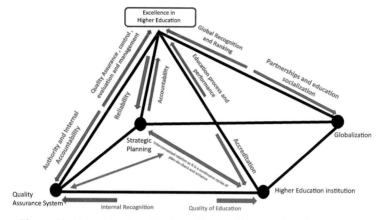

Figure 11.6 Building blocks of excellence of higher education pyramid.

The model can be used by an entire college or university and also by individual administrative, service, and student life organizations. It can also be used at the level of academic departments and among programs within the institution. The model places particular emphasis on the impact of quality as a means for assessing all these arguments as we have detailed in the last chapters. The methodology is based on a defined set of impost:

- Quality that is world-leading concept of education systems, quality of managing and teaching, learning, planning, evaluation, national, and globalization aims in terms of originality, connotation, and inflexibility.
- Quality procedures are excellence early stage standards performing nationally and internationally.
- Quality is the start-up setting for accountability and reliability of a higher education system.

The model is a quality-oriented model but as the higher education itself is submitting to changes, the model as well should also be flexible for adopting and enhancing. Following the five significates concerning promoting quality and attaining excellence in higher education:

1. Enriching quality orientation and international comparability of higher education.
2. Promoting quality teaching, teaching–learning process.
3. Amplifying research and extension capability.
4. Amending student selection/admittance system.
5. Value-added performance of graduates.

When we talk or mention an educational system in terms of quality, we do emphasize the quality of teaching–learning processes that are the backbone of an educational system and actually the criteria we would like to emphasize that quality plays a great role with. And it is a great chance to mention some details in specific criteria in terms of educational system performance leading to quality-oriented system that is expected at the end to end the journey attaining the excellence.

11.3.2 Excellence in Teaching

Elton (1998) presents teaching excellence in five parts, considering them inter-related and thus anyone without the other four is useless:

1. Teaching excellence is a multidimensional concept and its different dimensions call for different forms of recognition and reward.
2. If teaching quality is to be maintained and enhanced, teaching excellence must be recognized and rewarded.
3. The criteria for individual teaching excellence are no more difficult to enunciate and to evaluate than those for research excellence. They are, however, considerably more sophisticated than are appreciated by traditional academics, and they cannot be applied fairly as long as those who judge excellence lack the training for their task.
4. A prerequisite for real teaching excellence at the individual level is a trained teaching profession. A way forward, which links staff development directly to the process of analyzing teaching excellence, has been indicated.
5. Individual teaching excellence is a necessary, but not a sufficient condition for an excellent student learning experience. In addition, there must be excellences at departmental and institutional levels. They can, however, be developed on the foundation of individual excellence.

11.3.3 Excellence in Student Performance

The purpose is to promote students' learning and develop them to a level of professional performance that compares not only with national but also with international standards.

Universities should offer a high percentage of practical training and personal instruction. Student achievement is assessed primarily on the quality of performance. The development of technical skill is combined with academic study and professional development.

11.3.4 Excellence in Research

The Research Excellence Framework in the educational system, the concept of excellence, has been applied to the evaluation of the quality and value of research in higher education. The Research Excellence Framework aims to produce indicators of research excellence and to use these to benchmark the performance of higher education institutions (HEIs) against international standards.

11.3.5 International Peer-review Method

This approach encourages competition between institutions for recognition as centers of excellence, which may enhance higher education institution reputation and recognition wide-world.

While different HEI stakeholders like government, HEIs, students, and accrediting bodies, they continuously accept well-being lattices to ensure quality provision among HEIs, and still, it is essential to consider strategies written by known governmental organizations on the matter to provide and guide us with best practices. An example of this is the guidelines for quality provision in cross-border higher education which is a timely document that addresses key higher education issues in a more globalized society.

Finally, at the end of this chapter, we would like to promote the definition of excellence for higher education and re-define notions of the pyramid that have been made to provide a frame of mechanisms on quality assurance and excellence that depends on four main factors. We would say that: Excellence is not a specific procedure to follow or a check list to mark. Excellence is an architecture of building blocks that comes with process performance, effectiveness, harmony, and collaboration that should be with a quality-oriented concept toward a sustainable excellency of higher education.

And therefore, the journey to excellence is an integral operation of earlier stages contributing and aiming to reach. We also the integral at least should be including:

a. Excellency of educational system

Excellency of education is a two-way path of a continuous action of quality assurance and control to exploit the quality of education in order to assemble the education holding many factors as an affective group of teachers, institutions, and students to cope with the internal requirements of recognition and desired reputation.

b. Excellency of quality
Excellency of quality is a multi-task and permanent requests and calls to standardization and control the quality of education in order to meet the requirements of higher education supported with authorities that go beyond usual performance.

c. Excellency of strategic planning
Excellency of strategic plans is the base of collaboration between education and quality assurance agencies that harmonize quality on many levels in initiation and cope with community needs.

d. Excellency of higher education globalization
Excellency of globalization is the fourth corner of building block of higher education excellency as it is a must and expected happy end for the collaboration toward the external recognition.

The higher education sector needs to be willing to promote this kind of significant learning and help teachers learn how. It must create environments in which student learning of this kind can take place and in which teachers are rewarded for engaging in such practices. It will not happen by itself. Some steps should be taken to (Floud, 2010):

- Redefine professional standards for higher education teachers;
- Provide "Teaching Excellence Award" to deserving ones;
- Strive for "Quality" and then "Excellence" rather than "Quantity" in higher education;
- Measure teaching effectiveness objectively and provide constructive feedback to academics;
- Establish the institutional support base for educational development locally;
- Recognize "teaching excellence" in hiring and promoting decisions on a priority basis;
- Recognize the "teacher researcher" and provide due incentives;
- Promote researches on teaching and learning, and implementation of their relevant findings;
- Allocate enough funding for educational development and educational researches;
- Establish an "excellence forum" within each HEI that pools and shares resources and existing expertise on educational development in general and teaching pedagogy development in particular across borders.

References and Suggested Literature

[1] The Concept of Excellence in Higher Education, © *European Associa-tion for Quality Assurance in Higher Education AISBL*, 2014, available at: http://www.enqa.eu/index.php/publications/

[2] Sabio, R. A., Junio-Sabio, C., "Concerns for Quality Assurance and Excellence in Higher Education." In *International Journal of Information Technology and Business Management*, 2014. Available at: http://www.jitbm.com/JITBM%2023%20volume/4%20Quality%20 Assurance.pdf

[3] Bishop, J. H., *A Strategy for Achieving Excellence in Secondary Edu-cation: The Role of State Government*, 1991, available at: http://digital commons.ilr.cornell.edu/cgi/viewcontent.cgi?article=1355&context=ca hrswp

[4] Nadaf, Z. A., *Teaching Excellence in Higher Education*, 2016, available at: https://www.researchgate.net/publication/318723583_Teaching_Exce llence_in_Higher_Education.

[5] Kauppila, O., *Integrated Quality Evaluation in Higher Education*, University of Oulu, 2016.

[6] Available at: https://www.kodolanyi.hu/images/tartalom/File/hefop/qual ity2_mike_pupius.pdf

[7] Middle States Commission on Higher Education, *Characteristics of excellence in higher education, requirements of affiliation and standards for accreditation*, available at: https://www.msche.org/publications/ CHX-2011-WEB.pdf

[8] Middle States Commission on Higher Education, *Characteristics of Excellence in Higher Education, Eligibility Requirements and Standards for Accreditation*, 2002. Available at: https://www.qc.cuny.edu/about/ administration/Provost/Academic%20Program%20Review/Documents/ charac02.pdf

[9] Little, B., Locke, W., Parker, J., and Richardson, J., *Excellence in teaching and learning: a review of the literature for the Higher Edu-cation Academy, Centre for Higher Education Research and Informa-tion. The Open University*, 2007, available at: https://www.open.ac.uk/ cheri/documents/excellence_in_tl_litrev.pdf

[10] Ruben, B. D., *Excellence in Higher Education Workbook and Scoring Guide,* NACUBO.

[11] Council for the Advancement of Standards. *CAS Characteristics of Individual Excellence for Professional Practice in Higher Education, CAS Contextual Statement*, 6th edn. Washington, DC, 2006.

[12] *Teaching Excellence in Higher Education*. Available at: https://www.re searchgate.net/publication/318723583_Teaching_Excellence_in_Higher_ Education [accessed Nov 05 2017].

[13] Anand, A., *RUSA: Plan to Revamp Indian Higher Education*, 2013. Available at: http:// www.indiaeducationreview.com/features/rusa-planre vamp-indian-higher-education.

[14] Elton, L., Dimensions of Excellence in University Teaching. *Int. J. Acad. Dev.* 3, 3–11, 1998.

[15] Floud, S.R., *The Professionalization of Academics as Teachers in Higher Education*, 2010. Available at: http://archives.esf.org/fileadmin/Public_ documents/Publications/professionalisation_academics.pdf

[16] George, F. D., Excellence and Quality Driven Expansion: XII Plan Perspectives, 2011. Available at: http://hpuniv.nic.in/pdf/quality_excellence. pdf

[17] *Teaching Excellence in Higher Education*. Available at: https://www.res earchgate.net/publication/318723583_Teaching_Excellence_in_Higher_ Education [accessed Nov 05 2017].

[18] Dr. RALPH A. SABIO1, Dr. CECILIA JUNIO-SABIO2, CONCERNS FOR QUALITY ASSURANCE AND EXCELLENCE IN HIGHER EDUCATION, International Journal of Information Technology and Business Management 29th March 2014. Vol. 23 No. 1, https://www.jitbm.com/JITBM%2023%20volume/4%20Quality%20 Assurance.pdf

[19] BSI (1991) Quality Vocabulary Part 2: Quality Concepts and Related Definitions. London: BSI.

[20] Hosein N. Rad and Farzad Khosrowshahi, QUALITY MEASUREMENT IN CONSTRUCTION PROJECTS, http://laaturakentaminen.fi/ attachments/article/294/QualityMeasurements.pdf

[21] Rad, H., N. and Khosrowshahi, F. (1998). Quality measurement in construction projects. In: Hughes, W. (Ed.), 14th Annual ARCOM Conference, 9–11 September 1998, University of Reading. Association of Researchers in Construction Management, Vol. 2, 389–97.

[22] Mike Pupius Director, Centre for Integral Excellence Sheffield Hallam University Quality and Excellence in Higher Education, https://www.kodolanyi.hu/images/tartalom/File/hefop/quality2_mike _pupius.pdf

12

Quality Assurance for Higher Education in Japan

12.1 Quality Assurance in Japan Higher Education

Since Japan is a country with very high population that goes above 120 million, the Japanese educational system is expected to be a complicated and massive one. The higher education of Japan is a very bulky with millions of students enrolled in Japanese universities and colleges. Regardless of being public or private, Japanese academic institutions run their academic systems under the umbrella of Japan higher education. The Japan higher education runs in the framework of the Ministry of Education, Culture, Sports, Science, and Technology (MEXT) which is responsible for all education systems and practices in Japan. The ministry makes decisions to accept a proposal to found a new academic establishment or deny it. It plays an important role in budget determination for all national education institutions and grants for the private institutions. The ministry is directly responsible for overseeing and evaluating a number of research organizations. Figure 12.1 shows the organization of universities in Japan.

We have presented an excellence model in Chapter 11, where excellence of the higher education pyramid is a quality-oriented with four key components to be integrated toward achieving the ultimate of a quality system, in other words, the excellence. This chapter is written to discuss the case of Japanese higher education within the operative integrity of the four key components.

12.1.1 Quality-oriented Education System in Higher Education of Japan

The Japanese education system is a consolidated pyramid structure with MEXT at the top. The Ministry of Education has controlled the establishment

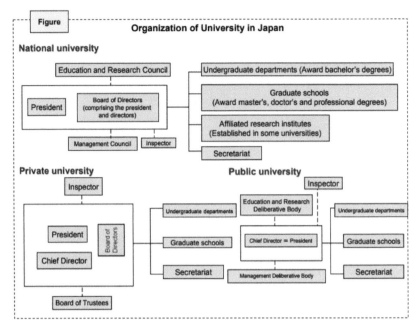

Figure 12.1   Organization of universities in Japan.

Source: http://www.mext.go.jp/en/policy/education/highered

of universities since the 1950s (MEXT, main website). This control has a quality assurance system that functions in higher education institutions. Quality assurance procedures are launched even before opening a new private university, when the MEXT examines the proposal for establishment in terms of the quality of teaching, facilities, and equipment, and other proper standards and environments to ensure the appropriation of the new education system. If the Ministry judges the level of quality inadequate, it declines the proposal. The quality assurance of universities is under the supervision and control of MEXT control for a long time, but once opened, private universities have some considerable freedom to manage the endowment of their higher education opportunities.

The proper implementation of the quality assurance system and the continuous cycles of evaluations are the best practices to attain higher quality and play an important role in the university improvement attempts. Japanese universities suffered of a huge number of university applicants, much more than available seats which means that the capacity of universities and junior colleges in the higher education was limited, so the quality of students had, to some extent, been assured by executing selective entrance examinations. Now the source exceeds the demand so unfortunately, some students are being allowed to register without adequate basic knowledge and learning abilities to study there effectively. The quality assurance of Japanese higher education is also an imperative dispute for institutions privileging them to international collaboration and competition such as increases in the number of foreign students studying in Japan, and vice versa, Japanese who study abroad in addition to students on the exchange programs.

Quality assurance framework is based on preceding principles, and the quality assurance framework is considered as prior regulations, through applying the Standards for Establishing University (SEU) and operating the establishment-approval system (EAS). The purpose of this strategy is to respect the independence and sovereignty of universities, and thus, accept that universities which are once approved will be ready to freely assure their own quality system. Since the SEU and the EAS were announced, the integral system has played a very significant role in guaranteeing appropriate stages of quality in all universities. On the other hand, there have been anxieties about assuring the quality of every educational activity. This kind of prior regulation, whose function is limited to indicate only what is needed to conduct various activities without any successor regulations to check the outcome afterward. This is not a completed cycle of a "quality" concept.

As higher education expanded more, sparkly the rise in enrolment rates and the development of society, more people criticized the total dependence on the prior regulations that would cause over-regulation and limit against pioneering projects. Thus, the government considered a new system that would be inspecting afterward, while sustaining the independence of universities. The new system is integrating preceding protocols with checking subsequently. The government modified the old procedures and formed extra litheness in applying the SEU, reviewing the EAS process, and announcing self-examination and assessment as obligatory in all universities.

Still, as the need for revolution from previous rules to checking afterward through probing current regulations became more widespread throughout the Japanese administrative system, the School Education Law was amended in 2002 to introduce the Quality Assurance and Accreditation System (QAAS), which derived into effect from April 2004. Concurrently, the EAS was renewed in 2003 so that all universities would be permitted as long as they meet the conditions stated in the apprehensive law. At this stage, requirements were condensed and standards were streamlined, while a new notification system in which universities could make minor organizational changes without approval was introduced.

The QAAS requires that all universities in Japan are to submit to an accreditation procedure, once every 7 years by any of the certified agencies. The objective of this system is to develop the mechanism to check afterward and complete the cycle of the quality by evaluating the outcomes. This structure plays a role as accreditation, confirming that qualified universities that satisfy the standards specified by the quality assurance agencies are automatically certified by the Minister of Education. The 2002 report by the Central Council of Education declares that this system should make it possible for universities to undergo the accreditation process from the outside world and thus to improve its quality. The result of the accreditation process is to be indicated by either all-purpose idea of "satisfactory in meeting the standards," "unsatisfactory in meeting the standards" or "pending."

From a legal standpoint, the Article 109(2) of the School Education Law stipulates that all universities need to undergo quality assurance and accreditation, and failure to do so this would be considered as an act in violation. On the other hand, the act of accreditation itself is to be considered a non-administrative measure, as the standards for accreditation are independently set by each certified agency, though the basic concept is expressed by the ministerial ordinance. As this new scheme of integrating prior regulations with checking afterward was introduced, the quality assurance context in Japan has both the benefit of the prior regulations that assure appropriate quality in advance and the inspection afterward that constantly assures quality convenience, while respecting the diversity of universities. Thus, it has been assumed that this incorporation of structures is the most operative and efficient for quality assurance. The structure of the SEU in this comprehension can be divided into four elements as follows:

1. Regulations concerning the basic framework such as qualifications for admission;
2. Regulations stipulating the minimum standards for human and material resources such as faculty, facilities, and equipment;
3. Regulations stipulating the norm for educational activities in university;
4. Regulations for taking courses and requirements for graduation; it should be noted that the SEU includes not only minimum standards but also certain desirable goals and duties that do not have to be specifically attained.

Though it should be noted that the legal frameworks of the ESU and the QAAS are different, it is important to consider a more coordinated application

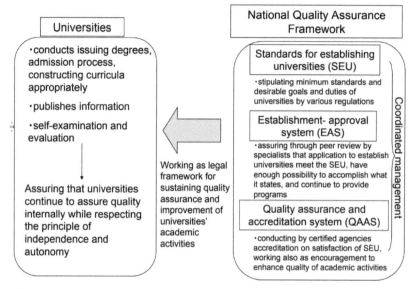

Figure 12.2   Illustration of quality assurance framework in Japan education system.

Source: http://www.mext.go.jp/component/english/__icsFiles/afieldfile/2011/06/20/1307397_1.pdf

of these systems, in order to systematically manage the quality assurance framework. Figure 12.2 shows the illustration of quality assurance framework in Japan education system.

12.1.1.1 System in order to realize quality assurance

With the aim of improving the standard for the international competitiveness of Japanese higher education, educational institutions and research efforts have been promoted to ensure and improve the quality of higher education. The establishment approval system is the combination of the quality assurance system and accreditation system to create a system that checks for quality in the education system with respect to the university's independency. In the next, we will go through the process of integration university systems' quality assurance:

1. Establishment Approval System

To assure the quality that can meet global demands and look after students' benefits, the approval system was designed and promoted by the Minister of Education, Culture, Sports, Science, and Technology to supervise foundation for a new university. Upon approval system's feedback and supervision, the Council for University Establishment can develop a system with experts for cycles of inspections that look for the minimal standards for university establishment. This system is a dynamic inspection of each standard and any change in the university. Since 2004, only prior notification to the Minister shall be required to change the organizational structure such as the types and areas of academic degrees awarded by the university. Checkpoints to be inspected upon university establishment (according to the higher education of Japan website) are as follows:

Purposes and objectives of establishment

Whether the aim for university education is clear, and whether that aim is sufficiently planned with considering the prospect of recruiting students and the expectation shared by the neighbors.

Curriculum

Whether the subjects required to accomplish the purposes and objectives of the establishment have been provided and the curriculum has been systematically organized. Whether teaching methods (lectures, seminars, experimentation, etc.) are carefully designed to accomplish the purposes and objectives of establishment.

Faculty

Whether professors required to accomplish the purposes and objectives of establishment have been sufficiently assigned. Whether full-time professors have been assigned to subjects deemed priority areas from an educational perspective.

Names

Names of universities and names of academic degrees.

Whether facilities and equipment, including school buildings, have met standards and verification that none of them are hampering education and research.

2. University Council

In Japan, the Central Council for Education, consisting of experts and leaders, is organized in order to make recommendations and provide suggestions for enhancing cycles in educational systems responding to the request of Minister of MEXT. University Council, which is a part of the Central Council, works for specific issues around higher education. Consulted by the Minister of

MEXT on September 11, 2008, the University Council had launched a debate on the future of university education in the mid- to long-terms, in order to assure the quality of education as well as to enhance social understanding of quality culture. Items for consultation can be as follows:

- Future of university system and its education responding to diverse needs of society and students

  - University education responding the needs of society and students
  - Reconstruction of the university system and its education which centers on degree programs
  - Human resource development in the fields where social demands are particularly high
  - Quality assurance system to realize whether the university education is responsive to diverse needs
  - Measures to support students in taking courses, in order to realize university education that is responsive to diverse needs.

- Future of university education in the progress of the globalization

  - Measures to improve the international competitiveness of universities
  - Introduction of international viewpoints in university quality assurance and accreditation, and response to university quality assurance and accreditation activities at the global level.
  - Facilitation of improvement in international mobility of students and faculty within the Asian and other regions.

- Overall picture of universities in Japan within its depopulating society

  - Perspectives for the sound development of university system as a whole within a depopulating society
  - Facilitation of functional differentiation of universities and establishment of networks among universities
  - Perspectives for university policy in response to human research development needs on both national and regional level.

3. The Central Council

The Central Council for Education is a body that carries out research and deliberations on significant substances related to the promotion of education, lifelong learning, sports, and other matters in response to requirements from the MEXT and provides feedbacks and suggestions to the Minister. The Central Council for Education has, under the all-purpose committee, five sectors in Education Systems, Lifelong Learning, Elementary and Lower Secondary Education, Universities, and Sports and Youth. In 2004, the Central Council for Education submitted three reports, regarding university entrance qualification tests, pre-school education, and higher education, respectively, and it does conduct energetic deliberations regularly concerning a variety of topics associated to quality in education systems. In February 2005, the Compulsory Education Special Working Group established under the general committee started intensive deliberations on the model form of compulsory education overall. MEXT is endeavoring to promote educational reform taking into account the reports submitted by the Central Council for Education. Central Council for education has subdivisions such as:

A. Council for Science and Technology

The Council for Science and Technology has six subdivisions under the general committee (Subdivision on Research Planning and Evaluation, Subdivision on Resource Study, Subdivision on Technology, Subdivision on Ocean Development, Subdivision on Geodesy, and Subdivision on Professional Engineers), two directly subordinate sections (Technology and Research Foundations Section and the Bioethics and Biosafety Commission), and three directly subordinate committees (International Committee, Committee on Human Resources, and Basic Plan Special Committee). The Council for Science and Technology makes full use of the respective special characteristics of science and technology and ensures their harmonization and a comprehensive approach to them, and is engaged in vigorous discussions designed at producing highly original and world-class research results.

B. Council for Cultural Affairs

The Council for Cultural Affairs is responsible for research and deliberations concerning important matters related to the promotion of culture and international cultural exchange, in response to requests from the Minister of Education, Culture, Sports, Science, and Technology or the Commissioner for Cultural Affairs.

Four subdivisions, the Subdivision on National Language, the Subdivision on Copyright, the Subdivision on Cultural Properties, and the Subdivision on the Selection of Cultural Awardees, and the Cultural Policy Committee have been created under this council. The four subdivisions conduct wide-ranging research and deliberations in specialized fields, for example, on matters concerning the improvement and dissemination of the

Japanese language, the copyright system, and the protection and utilization of cultural properties, respectively.

For example, the Council for Cultural Affairs published three major reports:

1. The report concerning the designation, etc., of cultural properties designated by the national government;
2. The report of the Subdivision on National Language;
3. The report of the Cultural Policy Committee "Let's Revitalize Japan through Regional Culture!" (Back February 2005).

These reports served as the foundation for the promotion of culture and the arts. The Agency for Cultural Affairs intends for further engagement in the promotion of culture and arts based on the recommendations contained in these and ones that have been released since then.

4. National Institution for Academic Degrees and University Evaluation (NIAD-UE)

NIAD-UE was established, based on the Act of General Rules for Incorporated Administrative Agency and the NIAD-UE law. NIAD-UE conducts accreditation processes on teaching conditions and research activities at universities. At the same time, in order to realize a society where the outcomes which learners earned at diversified environments are duly appreciated, NIAD-UE assesses the results of various learning provided at the higher education level and grants academic degrees to learners recognized as having fulfilled required academic standards.

Quality assurance and accreditation of national universities conduct accreditation processes of academic activities at national universities and inter-university research institutes, based on requests from the National University Corporation Evaluation Committee, held in MEXT, and provision of the results to the committee, each university and inter-university research institute evaluated, and the general public Granting Degrees Granting academic degrees (bachelor's, master's, and doctorate) in accordance with the guidelines prescribed by the School Education Law, for those recognized as having the same academic ability as those who have graduated from a university undergraduate college or graduate school, on evaluating various learning outcomes.

Researches on the evaluations of academic activities of universities and other institutions, and researches on the assessments of the learning outcomes are necessary to grant degrees' collection, filling, and dissemination of information concerning evaluations of academic activities of universities and other institutions, and information concerning various learning opportunities at universities.

5. Internationalization and Strategic Planning of Higher Education in Japan

The project entitled "Global 30 Project-Establishing University Network for Internationalization" is a funding project that aims to promote internationalization of academic environment of Japanese universities and acceptance of excellent international students to study in Japan. The selected core universities have been implementing a variety of approaches to internationalize academic systems and campuses such as developing degree programs conducted in English and enriching international student support, while they are expected to enhance inter-university network for sharing educational resource and other outputs including establishment of overseas office which can be jointly used by all Japanese universities.

An overview and survey of International Research Exchanges was released on April 19, 2016. In conducting this survey, the Ministry of Education (MEXT) has two objectives. The first one hopes to grasp what is happening with respect to the research exchanges that are undertaken annually between foreign nations and Japan's universities, its colleges of technology, its incorporated administrative agencies, etc. The second objective, from the results thus obtained, hopes to originate basic materials that link to the topic of policies for the promotion of international exchanges. With respect to the survey concerning the situation in 2014 which was conducted on behalf of MEXT by the Institute of Future Learning, on this occasion, notification is given as to a summary of the results.

1. Survey Content
   - Survey Subjects:
     Universities (national, public, and private institutions), colleges of technology, incorporated administrative agencies, etc. ($n = 902$ institutions) (valid responses: $n = 867$ institutions, response rate: 96.1%).
   - Survey Items:
     Yearly research exchange activities that were undertaken between the survey subjects and foreign nations in 2014. (Defined as the period of time from the start of April, 2014 until the end of March, 2015.)

2. Summary of Survey

Results concerning short-term accepted researchers from overseas, while throughout 2011 their numbers were down due to the impact of unexpected events such as the Great East Japan Earthquake, etc., a real ambition toward recovery has been witnessed more recently. With reverence to short-term dispatched researchers, since the commencement of this survey, the drift has been for their numbers to grow. Meanwhile, for both medium-/long-term accepted researchers and medium-/long-term dispatched researchers, the tendency in recent years has been by and large for their numbers to remain at similar levels. Regarding the numbers of research agreements concluded with overseas universities and research institutions, which is an analysis item that has been newly added to the survey, of the 867 institutions from whom a valid response was obtained on this occasion, some 474 indicated that they have concluded such agreements with overseas partners.

The emerge of strategic planning in Japanese higher education coincided with the difficulties experienced in all of education system in the 1970s and 1980s, as acceptances commenced to fluctuate, student demographics started to change, and funding became unreliable. At this point, future research and the rise of technology-enabled data collection and analysis pointed the way to strategic planning as one solution for developing a proactive stance in the environment of changing demands and declining resources. The difficulties with initial attempts to convert corporate excellence practicing and emergence of strategies to the culture of higher education were multitude. There was an urgent necessity to adjust a procedure intended to motivate a quality-based assessment for developing within a short timeframe, and that was frustrating at best, and disappointedly ineffective most often. While companies industrialized their planning processes based on market data and customer-driven production, academe was limited in the data it could collect and bring to bear on its issues and surely did not reflect itself as serving "customers." Up to date, MEXT has managed, regardless of obstacles, to complete determinations to enhance and develop various policies in accordance with the "Plan to Accept 100,000 Foreign Students," which was formulated in 1983.

Consequently and as a result, the number of foreign students studying at Japan's universities reached an all-time high of roughly 117,000 in May 2004 and the number of students learning at Japanese language educational institutions comes to about 35,000 as of July 2004. Meanwhile, the number of Japanese students studying at overseas universities has been noticeably increasing. An analysis and a close look at these statistics for each country show that in 2002, there were approximately 79,000 Japanese students studying in 33 major countries. In July 2008, in order to make Japan extra open-minded in or to maintain and improve our society, the anxious settled upon the context of the "300,000" International Students Plan," which sets the longer term goal of MEXT's strategic planning to accepting 300,000 international students by 2020.

Japanese strategic planning for higher education has positively impacted the strengthening the bridge among nations as we can tell today that the boundaries between nations are more frequently being crossed. In 2004, over 21,852 (19,059 China) Japanese students went to Asia and over 100,000 Asian students came to Japan. In 2006, 768 went to Germany (EU 5,607 in total) and 450 German students came to Japan. The integral and formation of personal and educational groups which transcend national boundaries are becoming increasingly popular. Where education once focused on developing skills that would be used within the nation, the education of the future will have bridge the gap and prepare students to be successful in a workforce which transcends national boundaries. Students will have to be trained for self-developing a transnational competence to get them into smooth transform and help them to cope. What we mean by transnational competence is the ability to imagine, analyze, and creatively address the potential of other local economies and cultures. What we expect from this trend is building a knowledge of commercial, technical, and cultural developments in these other locales. The awareness of who the key leaders of these locales and how to strategically plan for the best outcomes. The quality-oriented promotion includes understanding of local customs and negotiating strategies by cycle of improvement and enhancing using the feedback. All this is to attain general goals about having skills in business, law, public affairs, or technology as well having a faculty in English, at least one other major language, and computers (from William Cummings et al., "Transnational Competence").

Students are in the cycles of the quality-oriented strategic planning where they are with a priority to promote the understanding customs and communication strategies. And they are trained and knowledgably based promoted to understand and analyze each society and its culture. These skills will be necessary for both co-existence and co-development. As well as, learning the local language and customs of both countries will translate into mutual advantage. These skills will help preserve the dignity of nations. Japan higher education had also stepped toward the international Collaboration through the "towards Strategic Policy Planning" theory on Higher Education for the Global Age. This means that higher education institutions worldwide are paying efforts for re-evaluating their goals and objectives, especially in its governance for more integrating and incorporating opportunities of "excellence strategic plans." The challenges are elusive, but commonly known as words: access, expansion, privatization, distance education, technology, decentralization, globalization, and etc. Higher education in East Asia is still confronted with sustaining conservative values (autonomy, elite status, and other terms) while adjusting to the innovative environments of globalization at the same time. This globalization comes with more challenges and questions such as:

- Differentiation and expanded access: This raises the following arguments

  o How will traditional universities respond to policies focused on expanding access?
  o What are some of the tertiary models that are emerging?
  o The globalization of research: What coordination will need to be made in faculties, with respect to the reward?
  o Structure for promotion and retention, with respect to funding for research, copyright, and patent regulations as knowledge becomes more global?

- Privatization: Private higher education has grown rapidly. And the privatization of universities is quite accelerated. It is also true among many nations.
- Social Mobility: What will be the impact on issues of access and equity for groups such as minorities and women who had less access to higher education in many countries as universities adapt to global pressures?
- Internationalization: Knowledge production and dissemination is international in scope and widely available through new media and technology.

  Suggestions have been made and shared with those who are interested in promoting the strategic collaboration as (ToyoshiSatow, 2010):

  o Making strategic global alliances
  o Establishing the common transfer credit system
  o Creating the reward system of the collaboration
  o Writing the future vision

6. Providing Lifelong Learning Opportunities

In order to expand chances for lifelong learning, access to assorted types of learning, such as school education, social education, and home education, is necessary. Therefore, MEXT, working to enhance the purposes of facilities at the Open University of Japan, through which people can receive university education at home, indorses specialized training colleges that afford real-world vocational education and enhance libraries, museums, and community learning centers as situates for learning. MEXT administers the upper secondary school equivalency examinations that permit appropriate evaluation and employ effectively what has been learned and holds the national lifelong learning commemorations in order to stimulate individual attentiveness in lifelong learning (ToyoshiSatow, 2010).

Standards for Establishing Universities (Ordinance of the Ministry of Education, Science and Culture No. 28 of October 22, 1956) (Provisional Translation). Final Revision: Ordinance of the Ministry of Education, Culture, Sports, Science and Technology No. 40 of December 25, 2007.

12.1.2 The Future of Higher Education in Japan

On January 28, 2005, the Central Council for Education report "The Future of Higher Education in Japan" was released. The major content of the report is as follows.

1. Trends in quantitative changes to higher education
   It is forecast that the capacity (number of enrollees/number of applicants) of universities and junior colleges will reach 100% in 2007. The important issue in the future is development of higher education in which anyone can study the field they choose at the level they choose at any time.
2. Clarification of the diverse functions, individuality, and distinctiveness of higher education
   While, as a whole, higher education diversifies, it is necessary to further make clear the individuality and distinctiveness of each school in order to accurately respond to the various needs of learners.
3. Guaranteeing the quality of higher education
   Guaranteeing the quality of higher education in order to protect learners and maintain international validity will be an important issue. For this reason, an accurate operation of the approval system for establishing new departments and enhancement of the third-person evaluation system are necessary.
4. The Ideal Form of Higher Education Institutions
   In order to enhance education, it is important to review and reconstruct the form of "liberal arts education" in undergraduate courses and

strengthen the organizational development of courses of education at graduate schools.
5. The Role of Society Aiming for the Development of Higher Education
   Efforts to expand public funding of higher education and the building of a multifaceted funding system tailored to the diverse functions of each institution are necessary.

   The report also describes policies (the "twelve proposals," etc.) that must be worked on to realize the future of higher education.

12.1.3 Challenges of Higher Education in Japan

Japanese universities are not exception in terms of getting impacted and affected by the rapid change and transform in technology inside and outside Japan. General world educational system is heading for a world-class education system that integrates all systems into a universal one with a unified academic standard and quality-oriented scheme. Japanese universities should submit to the customer-based transform and changing needs, and in fact, they are achieving at this front. Stakeholders' expectations and demanding are surrounding the universities and they are endorsing to cope with technology and globalization. The change includes crucial aspects that cannot be skipped or left behind such as the necessity to develop a quality-oriented strategic plan to development of home culture and awareness of human resources with a solid knowledge-based process that helps to contribute and find solutions of different social issues.

To cope, universities have already started integrate strategies and develop quality assurance procedures to originate and characterize principles, objectives, and missions. They aim to qualitatively improve and enhance their educational systems with the quality-oriented processes they develop to "Fit for Purpose" goal. Japan higher education plays the main important and exclusive role to support universities' own education processes and activities. Higher education has participated with the following so far:

1. Assuring the quality of education system by incorporating and integrating the three systems: The establishment approval system, the quality assurance system and corresponding agencies, and the accreditation system.
2. Endorsing quality-oriented strategic plans and action processes to enhance undergraduate and graduate school courses.
3. Promoting the concept of international competitive.

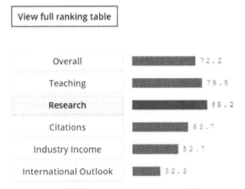

Figure 12.3   World university ranking.

Breakdown via year: RESEARCH

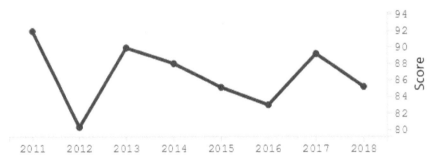

Note: Updated WUR methodology 2015 onwards

Figure 12.4   Research criteria.

Ranking position 2011 to 2018:

2011	2012	2013	2014	2015	2016	2017	2018
26	30	27	23	23	43	39	46

Figure 12.5   History of University of Tokyo ranking.

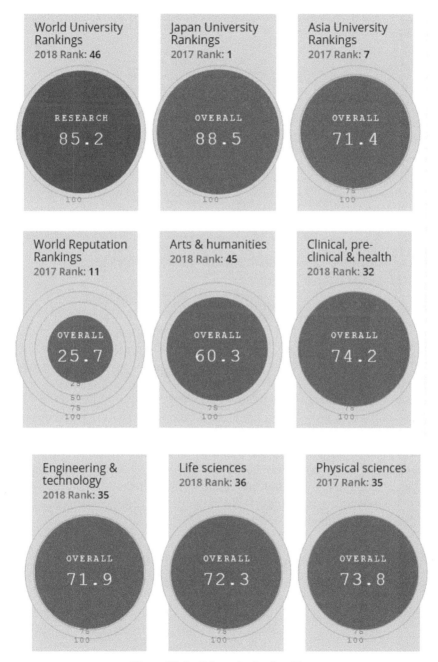

Figure 12.6 Other criteria of ranking.

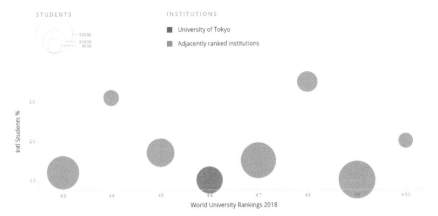

Figure 12.7 Adjacently ranked institutions against international students.

12.2 Japanese Universities and Ranking System: University of Tokyo

12.2.1 About the University of Tokyo

The University of Tokyo was formed in 1877 when the Tokyo Kaisei School and Tokyo Medical School merged to create the faculties of law, science, letters, and medicine, which came together with a university preparatory school.

Since then, the University of Tokyo has merged with a number of schools and institutions to grow into a large research university. It is made up of 10 faculties, 15 graduate schools, 11 affiliated research institutes, 13 university-wide centers, 3 affiliated libraries, and 2 institutes for advanced study and the University of Tokyo Hospital. The university has just over 450 international exchange agreements with universities from all over the world. The University of Tokyo can count 8 Nobel Prize winners, 15 Japanese Prime Ministers, and 5 Astronauts among its alumni.

12.2.2 Explore Rankings' Data for University of Tokyo

Figures 12.3 to 12.7 are related to the website of University of Tokyo.

References and Suggested Literature

[1] Available at: http://www.mext.go.jp/en/news/topics/detail/1381685.htm

[2] *Higher Education Bureau Ministry of Education, Culture, Sports, Science and Technology, Quality Assurance Framework of Higher Education in Japan*, 2009, available at: http://www.mext.go.jp/component/english/_icsFiles/afieldfile/2011/06/20/1307397_1.pdf

[3] Nováky, G., *Quality assurance and benchmarking in Europe and Japan: A shift to focus on teaching and learning, Open Seminar, RIHE*, Hiroshima University, 2015. Available at: http://arinori.hit-u.ac.jp/TuningJapan/pdf/Hiroshima_2015June19.pdf

[4] Maruyama, F., *An Overview of the Higher Education System in Japan*, available at: http://www.zam.go.jp/n00/pdf/nk001001.pdf

[5] *University of Tokyo*, available at: http://www.u-tokyo.ac.jp/en/about/history.html

[6] Available at: https://www.timeshighereducation.com/world-university-rankings/university-tokyo

[7] Hinton, K. E., *A Practical Guide to Strategic Planning in Higher Education*, 2012, available at: https://oira.cortland.edu/webpage/planningandassessmentresources/planningresources/SCPGuideonPlanning.pdf

[8] Newby, H., Weko, T., Breneman, D., Johanneson, T., and Maassen, P., *OECD Reviews of Tertiary Education*, 2009, available at: https://www.oecd.org/japan/42280329.pdf

[9] ToyoshiSatow, J. F., *Internationalization and strategic planning of Higher Education in Japan*, 2010, at Science and Conference Center, Berlin, available at: https://www.hrk.de/fileadmin/redaktion/hrk/02-Dokumente/02-07-Internationales/02-07-15-Asien/02-07-15-3-Japan/Konferenz_Hochschulreformen/6-Presentation_Professor_Satow_J_F_Oberlin_U.pdf

[10] Available at: http://www.mext.go.jp/en/news/topics/detail/_icsFiles/afieldfile/2017/01/31/1381685_001.pdf

13

Quality Assurance for Higher Education in Finland

13.1 The European Education System

European higher education used to be acknowledged for its very assorted national systems. Nowadays, it is characterized by the Bologna Process, the process that established the European Higher Education Area (EHEA). The EHEA is the consequence of the dogmatic will of 48 countries which, step by step throughout the last 18 years, erected an area using mutual apparatuses. These 48 countries instrument reforms on higher education on the basis of common key values – such as autonomy of expression, independence for institutions, independent students' unions, academic freedom, and free movement of students and staff. Over this process, countries, institutions, and stakeholders of the European area constantly acclimate their higher education structures assembling them more companionable and strengthening their quality assurance contrivances. So, in order to achieve the quality assurance, all European countries have presented or are introducing procedures to comfort the quality of the education delivered by their higher education institutions. They have expanded and decided upon Standards and Guidelines for Quality Assurance in the EHEA. These European Standards and Guidelines regard:

- Internal quality assurance within higher education institutions
- External quality assurance of higher education
- External quality assurance agencies
- A peer review system for quality assurance agencies.

European quality assurance agencies have organized themselves in the European Association for Quality Assurance in Higher Education (ENQA). In order to become or remain a full member of ENQA, quality assurance agencies have to undergo an external review. For all these countries, the focal objective is to rise staff and students' mobility and to facilitate employability. This has led to a proliferation in curriculums taught in English, to globally recognizable higher education assemblies and to transparent quality assurance measures. The EHEA should in fact be identifiable by its transparency, by its comparable degrees organized in a three-cycle structure, by its cooperation in quality assurance, and by its mutual recognition of degrees. Thus, in order to allow unrestricted mobility to students, graduates, and higher education staff, the official website of EHEA offers both general information on this procedure and exhaustive information for connoisseurs. The EHEA was launched along with the Bologna Process' decade centenary, in March 2010, during the Budapest-Vienna Ministerial Conference. As the foremost objective of the Bologna Development since its inception in 1999, the EHEA was destined to certify more comparable, compatible, and coherent systems of higher education in Europe. It is well know that between 1999 and 2010, all the exertions of the Bologna Process members were besieged to forming the EHEA, which developed into reality with the Budapest-Vienna Declaration of March 2010. One of the dedications of the Bologna Declaration (1999) was to embolden European cooperation in quality assurance of higher education with an interpretation to developing analogous criteria and practices. The European Ministers of Education embraced in 2005 the "Standards and Guidelines for Quality Assurance in the European Higher Education Area (ESG)" enlisted by the European Association for Quality Assurance in Higher Education (ENQA) in co-operation and consultation with its member agencies and the other members of the "E4 Group" (ENQA, EUA, EURASHE, and ESU). A new version was adopted in 2015 at Yerevan. In 2007, the European Ministers of Education proceeded to setting up the European Quality Assurance Register for Higher Education (EQAR). The Register was set up on 4 March 2008 as the first legal entity to emerge from the Bologna Process. EQAR citation has the ESG as criteria and thus provides statistics on quality assurance agencies that are in significant amenability with this common European framework. In addition, they have inaugurated a peer review system for quality assurance agencies. And as a consequence of the peer review system for quality assurance agencies, the European Ministers have agreed to set up a European register of quality assurance agencies (EQAR). Agencies are admitted to the register after a satisfactory peer review of the agency anxious. This peer review should be in line with the European Standards and Guidelines.

The standards and guidelines for quality assurance in the EHEA are often referred to as the European Standards and Guidelines or ESG. As we mentioned, they were proposed at the ministerial conference of the Bologna Process (2005) by the four focal European higher education stakeholders: ENQA, EUA, EURASHE, and ESU. The ESG aim to clarify the organization of quality assurance and lay down the standards and guidelines:

- For internal quality assurance within higher education institutions,
- For external quality assurance of higher education,
- For external quality assurance agencies.

Quality, whereas not easy to outline, is mostly a result of the collaboration between teachers, students, and the institutional learning environment. Quality assurance should ensure a learning environment in which the content of programs, learning openings, and amenities are fit for purpose. Quality assurance accomplishments are the matching purposes of accountability and enhancement. Taken together, these create trust in the higher education institution's performance. An effectively implemented quality assurance system will deliver materials to assure the higher education institution and the municipal of the quality of the higher education institution's activities (accountability) as well as deliver advice and endorsements on how it might expand what it is doing (enhancement). Quality assurance and excellence enhancement are thus inter-related. They can support the progress of a quality culture that is incorporated by all: from the students and academic staff to the institutional leadership and management. A variety of procedures can construct layers of quality-oriented education systems, but with ESQ, they need to state one framework and this is their main purpose and dedication. Setting a common framework for quality assurance systems for learning and teaching at European, national, and institutional level will help enabling the assurance and enhancement of quality of higher education in the EHEA. The frame supports a mutual trust, aiding recognition and mobility within and across national borders. It for sure reflects the quality assurance in the EHEA.

These drives provide a framework within which the ESG may be used and applied in different ways by different organizations, agencies, and countries. The EHEA is branded by its diversity of political systems, higher education systems, socio-cultural and educational traditions, languages, ambitions, and expectations. This makes a single colossal approach to quality and quality assurance in higher education unsuitable. Broad reception of all standards is a precondition for creating common understanding of quality assurance in Europe.

All these requirements necessitates, that the ESG prerequisite to be at a rationally generic glassy in order to ensure that they are applicable to all arrangements of endowment. The ESG provide the criteria at European level alongside quality assurance agencies and their accomplishments are evaluated. This guarantees that the quality assurance agencies in the EHEA follow the identical set of ideologies and the methods and measures are demonstrated to fit the purposes and necessities of their contexts. ESQ standards for internal and external quality are (EQNA, 2017):

A. European standards and guidelines for internal quality assurance within higher education institutions

1. Policy and procedures for quality assurance: Institutions should have a policy and associated procedures for the assurance of the quality and standards of their programs and awards.
2. Approval, monitoring, and periodic review of programs and awards: Institutions should have formal mechanisms for the approval, periodic review, and monitoring of their programs and awards.
3. Assessment of students: Students should be assessed using published criteria, regulations, and procedures which are applied consistently.
4. Quality assurance of teaching staff: Institutions should have ways of satisfying themselves that staff involved with the teaching of students are qualified and competent to do so. They should be available to those undertaking external reviews and commented upon in reports.
5. Learning resources and student support: Institutions should ensure that the resources available for the support of student learning are adequate and appropriate for each program offered.
6. Information systems: Institutions should ensure that they collect, analyze, and use relevant information for the effective management of their programs of study and other activities.
7. Public information: Institutions should regularly publish up-to-date, impartial, and objective information, both quantitative and qualitative, about the programs and awards they are offering.

B. Standards for the external quality assurance of higher education

External quality assurance procedures should take into account the effectiveness of the internal quality assurance processes described before of the European Standards and Guidelines.

1. **Development of external quality assurance processes**: The aims and objectives of quality assurance processes should be determined

before the processes themselves are developed, by all those responsible (including higher education institutions) and should be published with a description of the procedures to be used.

2. **Criteria for decisions**: Any formal decisions made as a result of an external quality assurance activity should be based on explicit published criteria that are applied consistently.

3. **Processes fit for purpose**: All external quality assurance processes should be designed specifically to ensure their fitness to achieve the aims and objectives set for them.

4. **Reporting:** Reports should be published and should be written in a style, which is clear and readily accessible to its intended readership. Any decisions, commendations, or recommendations contained in reports should be easy for a reader to find.

5. **Follow-up procedures:** Quality assurance processes which contain recommendations for action or which require a subsequent action plan should have a predetermined follow-up procedure which is implemented constantly.

6. **Periodic reviews:** External quality assurance of institutions and/or programs should be undertaken on a cyclical basis. The length of the cycle and the review procedures to be used should be clearly defined and published in advance.

7. **System-wide analyses:** Quality assurance agencies should produce from time to time summary reports describing and analyzing the general findings of their reviews, evaluations, assessments, etc.

The inspiration of the European Standards and Guidelines for quality assurance in higher education (ESG) is spreading and they are gaining acceptance as a shared reference point for all actors in European higher education. Currently, EQAR is listing 24 agencies in 23 countries, compliant with the ESG, which can carry out evaluations in any country of the EHEA. Though the essential responsibility for quality endures to rest within the higher education institutions, internal quality assurance is a duty of the institution, and the development of an effective "quality culture" is clearly linked with their degree of operational autonomy. The focus of the ESG is on quality assurance related to learning and teaching in higher education, comprising the learning atmosphere and related links to research and innovation.

In addition, establishments have policies and procedures to confirm and improve the quality of their other undertakings, such as research and governance. The ESG relate to all higher education offered in the EHEA regardless of the approach of education or place of delivery. ESG are also valid to all higher education comprising transnational and cross-border delivery. ESQ assures the vision of higher education institutions which in general aims to fulfill multiple purposes, including preparing students for active citizenship, for their future vocations, contributing to their employability, supporting their personal development, creating a broad innovative knowledge base, and encouraging research and innovation. Therefore, stakeholders, who may prioritize different purposes, can view quality in higher education differently and quality assurance needs to take into account these different perspectives.

Institutions on the other side have rights and space of applying the quality procedures. Therefore, ESQ set out good practice in the relevant area for consideration by the actors involved in quality assurance per country. Implementation will vary depending on different contexts.

- Higher education institutions have primary responsibility for the quality of their provision and its assurance.
- Quality assurance responds to the diversity of higher education systems, institutions, programs, and students.
- Quality assurance supports the development of a quality culture.
- Quality assurance takes into account the needs and expectations of students, all other stakeholders, and society.

Finland is a country in Europe that submits to the ESQ. In the following section, we will take an example of Finnish higher education and try to enlighten the four main factors of excellence for higher education: Education system, quality assurance, strategic planning, and internationalization.

13.2 Quality Assurance in Finnish Higher Education

Teaching and learning process in Finland is an education system with fully endowed features to students: Vocational education; higher education (University and University of Applied Sciences); and adult (lifelong, continuing) education. The Finnish strategy for achieving equality and excellence in education has been based on constructing a publicly funded comprehensive school system without selecting, tracking, or streaming students during their common basic education. Part of the strategy has been to spread the school network so that pupils have a school near their homes whenever possible or, if this is not feasible, e.g., in rural areas, to provide free transportation to more widely dispersed schools. Inclusive special education within the classroom and instructional efforts to minimize low achievement are also typical of

Nordic educational systems. Finland has a dual structure. Higher education is provided by universities and universities of applied sciences (UAS). Both sectors have their own profiles.

Universities emphasize scientific research, whereas universities of applied sciences adopt a more practical approach. Higher education institutions are very autonomous in organizing their instruction and academic year. There is restricted entry to all fields of study. The applicant volumes outweigh the number of places available. Therefore, universities and UAS use different kinds of student selection criteria. Most commonly, these include success in matriculation examination and entrance tests.

Equal access to higher education is ensured by the wide institutional network, the free education, student financial aid. as well as the flexible pathways to higher education. Efforts have also been made to lower the threshold to apply to higher education by developing an online joint application system.

This Joint application system is well-designed and presented to afford all required Information about higher education degrees and recognition of foreign qualifications in Finland.

13.2.1 Education System

There are 14 universities in the Ministry of Education and Culture sector; two of them are foundation universities and the rest are public corporations. University-level education is also provided by a military institution of higher education, the National Defence University, which is part of the Defense Forces. Altogether there were nearly 170,000 university students in 2012. From the beginning of 2010, universities have had the status of independent legal entities and been separated from the state. However, the state continues to be the primary financier of the universities. Direct government funding covers about 64% of university budgets. In addition, universities are encouraged to acquire private donations.

There are 24 polytechnics in the Ministry of Education and Culture sector. From the beginning of the year 2015, they have had the status of independent legal entities and been operated as limited companies. The State will be the primary financier of the polytechnics. In addition, there is the Åland University of Applied Sciences in the self-governing Province of Åland and the Police College of Finland subordinate to the Ministry of the Interior.

The main objective of Finnish education policy is to offer all citizens equal opportunities to receive education. The structure of the education system reflects these principles. The system is highly permeable, that is, there are no dead-ends preventing progression to higher levels of education.

The focus in education is on learning rather than testing. There are no national tests for pupils in basic education in Finland. Instead, teachers are responsible for assessment in their respective subjects on the basis of the objectives included in the curriculum.

The only national examination, the matriculation examination, is held at the end of general upper secondary education. Commonly admission to higher education is based on the results in the matriculation examination and entrance tests.

Governance has been based on the principle of decentralization since the early 1990s. Education providers are responsible for practical teaching arrangements as well as the effectiveness and quality of the education provided. Local authorities also determine how much autonomy is passed on to schools. For example, budget management, acquisitions, and recruitment are often the responsibility of the schools.

The European education policy is outlined in the Bologna Process, where the European ministers responsible for higher education have agreed on joint objectives for the development of the EHEA by 2010. The European higher education policy emphasizes quality assurance and aims to enhance the competitiveness of the Europe increasing the mobility of students, staff, and labor force. Increased student mobility and international degree programs are challenges for the future of HEIs.

To ensure that everyone has educational opportunities, student financial aid is designed to benefit a large proportion of students. Financial aid includes mainly study grants and housing supplements. Scholarships and/or other grants to households amount to 14.9% of public expenditure for tertiary education, above the OECD average of 11.4%. More than half of national students in first degree programs in tertiary-type A education (54%) benefit from scholarships and/or grants. Figure 13.1 presents the Specificities of the Finnish education system as it shows below.

In Finland, education providers have the main responsibility for quality

1. Skills demonstrations of competence-based qualifications
2. Self-evaluation
3. National evaluations of learning outcomes
4. System and thematic evaluations
5. International assessments
6. And most significance, teachers education and training.

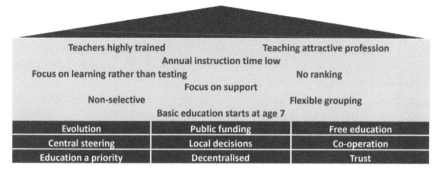

Figure 13.1 Specificities of the Finnish education system.

Source: http://www.oph.fi/download/175015_education_in_Finland.pdf

1. Teacher education

Teachers in Finland obtain a strong theoretical and practical teacher education and are highly respected. Pre-service teacher-training requirements in higher education institutions comprise a competitive inspection to enter training and a teaching workshop as part of pre-service training. As trusted professionals, teachers have considerable pedagogical autonomy to interpret the curricula and to choose teaching methods and materials, as well as resources and methods to continuously assess students' progress. Teachers at universities of applied sciences are obligatory to have either a Master's or a post-graduate Licentiate's degree, depending on their position. They must also complete pedagogical studies.

University teachers are generally compulsory to hold a Doctoral or other postgraduate degree. Teacher training can be either synchronized, with pedagogical training integrated into the Master's program, or uninterrupted, with the pedagogical training completed after the initial degree. At most levels of education, teachers are requisite to contribute in in-service training every year. Finnish teachers consider in-service training to be a privilege and therefore partake aggressively.

In-service training is presented by different providers. The state funds in-service training programs, principally in areas important for implementing education policy and reforms. Education providers can also apply for funding to improve the professional competence of their teaching personnel.

2. Progression of students

University student goes through lot of evolution and rapid development in their studies by finishing individual courses and study modules. The liberty of choice concerning the order of studies varies between different subjects: in some fields, students are free to plan the sequence of their studies, while the order of courses is defined in more detail in other fields. For some courses, the student may be required to have completed certain introductory studies or received, for example, the grade "good" from earlier studies. University students may take studies included in the degree programs or other conceivable studies offered by the university, or complete studies at other Finnish or foreign universities and institutions of higher education. Universities have discrete covenants on the right to study at these institutions. The system of individual study plans facilitates the planning of studies and the observing of progress in studies and supports student guidance and counseling. Polytechnic students progress in their studies by completing courses. Each degree program consists of central studies and professional studies, optional studies, practical training to promote professional skills, and a final project which also includes a maturity test. Studies are compulsory, optional, or free choice. After completion of a polytechnic or another appropriate higher education degree and at least three years of work experience in their field, students may apply for the right to complete a polytechnic Master's degree.

3. Employability

Higher education institutions have been fortified to collaborate with the labor market and the initiatives. One objective of this collaboration is to ensure that studies and degrees are relevant to the labor market. Many degree programs further include compulsory practice in enterprises in their study programs. In addition, higher education institutions have recruitment services. There are joint recruitment services for both universities and polytechnics. Both of these have portals where users can find information on degrees and qualifications, career planning, and writing applications. These services are meant for students who seek placements for training periods during their studies or vacancies after they have graduated. The services are also used by employers for their recruitment purposes.

4. Student assessment

At polytechnics and universities, student assessment is based on continuous assessment. In most cases, students are assessed on the basis of written examinations at the end of lecture series or larger study units, but there are also oral examinations. In addition, students write papers for seminars and other papers. For the university Bachelor's degree, students write a thesis. At art academies, the thesis may take the form of an artistic production, such as a concert, a play, or some other performance, which also includes a written part. At polytechnics, students do an individual final year project that commonly includes a thesis. The examiners of coursework are usually the course lecturers or the teachers responsible for the study unit or module.

5. ESIB – the National Unions of Students in Europe

ESIB is the umbrella organization of 50 national unions of students from 37 countries and through these members. The aim of ESIB is to represent and promote the educational, social, economic, and cultural interests of students at a European level toward all relevant bodies and in particular the European Union, Council of Europe, and UNESCO. Aims and objectives: ESIB – the National Unions of Students in Europe has the following goals:

- To promote the views of students on the educational system as a whole.
- To promote the social, economic, political, and cultural interests of students and the human rights of students which have a direct effect or an indirect effect on education and on the status and welfare of students in society.
- To promote equal opportunities for all students regardless of their political belief, religion, ethnic or cultural origin, gender, sexual orientation, social standing, or any disability they may have.
- To promote equal chances of access to higher education for all people.
- To promote European and global co-operation and to facilitate information exchange with students and students' organizations.
- To promote co-operation with other organized groups in matters pertaining to education and student life.
- To provide assistance and support to National Unions of Students across Europe – hereafter referred to as NUSs – in their work to protect student interests.

About the QA project, The ESIB Quality Assurance project was initiated in June 2001 and was funded by the European Commission, the Council of Europe, and the Dutch Government. The project lasted for slightly more than a year and had the following key objectives: Improving quality assurance processes and students' involvement in them by:

- Collecting, analyzing, and disseminating theory, good practices, and experiences of (student involvement in) quality assurance in Europe, focusing on exchanges between well-developed quality assurance systems and less developed.

- Raising awareness of the importance of (student involvement in) quality assurance processes.
- Identifying and promoting European-wide strategies to involve students and student organizations in quality assurance.
- Promoting co-operation of European student organizations on one of the key themes of the Bologna Process. The project had several target groups which it sought to involve in its work and also affect with its outcomes. These can be outlined as:
- Primary: Students, especially those who are actively involved in organizations and bodies dealing with quality of education.
- Secondary: Other parties dealing with quality assurance and student involvement in European higher education, such as educational staff and policy makers.
- Throughout the time of the project, a number of activities were undertaken which resulted in both direct and indirect outcomes toward the objectives of the project. The main activities are outlined below:
- October 2001 – European Student Seminar on Quality Assurance. This seminar brought together student representative and QA experts from throughout Europe to examine developments in QA.
- November/December 2001 – A set of checklists and guidelines of best practices arising from the seminar was prepared and distributed to member organizations.
- January/June 2002 – The production of the "European Student Handbook on Quality Assurance" for local and national student representatives was started with the completion of the first edition in June 2002.
- May 2002 – Training for national student representatives on involving students in quality assurance mechanisms.

6. High quality and research

Excellent basic education and competitive student admission underpin the high standards in Finnish higher education.

The Finnish attitude to studying follows an approach that benefits the individual: students are confronted and encouraged to think for themselves. Teaching in universities is based on research and the institutions' precise expertise. These high-quality study programs combine in-depth research with the needs of your future working life. At universities of applied sciences (UAS), Research and Development is the foundation for teaching. Co-operation with local industries boosts learning outcomes as the education is shaped to answer the requirements and expectations of both industry and commerce.

13.2.2 The Quality Assurance of Finnish Higher Education

FINEEC, the Finnish Education Evaluation Centre, was established on 1 May 2014 by merging FINHEEC, the Finnish Higher Education Evaluation Council, the Finnish Education Evaluation Council, and the evaluation of education undertaken by the Ministry of Education and Culture.

The Finnish Education Evaluation Centre (FINEEC) is the national quality assurance agencyresponsible for evaluations of higher education in Finland. FINEEC is a full member of the EuropeanAssociation for Quality Assurance in Higher Education (ENQA) and is included in the European QualityAssurance Register for Higher Education (EQAR). An audit is an independent and systematic external evaluation. It assesses whether the quality system of a higher education institution is fit for purpose and functioning and whether it complies with the agreed criteria.

One of the main principles of the FINEEC audit is the autonomy of higher education institutions (HEI), also stated in the Finnish Universities Act and Polytechnics Act: the HEIs are responsible for the quality and continuous development of their education and other operations. Another guiding principle of the audits is enhancement-led evaluation. The aim is to help HEIs to identify strengths, good practices, and areas in need of development in their own operations.

Finnish quality assurance system is enhancement-oriented, aiming at refining the study and learning experience for all students. Independent audits of the quality control systems of higher education institutions are conducted by FINEEC, The Finnish Education Evaluation Centre. The criteria used in auditing these quality assurance systems are made public, as are all the final reports. The quality audit evaluates the effectiveness of the system, issuing recommendations for its improvement and highlighting best practices. A successful higher education institution is awarded an audit certificate indicating that it has been successfully audited. The quality developments outlined two trends, which could be labeled as:

1. Quality assurance of internationalization

Based on the lack of attention that was paid to internationalization of higher education in most quality assurance systems, it was concluded that such approaches should be promoted. In order to do so, initiatives were taken to develop suitable methodologies for quality assurance of internationalization and efforts were made to enhance the dialog with quality assurance actors and agencies.

2. Internationalization of quality assurance

The increasing international networking between quality assurance actors and agencies includes the exchange of information and experiences, inclusion of foreign experts in review panels, various forms of cooperation, EU projects, and international and European networks.

3. Strategic planning

FINEEC's evaluation activities have four strategic focus areas that are determined on the one hand by the analysis of the changes in the national and international operating environment and on the other hand by foresight and effective evaluation:

1. Developing learning and competence with evaluation. Evaluations implemented with different enhancement led methods aim at improving learning outcomes and competence at all educational levels in both official language groups.
2. Functionality and development of the educational system. Evaluation activities that cover all educational levels provide information on the functionality of the entire educational system and policy. The evidence-based evaluation information forms a basis for development work. Evaluations are also targeted at the educational level boundaries and various transition phases.
3. Themes which are central and critical in the society. Evaluations are targeted at societally important and critical themes. Based on an analysis of the changes in the operating environment, significant development targets in education, which are not included in the evaluation plan, may be raised for evaluation.
4. Supporting education providers in quality management and in strengthening an operating culture based on enhancement-led evaluation. FINEEC supports education providers in developing quality management by evaluating their quality systems and producing information on good practices in quality management and development, as well as by spreading the information across different educational levels.

13.2.3 Strategic planning in Finnish Higher Education

Investigate the integration of quality assurance and strategic management in higher education. The integration concept presents the quality map, which is a graphical representation of the quality assurance system. The quality map explicitly takes into account the environment, strategic planning, and the internal processes of the organization. The quality map helps the management of the higher education institution to present an overview of the quality assurance system to the external evaluators, members of the organization, students, and other stakeholders. The framework developed in this study is used to analyze the engagement of HEIs in regional development. The HEIs are accountable for the quality of their education and other activities to the quality assurance agencies, the ministries of education, and other stakeholders. The institutions are obliged to regularly evaluate their activities and participate in external evaluations (Ministry of Education, Finland, 2005).

Finnish higher education Strategic plans have the main as follows (https://karvi.fi/app/uploads/2014/10/KARVI_Strategia_2015_EN_WWW.pdf):

Vision
Finland develops education based on the versatile and up-to-date evaluation information produced by the Finnish Education Evaluation Centre.

Mission
FINEEC is a nationally significant and internationally desired evaluation partner in the field of education and an inspiring developer that produces evidence-based evaluation information that has an impact on the development of education.

Operating principle
FINEEC develops operating principle that can be visualized in the following Figure 13.2 and represents steps of how they perform.

13.2.4 Internationalization of Finnish Higher Education

Internationalization is not a new phenomenon as previous scholars traveled throughout Europe and in this contemporary world where the high-tech utensils and innovative knowledge of global communication and attentiveness are considered as the main possessions. Due to the on-going labor marketplace, there is an essential for students to have international, foreign communications and intercultural abilities to interrelate worldwide, and therefore, institutions are focusing more importance on internationalization. Bruch and

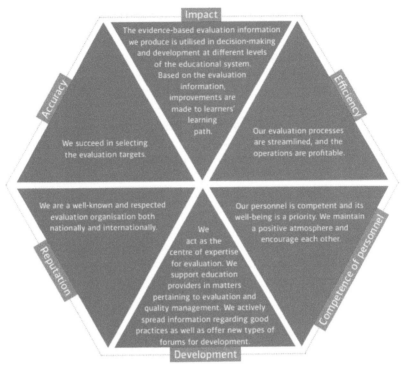

Figure 13.2   FINEEC develops, experiments, reforms, involves, and serves.

Source: https://karvi.fi/app/uploads/2014/10/KARVI_Strategia_2015_EN_WWW.pdf)

Barty (1998) write that internationalization in the higher education in the present context is not a new incidence as traveling to places has been an aspect in education for centuries, ever since "wondering scholars" ventured from place to place to expand their knowledge (Cited in Montgomery, 2010). As mentioned by Davies et al. (2009), the higher education internationalization strategy aiming to increase the number of international students is required as to satisfy the rapidly growing labor market needs with an increased input of foreign talent (Cited in Shumilova et al., 2012).

In favor of supporting the internationalization of the Finnish HEIs, the Ministry of Education has recruited a strategy for the internationalization of higher education in Finland. Furthermore, as one of the central features of the internationalization process is the involvement of the international students pursuing their higher education in Finnish universities of Applied Sciences and the Universities, it is of critical importance to shed light on the life circumstances of the international students living in Finland. FINEEC is active in the main international evaluation networks. Being vigorous in the networks strengthens the conspicuousness of Finnish evaluation activities and enables influencing the development of European evaluation activities. Moreover, the information derived through the international networks on, for instance, the schedules for multinational evaluations (e.g., PISA, TIMMS, and PIRLS), is important for the scheduling of the national assessments of learning outcomes.

FINEEC reinforces its role as an expert in various international projects and networks. Cooperation structures within the EU and OECD will be in particular focus as instruments for collaboration and influence. During the validity of the evaluation plan, FINEEC pledges international peer review and benchmarking activities with evaluation and quality assurance organizations within general education and professional education and training. FINEEC continues to contribute in European network projects for comparing evaluation models for higher education institutions. In particular, FINEEC participates in support that produces added value to the expansion of the quality of education, learning, and the Finnish educational system as well as promoting the export of Finnish evaluation expertise. Such projects include the EU's twinning projects.

In 2016, FINEEC implemented the Common Assessment Framework (CAF). With respect to its evaluations of higher education institutions, FINEEC participated in an external ENQA review in 2016 to maintain its full ENQA membership and EQAR listing. FINEEC is also preparing for a full external review of its activities during the period covered by the plan.

13.3 Examples of Finnish Universities: University of Helsinki

13.3.1 University of Helsinki

(https://www.helsinki.fi/en)

The University of Helsinki is a university located in Helsinki, Finland, since 1829, but was founded in the city of Turku (in Swedish Åbo) in 1640 as the Royal Academy of Åbo, at that time part of the Swedish Empire. It is the primogenital and largest university in Finland with the broadest range of disciplines available. Around 36,500 students are currently enrolled in the degree programs of the university spread across 11 faculties and 11 research institutes.

13.3.2 University of Helsinki Rankings

http://www.shanghairanking.com/World-University-Rankings/University-of-Helsinki.html

The University of Helsinki is ranked at 56th in the world by the 2016 Academic Ranking of World Universities published by Shanghai Jiao Tong University. According to the Times Higher Education World University Rankings for 2016, the University of Helsinki is ranked at 76th overall in the world. In the 2014 QS World University Rankings' list, the University of Helsinki was ranked 67th. Figures 13.3 to 13.5 are from ranking systems.

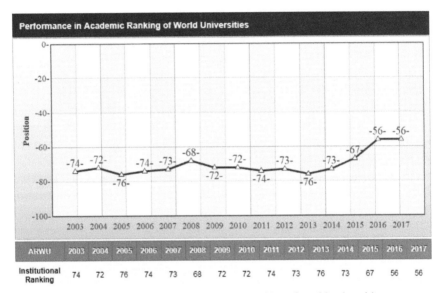

ARWU	2003	2004	2005	2006	2007	2008	2009	2010	2011	2012	2013	2014	2015	2016	2017
Institutional Ranking	74	72	76	74	73	68	72	72	74	73	76	73	67	56	56

Figure 13.3   Performance in academic ranking of world universities.

Performance in Academic Ranking of World Universities by Broad Subject Fields										
Broad Subject Fields	2007	2008	2009	2010	2011	2012	2013	2014	2015	2016
SCI	/	/	/	/	/	101-150	76-100	76-100	76-100	76-100
ENG	/	/	/	/	/	/	/	/	/	/
LIFE	77-106	76-107	76-100	76-100	51-75	76-100	76-100	51-75	51-75	51-75
MED	51-75	52-75	51-76	76-100	51-75	76-100	51-75	51-75	51-75	51-75
SOC	/	/	/	/	/	/	/	/	/	/

Figure 13.4   Performance in academic ranking of world universities by broad subject fields.

Performance in Academic Ranking of World Universities by Subject Fields							
Subject Fields	2009	2010	2011	2012	2013	2014	2015
Mathematics	/	/	/	151-200	/	151-200	/
Physics	78-104	76-100	76-100	76-100	51-75	101-150	101-150
Chemistry	/	/	/	/	/	/	/
Computer Science	/	/	/	/	/	/	/
Economics/Business	/	/	/	/	/	/	/

Figure 13.5 Performance in academic ranking of world universities by subject fields.

References and Suggested Literature

[1] European Association for Quality Assurance in Higher Education, *Standards and Guidelines for Quality Assurance in the European Higher Education Area*, 2009, Helsinki, 3rd edn. Available at: http://www.enqa.eu/wp-content/uploads/2013/06/ESG_3edition-2.pdf

[2] *Education in Finland, Finnish National Agency for Education*, 2017. Available at: http://www.oph.fi/download/175015_education_in_Finland.pdf

[3] Available at: http://www.oph.fi/english/education_system/higher_education

[4] Available at: http://www.oph.fi/english/education_system/international_perspectives

[5] Hanhijoki, I., Katajisto, J., Kimari, M., and Saviojam, H., *Education, training and demand for labour in Finland by 2025*, 2012. Available at: http://www.oph.fi/english/publications/2012/education_training_and_demand_for_labour_in_finland_by_2025, http://www.oph.fi/download/144754_Education_training_and_demand_for_labour_in_Finland_by_2025_2.pdf

[6] Kauppinen, J., *Curriculum in Finland*, 2016. Available at: http://dge.mec.pt/sites/default/files/Noticias_Imagens/1_curriculum_in_finland.pdf

[7] Available at: http://ncee.org/what-we-do/center-on-international-education-benchmarking/top-performing-countries/finland-overview/finland-teacher-and-principal-quality/

[8] *Finnish Education Evaluation Center*, available at: https://karvi.fi/en/fineec/

[9] OECD, *Education Policy Outlook Finland*, 2013. Available at: http://www.oecd.org/edu/EDUCATION%20POLICY%20OUTLOOK%20FINLAND_EN.pdf

[10] Finnish Education Evaluation Centre, *National Plan for Education Evaluation 2016–2019*. Available at: https://karvi.fi/app/uploads/2016/06/National-Plan-for-Education-Evaluations-2016-2019.pdf

[11] Chand, S., *Internationalization of Finnish Higher Education, A Literature Review on the Living Scenarios and Employability of International Students in Finland*. Available at: https://www.theseus.fi/bitstream/handle/10024/83950/chand_suraj.pdf?sequence=1

[12] Finnish Education Evaluation Center, *Foresight and Effective Evaluation 2020, The Strategy of Finnish Education Evaluation Centre*.

[13] Haukijärvi, I., *Strategizing Digitalization in a Finnish Higher Education Institution Towards a Thorough Strategic Transformation*, 2016. Available at: https://tampub.uta.fi/bitstream/handle/10024/99877/978-952-03-0156-9.pdf?sequence=3

[14] Kettunen, J., *Integration of Strategic Management and Quality Assurance*. Available at: http://citeseerx.ist.psu.edu/viewdoc/download;jsessionid=4BAC0A4E7A314479D63272F37FF6ABA4?doi=10.1.1.626.3185&rep=rep1&type=pdf

[15] Available at: *Five Finnish Universities Ranked In The Academic Ranking Of World Universitie*. Available at: http://www.studyinfinland.fi/frontpage_news/101/0/five_finnish_universities_ranked_in_the_academic_ranking_of_world_universities

[16] Available at: http://www.euroeducation.net/prof/finco.htm

[17] Available at: https://en.wikipedia.org/wiki/University_of_Helsinki

[18] Available at: http://www.uta.fi/laatujarjestelma/en/in_Finland.html

[19] Available at: http://www.shanghairanking.com/World-University-Rankings/University-of-Helsinki.html

[20] Bernhard, A., *Quality Assurance in an International Higher Education Area: A Case Study*, Springer Science and Business Media.

[21] Vlâsceanu, L., Grünberg, L., and Pârlea, D., *Quality Assurance and Accreditation: A Glossary of Basic Terms and Definitions*. Available at: http://unesdoc.unesco.org/images/0013/001346/134621e.pdf

[22] European Association for Quality Assurance in Higher Education, Helsinki, *Terminology of Quality Assurance: Towards Shared European Values?*, 2006 Available at: http://www.enqa.eu/indirme/papers-and-reports/occasional-papers/terminology_v01.pdf

[23] Lim, D., *Quality Assurance in Higher Education in Developing Countries*. Available at: http://www.tandfonline.com/doi/abs/10.1080/0260293990240402

[24] ESIB, The National Unions of Students of Europe, *European Student Handbook on Quality Assurance in Higher Education.* Available at: http://www.aic.lv/bolona/Bologna/contrib/ESIB/QAhandbook.pdf

[25] Available at: http://www.euroeducation.net/prof/finco.htm

[26] Campbell, C., van der Wende, M. *International Initiatives and Trends in Quality Assurance for European Higher Education. The European Network for Quality Assurance in Higher Education Helsinki*, Helsinki, 2000. Available at: http://www.enqa.eu/indirme/papers-and-reports/occasional-papers/initiatives.pdf

[27] Available at: https://en.wikipedia.org/wiki/European_Higher_Education_Area

[28] European Association for Quality Assurance in Higher Education (ENQA), European Students' Union (ESU), European University Association (EUA), European Association of Institutions in Higher Education (EURASHE), In cooperation with: Education International (EI), BUSINESSEUROPE, European Quality Assurance Register for Higher Education (EQAR), *ESG 2015, Standards and Guidelines for Quality Assurance in the European Higher Education Area (ESG).* Available at: http://media.ehea.info/file/ESG/00/2/ESG_2015_616002.pdf

[29] Kauppinen, J., *Curriculum in Finland*, 2016. Available at: http://dge.mec.pt/sites/default/files/Noticias_Imagens/1_curriculum_in_finland.pdf

[30] ENQA Agency Review: Agency for Quality Assurance in Higher Education, (AEQES), Norma Ryan, Oana Sarbu, Jean-Pierre Finance, Rok Primozic 13 February 2017, https://backend.deqar.eu/reports/EQAR/AEQES_External_Review_Report_2017.pdf

Index

A

academic collaboration 123
academic experience 61, 196
academic leadership 102
academic plans 42, 112, 133
academic programs
    committee 204
academic standards 39, 154, 385
accountability 4, 142, 399
accreditation 6, 139, 391
accumulative knowledge based
    system 48
active committee 54
adopting quality assurance 12
aggregation and disaggregation
    270, 272
alumni outcomes 332, 333
approval system 379, 390, 391
artificial intelligence-led quality
    assurance 180
artificial intelligence 29, 167, 182
ARWU 279, 298, 344
assessment conceptual model
    232
assessment for excellence 141
assessments 6, 229, 413
assessments committee 72

B

benchmarks 118, 129, 357
blended learning 23, 198, 252
broadening 122, 217, 320

C

central council 380, 383, 390
characteristics of excellence
    359, 360
checklist 236, 408
cloud 156, 258
cloud computing 57, 166
combined learning 192
community_based
    organization 72
components of strategic
    planning 67
consultation 26, 269, 398
continuous process 51, 94, 223
conventional strategic
    planning 76
coordination 19, 150, 389
core design principles 263, 267
core pedagogic principles
    263, 267
course – coordination
    system 237, 240
course content 21, 154, 207
criterion references
    approach 360
customer driven focus 109

D

development procedures 67
diagnostic assessments 233,
    265
digital divide 26, 31, 125

digital resources 262
domain model 171, 173, 178

E
economic engines 39
education process 27, 176, 391
education pyramid 4, 95, 377
educational assessments 218
educational metadata 250
educational resources 251, 258
educator perspective 26
e-government 28, 33
e-learning 1, 252, 344
embedded learning
    resources 250
employment opportunities
    130, 206
encapsulation 47
ESIB 407
evaluation 4, 390, 413
excellence 6, 91, 353
excellence model 10, 355, 377
excellence strategies 65, 67
expansion of higher
    education 55, 122
external environment 64, 76, 81
external quality
    assurance 39, 134, 401

F
faculty affairs 42
faculty performance 66
feedback 19, 174, 388
fit for purpose 7, 391, 409
foundation program 130

G
global academic ranking
    system 292

global realization 39
global university rankings
    279, 316
globalization 9, 24, 121
grade correlation 284
graduate employment rate 333
graduates 2, 331, 398
granularity 270, 271

H
heterogeneous learning
    resources 251
high class 351
higher education 3, 52, 318
higher education council 72, 140
higher education
    framework 39, 91, 359
higher education standards
    42, 99, 220
householders 30

I
ICT 21, 123, 261
impact of ranking 319
impact of ranking
    systems 323
improvements cycles
    75, 111, 356
incorporated quality
    system 48
individual level 193, 208, 372
innovation & excellence 45
inspirational strategic
    planning 77
institutional effectiveness 86, 147
institutional audit 116, 209
institutional level 67, 119, 316
intelligent tutoring system
    171, 179

interactive competencies
360, 361
internal quality assurance
10, 140, 400
international partners 125, 314
international ranking 296, 344
internationalization 12, 125, 407
interpretation 79, 286, 398
investigation and research
methodologies 192
ISO 10, 103, 363
issue – based strategic
planning 76
iterative knowledge based
system 199

K
Kerby simple difference
formula 291, 292
know-how 27, 48
knowledge – transfer 111, 149
knowledge based learning
process 336

L
landmark of development 47
learning bodies 48
learning environment 28,
262, 399
learning frameworks 192
learning management
system 21, 177, 269
learning objectives
145, 217, 366
learning outcomes 10, 143,
385, 408
learning practices 135, 204, 326
learning process 1, 116, 204, 402
learning resources 11, 221, 273

learning strategies 27, 101, 256
learning-teaching units 195
lifelong learning 39, 112, 390

M
meta quality assurance
system 140
metamorphoses 54, 177
MEXT 377, 382, 390
mission 8, 366, 411
monitor standards 72
multi-campus universities 223
multi-layer of quality-oriented
building blocks 47
multination education 122

N
national framework 40, 123
NIAD-UE 385
non-parametric approach 281

O
on-going process 222
operational system 47
organic strategic planning 76
outcomes of learning 49

P
passive motivator 334
pedagogical scheme 27
peer-reviews 140, 318, 373
performance-based
assessments 233
performance indicators
95, 206, 306
portfolio 118, 210, 236
privatization 122, 387
problem solving 14, 123, 173
program delivery 19, 154

program level 119, 144, 208
project based learning 192
protocols 40, 257, 379
provision 18, 156, 385
public universities 40, 55, 330

Q

QS 279, 304, 414
qualitive and quantitive
    standards 192
quality agents 44
quality assurance 42, 145,
    211, 407
quality assurance agencies
    10, 134, 411
quality assurance enhancement
    52
quality assurance in Finnish
    higher education 402
quality assurance in Japan higher
    education 378
quality-based melody 43
quality benchmarks 253
quality control 13, 96, 235, 409
quality culture 4, 132, 205, 401
quality enhancement cells 100
quality management 10, 97, 410
quality map 82, 411
quality of teaching 6, 49,
    211, 378
quality orientation 9, 124, 371
quality oriented
    assessments 230
quality-oriented concepts
    14, 193
quality-oriented curriculum 136
quality performance 109
quality-oriented procedures 43,
    116, 391

query dependent rank 278
query independent rank 278

R

rank-biserial 291
rank correlation 277, 292
ranking system 42, 280, 318
real-time strategic planning 76
research 2, 59, 161, 254
research quality 17, 343
reviewing 57, 196, 258, 379
risk management 218
robust academic
    management 54

S

scholarships 129, 334, 404
search engines 167, 278
selection procedures
    committee 204
self-assessment 9, 116, 244, 354
self-evaluation
    documents 107, 150
self-evaluation 12, 102, 404
self-mastery 360, 362
semantic connection of learning
    objects 268
site visit 117, 315
skill acquisition 218
social environment 259
social networks 176, 260, 337
social role 43, 54
spearman 285, 291
sponsors 43, 48
staff development 84, 204, 372
stakeholders 6, 91, 411
steered independent
    learning 192
strategic planning 48, 144, 337

student model 172, 178
students performance 8, 120,
     237, 264
students' learning
     experiences 137, 201, 372
study program 136, 325, 409
substantial link 44
summative assessment 221, 267
sustainability 42, 66, 133
SWOT 77, 79, 86
synergistic relationships 112
systematic failing 54

T
tactical planning 63
teacher quality 192, 200, 205
teaching and learning
     development committee 204
teaching-learning effectiveness
     206
teaching-learning process 42,
     133, 371
teaching materials 161, 252
teamwork 108, 112, 114

technology – based assessments
     233
the European education system
     397
times higher education 304,
     321, 414
toolkit 41, 258
top-down method 73
total-approach 109
total quality management 10,
     114, 210
transparency 42, 134, 398

U
university council 382
university quality system 56, 129
user interface 159, 172, 175
values 71, 203, 285, 398
virtual learning 15, 160, 237
vision 16, 70, 353, 410

W
web-distributed objects 250

About the Authors

Prof Dr. AL Jaber is the Founder/Member of the Board-Cross border for Excellency and Quality in Higher Educations, JORDAN (2013 and Founder/Chairman of the Board-Gazaniasolar, Jordan (2013) and he is the President of Gulf University, Bahrain. Since 2014, who has a PhD in Computer Science, Clarkson Uni. NY USA, 1984., MSc: Geometric Programming/Math, Bagdad Uni. Iraq, 1978. BSc: Math., Basra Uni., Basra, Iraq 1976. MSc: Geometric Programming/Math, Bagdad Uni. Iraq, 1978. BSc: Math., Basra Uni., Basra, Iraq 1976.

Prof. AL Jaber has long experience and remarkable teaching experience in computer science as he worked as an associated professor then a full professor for many Universities inside and out side Jordan. His achievements have been lasting as he has been Board of Trustees for many public and private universities in Jordan. He is also an External Reviewer for Quality Assurance-IT Program in Arab Universities, UNDP. As well as a Higher Education and IT Consultant. Prof. Al Jaber worked for as a Dean of graduate studies for **Amman Arab University for Graduate Study being** Responsible for the regulations of scientific research affairs at the university in addition to presenting encouragement and support for research as well being a President of the Council of Scientific Research. He worked for two years as a Secretary General of public Sector Development for **Ministry of Public sector Development - Amman, Jordan** as well being a vice President for Academic Planning and development affairs for more than three years for **Al Balqa' Applied University - Salt, Jordan; a** <u>Dean of King Abdullah II School for Information Technology</u> **and then a President** <u>Advisor for Planning and Development</u> for The university of Jordan from 2000–2005.

Prof. Al Jaber has been playing crucial roles in many scientific and academic committees as being a Coordinator of the IT curriculum committee of the ministry of higher education in (90–91); a Member and coordinator of tender committees; a Coordinator of the committee of His ham AI-Hijjawi Award (Jordan and Palestine); a committee member for the

King Abdullah II award of 2002/2003; a Committee member to select the best programmer in Jordanian universities; a Committee member of Quality Assurance for the Computer Science programs in the University of Jordan \Al-Hussein fund for Excellency; a Coordinator of the committee of Quality Assurance for the Computer Science programs in the University of Jordan United Nations Project; a Committee member of reviewer of Quality Assurance for the Computer Science Department in the University of Sanah - Yemen and University of Polytechnic-Hebron.; a Committee member in accreditation committees for the Computer Science programs in at different universities in Jordan; a Member of the Committee for curriculum design for computer science at different universities in Jordan; a Member of the Committee responsible for the development of Information Technology in Jordan; a Member of the steering committee for the British Council in the project "Jordan-UK get connected"; a Member of the editorial board of the Arab international Journal in Computer, as well as the editorial board of Abhath AI-Yarmouk Journal of AI-Yarmouk; a Reviewer for many Journals; a -Member in the council of Computer Center UJ; a Member in the council of Human Resources and Development-government of Jordan; a Member in the Association of Computing Machinery (ACM); a Member of the board of trustees Zeitonah University 2006–2009; a member of the board of trustees Philadelphia Private University 2009-now. And also a <u>Chairman of the Research Priorities ITC Committee</u> for the coming 10 years in Jordan-HCST.

Prof. Dr. AL Jaber is a Seasoned professional with experience in Sr. Management and administrative positions with academic institutions and public-sector organizations. Experienced professional with solid track record in designing and implementing university degree programs and required curriculums as well as developing research policies, programs and priorities. In-depth knowledge in educational quality assurance systems and strategic planning and hands-on experience in establishing QA programs, policies, procedures, reporting, organization, staffing and assessment. Excellent knowledge and solid experience in public sector developments and reforms, services improvement programs, change management, turn-around programs and human resource development plans. Computer Scientist with excellent research background and expertise in Computer Algorithms, Computer Security and other Computer Science related fields.

Sincerely,
Prof. Dr. Ahmad Odeh Al-Jaber
mmmsmkm@yahoo.com

Haifaa O. Elayyan is an Academic Researcher interested Quality Assurance for Higher education Institutions and has many recent publications in Re-forming and re-presenting Quality Assurance concept along with e-tutoring technologies. Haifaa O. Elayyan has BCS. In computer Science from The University of Jordan (2000), and Joined the Master Degree program/ Computer Science MEU, Jordan (2013).

Mrs. Elayyan has contributed with publications in the Quality Assurance concepts and has some of these papers are selected for Journal publications such as IEEE. Mrs. Elayyan has long experience lecturing, curriculum designing and course coordinating as being an academic staff for universities such as Arab Open University. KSA branch. Mrs. Elayyan has been a member in quality assurance committees for universities she worked for and contributed with preparing quality assurance procedures; requirements for universities internal and external accreditation.

Mrs. Elayyan research interests are also including Database implementation & techniques. in e-Learning, and e-Solutions technologies for quality assurance purposes. Also Interested in Data mining field.

Haifaa Omar Elayyan
Haelayyan@gmail.com